AF389201

Sustainable Structures and Construction in Civil Engineering

Sustainable Structures and Construction in Civil Engineering

Editors

Zhihua Chen
Qingshan Yang
Yue Wu
Yansheng Du

MDPI • Basel • Beijing • Wuhan • Barcelona • Belgrade • Manchester • Tokyo • Cluj • Tianjin

Editors
Zhihua Chen
Tianjin University
Tianjin, China

Qingshan Yang
Chongqing University
Chognqing, China

Yue Wu
Harbin Institute of
Technology
Harbin, China

Yansheng Du
Tianjin University
Tianjin, China

Editorial Office
MDPI
St. Alban-Anlage 66
4052 Basel, Switzerland

This is a reprint of articles from the Special Issue published online in the open access journal *Sustainability* (ISSN 2071-1050) (available at: https://www.mdpi.com/journal/sustainability/special_issues/Q92187OEGO).

For citation purposes, cite each article independently as indicated on the article page online and as indicated below:

LastName, A.A.; LastName, B.B.; LastName, C.C. Article Title. *Journal Name* **Year**, *Volume Number*, Page Range.

ISBN 978-3-0365-8248-1 (Hbk)
ISBN 978-3-0365-8249-8 (PDF)

Cover image courtesy of Ting Zhou

Contents

Preface to "Sustainable Structures and Construction in Civil Engineering"

With the development of society and the economy, sustainable structures and construction are becoming increasingly necessary in civil engineering. Sustainability in construction refers to eco-friendly and economical practices that occur throughout the life cycle of structures, including their design, construction, service and demolition. Sustainable structures include fabricated and modular structures, structures constructed using bio-based materials, as well as others. Specifically, timber, bamboo and light steel have become increasingly popular building materials. Additionally, wall panels made from eco-friendly materials can be used to reduce buildings' energy consumption and carbon emissions. Sustainable construction necessitates prefabricated construction and the recycling and reuse of materials, structures, etc. Key problems have been proposed and solved for sustainable structures and construction in practice.

This Special Issue book provides an international forum for the presentation and discussion of the latest developments in sustainable structures and construction in civil engineering.

We appreciate all the authors of the articles and contributors to this book. We also thank the Chinese National Key Research and Development Program (Grant No. 2019YFD1101000) for sponsoring some of the research in this book. We wish to thank the 23rd National Symposium on Modern Structural Engineering conference for supporting this special issue.

Zhihua Chen, Qingshan Yang, Yue Wu, and Yansheng Du
Editors

 *sustainability*

Article

Automated Optimum Design of Light Steel Frame Structures in Chinese Rural Areas Using Building Information Modeling and Simulated Annealing Algorithm

Ting Zhou [1], Kezhao Sun [2], Zhihua Chen [3,*], Zhexi Yang [2,4] and Hongbo Liu [3]

[1] School of Architecture, Tianjin University, Tianjin 300072, China; zhouting1126@tju.edu.cn
[2] International Engineering Institute, Tianjin University, Tianjin 300072, China
[3] School of Civil Engineering, Tianjin University, Tianjin 300072, China
[4] Department of Architecture and Civil Engineering, City University of Hong Kong, Hong Kong SAR, China
* Correspondence: zhchen@tju.edu.cn

Abstract: Many manual calculations and repeated modeling are required during the traditional structural design process. However, due to the high cost, rural buildings in China cannot be professionally designed and verified by designers as urban buildings, and their safety and economy cannot easily meet the requirements. Building Information Modeling (BIM) technology and intelligent optimization algorithms can effectively improve the structural design process and reduce design costs, but their applications in the field of rural residential buildings in China are limited. Therefore, this paper presents an innovative framework that realizes the structural design of rural light steel frame structures on the BIMBase platform (widely used BIM software in China, BIMBase 2023R1.3). Based on the parametric library of structural components built on standardized component coding, the framework completes the rapid modeling of rural light steel frame structures and the interaction between the BIMBase platform and structural analysis software, SATWE. The improved two-stage simulated annealing (SA) algorithm is applied to the structural design of rural buildings to obtain a design scheme that meets the design requirements and reduces the material consumption as much as possible. Two prefabricated rural light steel frame structures were analyzed to evaluate the efficiency of the proposed framework. The results show the feasibility of the proposed framework. Compared with traditional manual design methods, the design period can be reduced by six times while maintaining comparable levels of material consumption and structural design indicators.

Keywords: optimum design; rural light steel frame structure; building information modeling; simulated annealing algorithm

Citation: Zhou, T.; Sun, K.; Chen, Z.; Yang, Z.; Liu, H. Automated Optimum Design of Light Steel Frame Structures in Chinese Rural Areas Using Building Information Modeling and Simulated Annealing Algorithm. *Sustainability* **2023**, *15*, 9000. https://doi.org/10.3390/su15119000

Academic Editor: Paulo Santos

Received: 24 April 2023
Revised: 30 May 2023
Accepted: 31 May 2023
Published: 2 June 2023

1. Introduction

In recent years, the rural economy in China has grown by leaps and bounds. Under the impetus of the residential construction boom, the development of rural buildings has shifted from quantitative increase to qualitative improvement. The improvement of construction quality and living conditions has become an inevitable requirement for the next stage of the development of rural buildings [1,2].

However, the current design and construction of rural buildings in most parts of China are mostly carried out in accordance with experience, and there are common key problems such as outdated construction techniques, insufficient disaster resistance, and excessive resource and energy consumption [3–6]. At the same time, the prevailing structural design in China mainly depends on the experience of designers [7]. The designers give the primary scheme by referring to similar engineering designs, then check the structural layout, and optimize it repeatedly according to the design results until a reasonable design scheme is obtained. Given the economic constraints in rural areas, it is impractical to conduct precise and meticulous structural design for every rural building. Sufficient validation of the safety and economic viability of the structural design of rural buildings cannot be provided by relying solely on engineers' experience. Consequently, there is a need to explore a cost-effective and high-quality structural design process that is suitable for rural buildings in China.

In rural areas of China, it is common for households to reside in single-family homes with two stories. In comparison to the prevalent use of concrete and masonry structures, light steel frame structures offer several advantages, including simple structural forms, they are lightweight, have high strength, good ductility, and have a convenient construction [5,6]. In consequence, it is necessary to promote their application in rural areas. The application of building information modeling (BIM) technology and optimization algorithms to realize the automatic design of rural light steel frame structures can effectively improve the efficiency of the traditional structural design while ensuring the safety and economy of the optimized structure [8]. To our best knowledge, these technologies have been extensively applied in urban buildings, but they have not been integrated with rural buildings. Thus, it is of great significance to study the automatic design method of light steel frame structures to ensure the rapid development of the rural construction industry.

BIM, the cornerstone of digital transformation in the architecture, engineering, and construction (AEC) industry, has been widely used in infrastructure construction in recent years [8]. A BIM model can be perceived as a data repository that consolidates all building information throughout various project lifecycle stages, emphasizing the sharing and transmission of model information [9–11]. Therefore, BIM can be integrated into the structural design of rural buildings, with its advantages of automation to improve the structural design process and reduce design costs.

Furthermore, the recent advancements in artificial intelligence have had a profound impact on various traditional industries. The resulting effects and challenges have also spurred the transformation of the AEC industries [12]. The intelligent optimization algorithm is called a modern heuristic algorithm. It is the result of the intersection, penetration and mutual promotion of artificial intelligence and engineering science [13]. By integrating the intelligent optimization algorithm with structural design, an intelligent design system and decision support system can be created. This integration not only reduces investment costs but also ensures the accuracy and rationality of each stage in the rural building construction process [14]. It is evident that this technology holds immense potential for extensive application within the AEC industry.

Therefore, the objective of this research is to develop a new framework that combines BIM and optimization algorithms for light steel frame structures in Chinese rural areas. According to the state standard [15,16], a set of standardized components suitable for rural buildings is established, which can be directly used for model assembly on the BIM platform. The standardized component coding and index are studied to facilitate the retrieval of components in the component library for rapid BIM modeling. The interaction between BIM and other structural analysis software is realized through the secondary development of BIM, with the component of the library serving as the basic unit. The improved two-stage simulated annealing algorithm is applied to the structural design and optimization of rural buildings to achieve greater material savings and economical design solutions. Illustrative examples of two prefabricated rural light steel frame structures show the effectiveness and efficiency of the proposed framework in providing structural design and optimization solutions. Through the framework proposed in this paper, a simple, economical and efficient automatic design method can be provided for rural buildings.

The remainder of this paper is organized as follows: Section 2 reviews several works on the interworking between BIM and other software and the research status of structural intelligent optimization algorithms. Section 3 discusses the preparatory steps taken for the proposed framework. Section 4 presents the proposed framework for the automated optimum design of the rural light steel frame structure. In Section 5, two illustrative examples of prefabricated rural light steel frame structures are presented to demonstrate the efficiency and performance of the proposed framework. Conclusions and future work are discussed in Section 6.

2. Research Background

In the current AEC industry, a shift toward BIM technology has been observed due to its capability to reduce errors, wastage, and cost and also improve efficiency [17]. BIM serves as an integrated database that represents the building and facilitates the exchange of information between different software tools. BIM can be used as the central model for analyzing multiple performance standards, including architectural, structural and lighting aspects [18]. The predominant focus of current research regarding BIM as a central data model for building performance analysis revolves around diverse energy simulation software or tools. The common approach is to convert the BIM models into energy input files to solve interoperability issues and create an automatic link between the energy simulation software and BIM authoring tools [19,20]. However, there are few studies on the integration between BIM and structural analysis software. Existing research has highlighted potential issues with information consistency and missing data when converting models between different BIM and structural analysis software. For instance, when transferring model data between Tekla Structure version 13.1 and Revit Architecture 2008, only geometric components could be successfully transmitted. However, important cross-sectional properties were missing and had to be manually set for models containing multiple elements [21]. Moreover, preliminary tests have revealed that in SAP2000, the imported material information from an industry foundation class file could not be successfully loaded into a structural model for performing structural analysis [22]. When transferred to ETABS and SAFE software (ETABS 2015 or later, and SAFE 2014 or later), various types of boundary conditions in Revit (such as pinned, roller and fixed) are uniformly interpreted as pinned conditions [23]. A limited number of researchers have investigated BIM interoperability workflows between BIM and structural analysis software. Ramaji and Memari [24] outlined three specific workflows connecting building information models with structural analysis models. Aldegeily [23] summarized three types of paths for data transfer between architectural and structural models. However, there remains a deficiency in fundamental methods that facilitate smooth interoperability between BIM and structural analysis software. Furthermore, the current software available for model conversion operations is of limited use in the Chinese construction market due to inadequate alignment with Chinese standards [25,26]. Thus, it is imperative to investigate the integration of mainstream domestic structural analysis software with BIM in order to facilitate structural design for rural buildings in China.

At present, structural engineers in China heavily rely on intuitive knowledge and experience to manually generate and optimize conceptual designs [27]. This is a cumbersome and iterative process during which structural engineers can only consider a few possible alternatives [28]. Restricted by construction costs, it is more difficult to perform manual iterative checking calculations for the structural design of rural buildings in China. Over the past few decades, a multitude of nature-inspired approaches have been developed to optimize engineering design problems. These approaches, referred to as "Meta-heuristic Algorithms," are formulated based on global search techniques. They are designed to overcome the limitations of traditional methods and tackle challenging engineering problems more effectively. Gan [29] proposed an improved genetic algorithm to design high-rise reinforced concrete structures based on structural topology and element optimization. Tafraout [30] developed a genetic algorithm for the automated generation of optimal structures, incorporating seismic stability criteria from relevant design specifications. Zhou [31] conducted an automatic optimization design of shear wall structures using an enhanced genetic algorithm and leveraging prior knowledge. Gholizadeh [32,33] employed intelligent algorithms, including a bat algorithm and dolphin echolocation algorithm for the optimization design of planar steel frame structures. The study focused on optimizing the structural dimensions, resulting in a successful reduction in the total mass of the structure. Furthermore, the study investigated the optimization of shear wall locations and obtained the optimal layout. Kaveh [34] utilized an advanced charging system search algorithm to optimize steel frame structures, resulting in a structure that met the required constraints and limits while significantly reducing the total material cost. Various optimization algorithms have been widely used in reinforcement design [35,36] and high-rise structure design [37–39] but few scholars have applied them to the structural design and optimization of rural buildings. In the context of structural design for urban buildings, optimization presents a highly complex problem involving non-convexity, non-linearity, variable discretization, and numerous local optima. Moreover, the optimization targets are characterized by significant structural diversity and strong interactions between components. The quality of the optimization outcomes and the speed of convergence depend heavily on the updating mechanisms and search capabilities of the employed algorithms. When compared to urban buildings, the structural design of rural buildings involves different constraints, with smaller-scale structures and relatively simpler component types. Given the economic limitations of rural areas, algorithms suitable for designing rural buildings must be simpler and more general. Further research is needed to assess the suitability of algorithms for achieving efficient optimization design of rural housing structures.

In view of the above discussion, the aim of this paper is to address the existing challenges in the structural design of rural buildings in China by integrating BIM technology and optimization algorithm. This framework is developed on the foundation of BIMBase, a BIM tool widely used in China, and its programming tool. It integrates the comprehensive information stored in parametric BIM models with structural analysis software, thereby enhancing the accessibility of structural optimization throughout the design process. SATWE is an open-source structural analysis software extensively used in China. In comparison with other structural calculation software, it demonstrates closer integration with China's current specifications and provides support for multi-functional software secondary development. By utilizing an application programming interface (API) to communicate with BIMBase, its structural analysis function can be extended to the BIMBase project level. The structural optimization of rural light steel frame structures is solved by using the SA optimization algorithm. Compared with particle swarm optimization and genetic algorithm, which are widely used in the field of structural optimization, simulated annealing algorithm does not require extensive parallel computing. The calculation speed of each iteration step is faster, which is more suitable for the optimization design of rural buildings with fewer variables and dimensions. The improved two-stage SA algorithm used in this framework is essentially based on the improvement of the technique as formulated by Tort et al. [40]. It has been shown to effectively solve the problem of poor convergence characteristics and inefficient search processes of the traditional simulated annealing algorithm. The significant potential values of this proposed framework lie in its ability to save time and cost through the automation of processes, which are derived from rigorous and reliable analysis. Furthermore, it aims to achieve better quality and higher performance in buildings, further enhancing its value proposition.

3. Preparation for the Proposed Framework

3.1. Parametric Structural Component Modeling

In the context of structural design for rural buildings, the geometry of the components plays a crucial role as a design parameter that requires iterative testing and optimization to achieve the most reasonable design results. BIMBase is based on the Python language and its own pyp3d database for parametric modeling. Compared with Revit, which uses "family" as the graphic element for "class", modeling in a programming language is more flexible and allows for the more accurate and efficient modeling of more complex geometric components, as well as the accurate expression and transmission of model information. Using the model operation commands provided in the pyp3d database, the parametric component code is written in the Python programming language to generate corresponding parametric components in the BIMBase platform. For a given type of component, the variable parameters should first be determined; afterward, the equation of the geometric logic is determined using the model basic body and the model basic command functions with the variable parameters as variables; finally, the initial values of each variable parameter are set to generate the parametric component in the BIMBase platform. It should be noted that although the various parameters of the component are variable, they still adhere to standard components, and the fixed modulus required by the state standard still must be met. Taking a parametric H-beam component as an example, the process of parametric component modeling is shown in Figure 1.

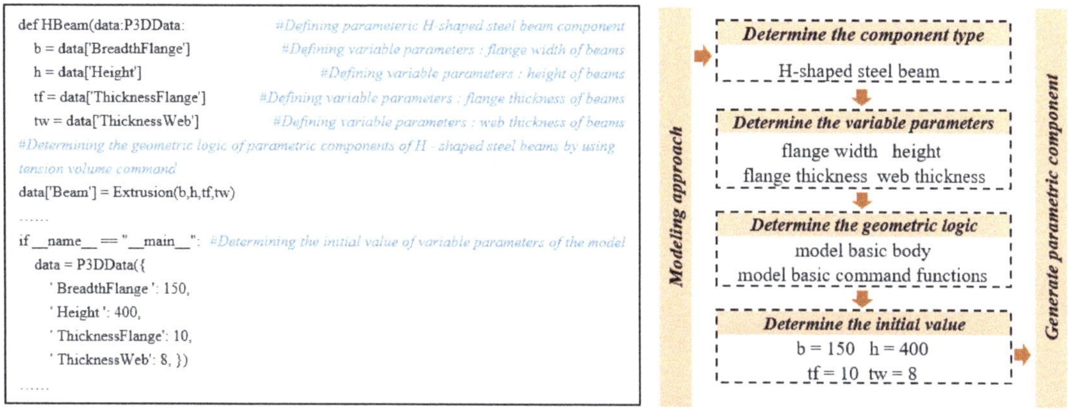
def HBeam(data:P3DData:
#Defining parameteric H-shaped steel beam component
b = data['BreadthFlange']
#Defining variable parameters : flange width of beams
h = data['Height']
#Defining variable parameters : height of beams
tf = data['ThicknessFlange']
#Defining variable parameters : flange thickness of beams
tw = data['ThicknessWeb']
#Defining variable parameters : web thickness of beams
#Determining the geometric logic of parametric components of H - shaped steel beams by using tension volume command
data['Beam'] = Extrusion(b,h,tf,tw)
if __name__ == "__main__": #Determining the initial value of variable parameters of the model
data = P3DData({
'BreadthFlange': 150,
'Height': 400,
'ThicknessFlange': 10,
'ThicknessWeb': 8, })
Modeling approach
Determine the component type
H-shaped steel beam
Determine the variable parameters
flange width height
flange thickness web thickness
Determine the geometric logic
model basic body
model basic command functions
Determine the initial value
b = 150 h = 400
tf = 10 tw = 8
Generate parametric component

Figure 1. Process of parametric component modeling.

3.2. Parametric Library of Structural Components for Rural Buildings

The integrated management of structural components through the establishment of a BIM component library can greatly improve the efficiency of BIM applications. The key to integrated management is the coding method of the components. Once the component modeling is completed, component coding is added based on a standardized classification, following a unified standard. This process results in the formation of component information coding in a specific format. Combined with the characteristics of rural buildings, the components in the component library are encoded at five levels with a total length of 12 digits:

- Two digits at the first level represent the professional type.
- Two digits at the second level represent the structural system.
- Two digits at the third level represent the component category.
- Three digits at the fourth level represent the sectional dimension.
- Three digits at the fifth level represent the component length.

Taking a column with a height of 3000 mm in the □150 × 6 component class as an example, its coding is 020101007007, as shown in Figure 2.

After determining the coding rules of the components, the components can be stored as model folders in sequence according to the coding sequence and saved in the component library folder. With a reasonable folder level and coding index, the location path of the corresponding component model file can be quickly and conveniently found. After the compilation of structural components of rural buildings is completed, it is packaged and imported into the component library management platform of BIMBase 2023R1.3, facilitating the creation of a parametric library for rural building structural components (Figure 3). Relying on the versatility of the BIMBase platform, the components can be imported for modeling through a user-friendly visual interface, or their model information can be directly accessed from the component library using Python code. This integrated component management greatly improves the efficiency of modeling rural building structural models while providing a solid foundation for subsequent structural design and optimization, which is the fundamental work of the framework proposed in this paper.

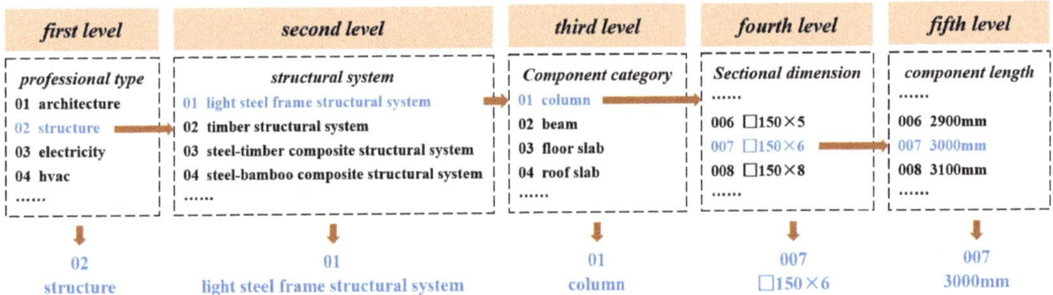

Figure 2. Example diagram of BIM component coding.

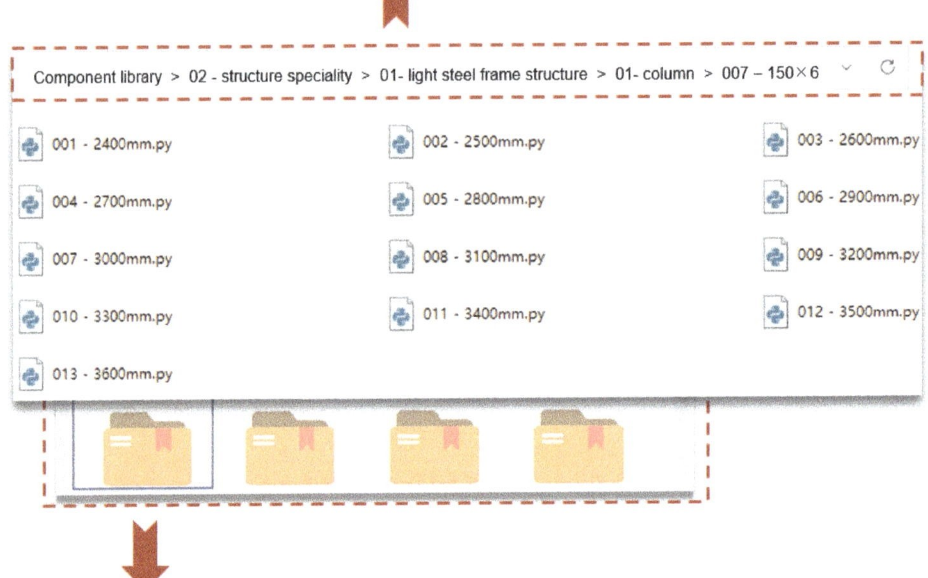

Figure 3. Construction of the parametric library of structural components for rural buildings.

4. The Proposed Framework for the Structural Design of the Rural Buildings

The structure of the proposed framework with BIM and optimization algorithm for the structural design of the rural buildings is explained in this section. As shown in Figure 4, four modules are consisted in the proposed framework: (1) Model generation, (2) Model transformation I, (3) Structural optimization, (4) and Model transformation II, which will be discussed in the following sections.

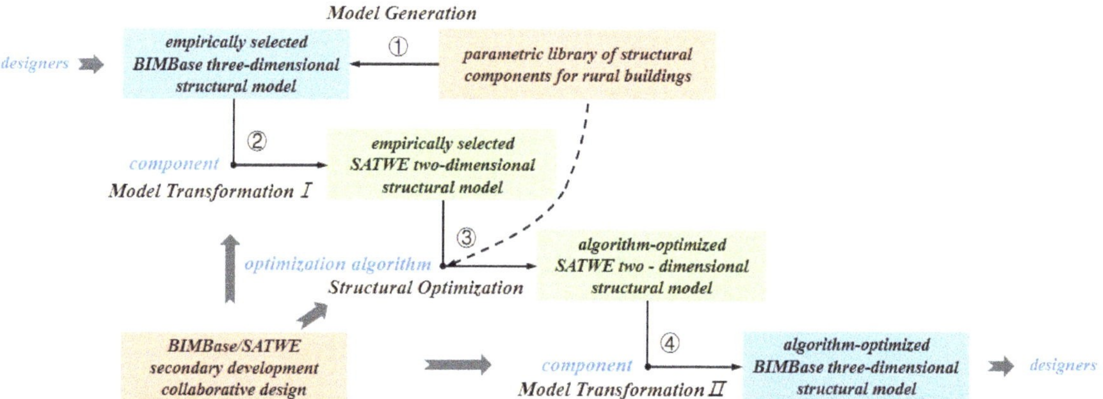

Figure 4. Flow chart of structural model intelligent design for rural buildings.

4.1. Model Generation Module

The structural designers can directly import components from the component library to complete the rapid modeling of a three-dimensional structural model of rural buildings in the BIMBase platform according to the previous design experience. The components are organized into model folders in a specific coding sequence and stored in the component library folders. These folders are then packaged and imported into the component library management platform of BIMBase. Designers can directly access these component models on the platform, which greatly improves the modeling efficiency of the model. It should be noted that each component in the component library is indexed by a specific coding sequence and stores its geometric and material information through Python code. Therefore, the three-dimensional structural model can be equivalent to a Python code containing a set of component information. Once the structural modeling of the rural buildings is completed, the model can be further refined by specifying the behavior and physical characteristics of the structure. The behavioral characteristics include information about the loadings (e.g., dead load, live load and wind load) acting on the component, while the physical characteristics include end-support information (e.g., fixed, pinned and roller end-support). An example of a structural frame model built by the component library is given in Figure 5.

Figure 5. The example frame structural model built by the component library.

4.2. Model Transformation I Module

SATWE software (SATWE 2022) is based on two-dimensional structural models for structural calculation and analysis, which requires the transformation of a three-dimensional structural model into a two-dimensional structural model. Many model transformation plug-ins and program interfaces are based on IFC (Industry Foundation Classes) data standards, with coordinate points, geometry and layer information as the basic object to achieve model transformation function. However, this approach often involves the extensive identification of coordinate information and frequently results in information loss and deviations, making the transformed structural model inaccurate and unable to perform the next structural calculation directly. In this framework, the components in the component library are served as the basic unit of model transformation. The integral BIMBase three-dimensional structural model is disassembled into component models included in the component library, and the three-dimensional model information of each component is extracted in turn. Through Python language, this three-dimensional model information is transformed into two-dimensional model information of components that can be identified by SATWE so as to generate the two-dimensional structural model of each component. The two-dimensional structural models of these components are assembled into a folder that can be recognized by the SATWE software (SATWE 2022), and a complete SATWE two-dimensional structural model is finally formed. This process enables the automatic transformation from BIMBase three-dimensional structural model to an SATWE two-dimensional structural model.

In the BIMBase three-dimensional structural model, two types of three-dimensional model information can be extracted from any component. The first type of three-dimensional information is standardized component coding. Through this coding, the key information, such as the cross-sectional parameters, material properties and length of the component, can be searched from the component library. The second type of three-dimensional model information is the three-dimensional position coordinates of the component in the BIMBase three-dimensional model, which determines the position of the component in the overall model. When generating components in the SATWE two-dimensional model, it is crucial to first determine the appropriate standard layer and axis network where the components will be situated. Subsequently, the components are positioned on the corresponding nodes or axes based on their type (section information) and two-dimensional position within the standard layer. In order to realize this component generation process, it is necessary to transform the three-dimensional model information extracted from the BIMBase model into the two-dimensional model information required by the SATWE model, as shown in Figure 6. The process of extracting and transforming model information through component-based methods avoids the loss of data and errors in data conversion. This enables smooth collaboration among different software systems during the design process. Furthermore, additional structural behavior and physical information extracted from BIMBase three-dimensional models are packaged into a compatible data format for structural analysis. Based on the above-mentioned model transformation method, in conjunction with the API interface of the BIMBase platform, a program for the mutual transformation of BIMBase and SATWE is developed to realize the collaborative design of the two pieces of software.

4.3. Structural Optimization Module

Considering that the structural design scheme selected by experience is often conservative, this excessive structural redundancy design will lead to a waste of resources and high construction costs. Therefore, further optimization design of the primary structural scheme is required. Due to the limited budget of rural buildings in China, it is difficult for designers to complete a large number of structural trials. Therefore, the optimization algorithm is applied in this framework to the structural design optimization of rural buildings. A number of variations and enhancements of the SA algorithm have been proposed in the literature to improve its performance [40–45]. The two-stage SA algorithm employed in this framework is based on the improvement of the technique, as formulated by Tort et al. [40]. The basic elements involved in this algorithm are outlined briefly in the following.

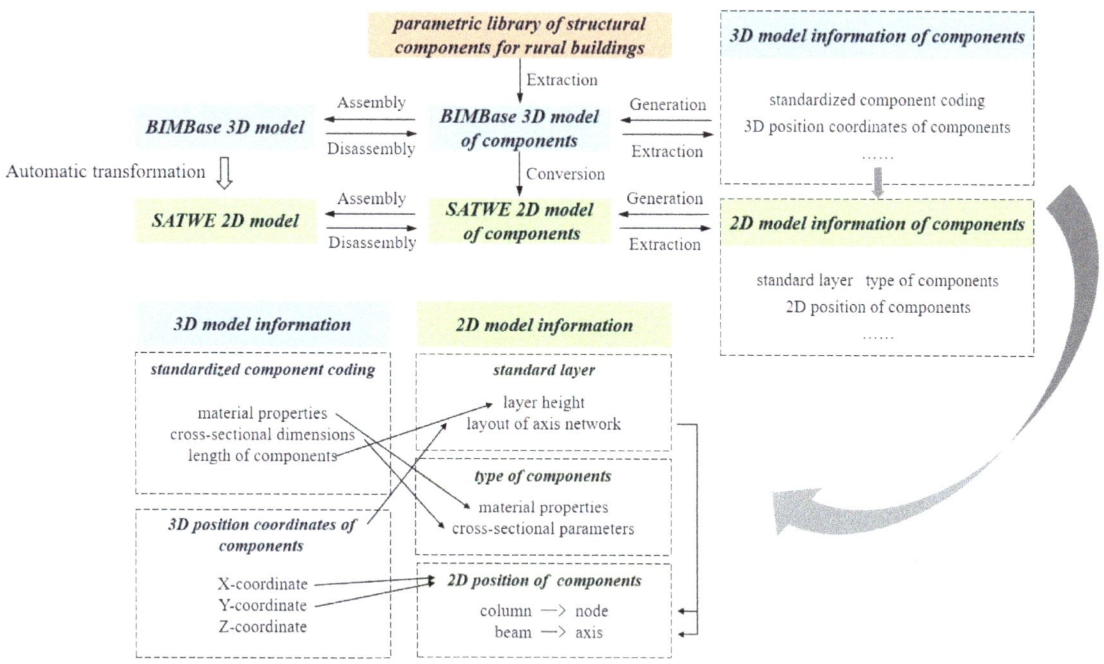

Figure 6. Model transformation diagram based on parametric library of structural components.

4.3.1. Mathematical Model and Boundary Conditions of Intelligent Structural Optimization

The objective of structural intelligent optimization is to ensure that the structural design scheme meets the limit state of bearing capacity and the limit state of normal use while the steel consumption of the structure reaches the lowest possible value. For any SATWE two-dimensional model of rural buildings, the structural design scheme can be simplified as the mathematical model represented by Equation (1):

$$\begin{cases} P = \left[C_1, C_2, C_3, \ldots, C_i, \ldots, C_m, B_1, B_2, B_3, \ldots, B_j, \ldots, B_n \right]^T \\ C_i = \left[F_{ci}, T_{ci}, S_{ci} \right] \\ B_j = \left[F_{bj}, T_{bj}, S_{bj} \right] \end{cases} \tag{1}$$

In Equation (1), P represents the structural design scheme imposed on m column components and n beam components. C_i represents the i-th column component and B_j represents the j-th beam component. Their structural two-dimensional model information is determined using three variables: standard layer number, F, type of component, T, and two-dimensional position of the components, S, which correspond to the three modules of the standard layer, type of component and two-dimensional position of the components in SATWE two-dimensional model information, respectively.

In general, for a given component, S, the standard layer number F and two-dimensional position will be determined during the design of the building scheme. That is, in this mathematical model, F and S are constants, and T is the design variable to be optimized. Then, the structural design scheme P can be simplified as a function related to the variable T:

$$P = g\left(T_{c1}, T_{c2}, T_{c3}, \ldots, T_{ci}, \ldots, T_{cm}, T_{b1}, T_{b2}, T_{b3}, \ldots, T_{bj}, \ldots, T_{bn} \right) = g(T) \tag{2}$$

For a specific standardized component, the material consumption of the component is constant, as its cross-sectional dimension and length of component are constant. For an overall structural design scheme, the total material consumption of the structure, Q, is related to the component type of all components. Therefore, the objective function of the algorithm can be organized as follows:

$$Q = h(T) = \sum_{i=1}^{m+n} h(T_i) = h(g^{-1}(P)) = f(P) \tag{3}$$

According to the state standard [15,16], the following structural constraints need to be considered in the optimal design of rural light steel frame structures.

Normal stress constraint:

$$g_{s,k} = \frac{|\sigma_k|}{\sigma_u} \le 1 (k = 1, 2, \ldots, m + n) \tag{4}$$

Overall stable stress constraint:

In-plane stability:

$$g_{stax,k} = \frac{N_k}{\varphi_{xk} A_k f} + \frac{\beta_{mxk} M_{xk}}{\gamma_{xk} W_{xk} \left(1 - \frac{0.8N_k}{N'_{Exk}}\right) f} + \frac{\eta_k \beta_{tyk} M_{yk}}{\varphi_{byk} W_{yk} f} \le 1 (k = 1, 2, \ldots, m + n) \tag{5}$$

$$N'_{Exk} = \pi^2 E_k A_k / 1.1 \lambda_{xk}^2 \tag{6}$$

Out-plane stability:

$$g_{stay,k} = \frac{N_k}{\varphi_{yk} A_k f} + \frac{\beta_{myk} M_{yk}}{\gamma_{yk} W_{yk} \left(1 - \frac{0.8N_k}{N'_{Eyk}}\right) f} + \frac{\eta_k \beta_{txk} M_{xk}}{\varphi_{bxk} W_{xk} f} \le 1 (k = 1, 2, \ldots, m + n) \tag{7}$$

$$N'_{Eyk} = \pi^2 E_k A_k / 1.1 \lambda_{yk}^2 \tag{8}$$

Maximum interlayer displacement angle constraint:

$$g_{dr,\chi} = \frac{|dr_\chi|}{dr_u} \le 1 (\chi = 1, 2, \ldots, \chi_s) \tag{9}$$

Period ratio constraint:

$$g_{rp} = \frac{|r_p|}{r_u} \le 1 \tag{10}$$

In Equations (4)–(10), σ_k represents the maximum stress of the k-th component, σ_u represents the yield strength of the steel. E_k, A_k, W_{xk}, W_{yk}, N_k, M_{xk}, M_{yk}, f represent elastic modulus, cross-sectional area, x-direction gross cross-section modulus, y-direction gross cross-section modulus, axial force, x-direction maximum bending moment, y-direction maximum bending moment and strength design value of k-th component, respectively. λ_{xk}, λ_{yk} represent the slenderness ratio of k-th component to x axis and y axis. γ_{xk}, γ_{yk} represent the overall stability coefficient of k-th component to x axis and y axis under axial compression. β_{mxk}, β_{myk}, β_{txk}, β_{tyk} represent the in-plane and out-plane equivalent bending moment coefficients of k-th component. η_k represents the sectional influence coefficient, dr_χ represents the interlayer displacement angle of χ-th layer, dr_u represents the limit of the interlayer displacement angle, r_p represents the period ratio, r_u represents the limit of the period ratio.

The external penalty function method is used to calculate the total steel consumption of the structure after punishment, and the auxiliary function is constructed to deal with the constraints. The expression is:

$$f(g_i) = \begin{cases} 0, & g_i \le 1 \\ 1, & g_i > 1 \end{cases} \tag{11}$$

In Equation (11), g_i represents the constraint value, which can be calculated by Equations (4)–(10).

Through the auxiliary function, the constraint conditions are taken into account in the objective function, and the pseudo-objective function after punishment is obtained. The mathematical expression is:

$$F = Q(1 + \lambda_1 \sum_{i=1}^{t} f(g_i))(i = 1, 2, \ldots, t) \tag{12}$$

In Equation (12), λ_1 represents the penalty coefficient, and t represents the total number of constraints.

4.3.2. Intelligent Structural Optimization Based on the Two-Stage Simulated Annealing Algorithm

A number of variations and enhancements of the annealing algorithm have been proposed in the literature to improve its search performance. The two-stage SA algorithm employed in the present study is based on the improvement of the technique as formulated by Tort et al. [40]. The generated flow chart of the initial solution is shown in Figure 7. In the first stage of this method, only the size of the frame columns in the primary structural scheme obtained from the engineer's experience is optimized by the annealing algorithm, while the frame beams are sized with a fully stressed design based on the heuristic approach. The initial design is rapidly improved in a relatively small number of iterations. In the second stage, the previously obtained optimal design is utilized as the initial solution, and under a set of new annealing parameters, the structural component size variables are iteratively optimized until the structural design scheme, satisfying both safety and economy, is obtained. The main process of the algorithm (Figure 8) is as follows.

Stage 1:

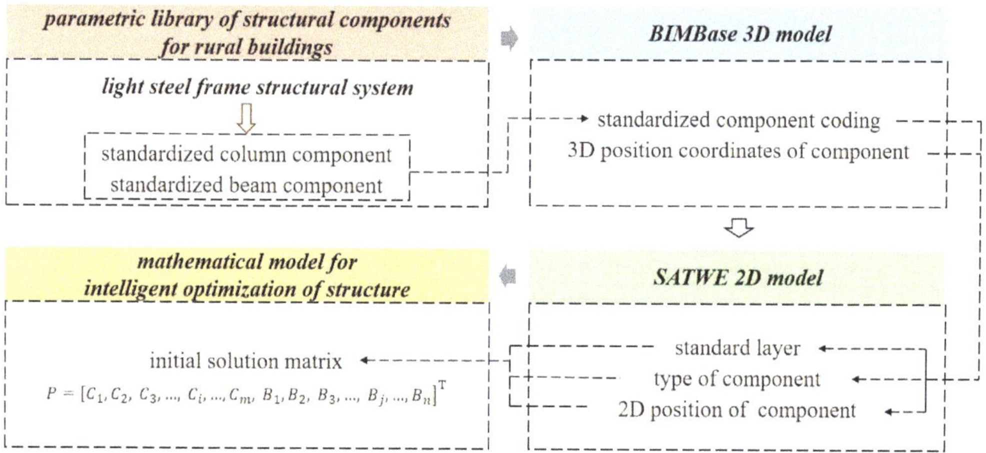

Figure 7. The generation process of initial solution.

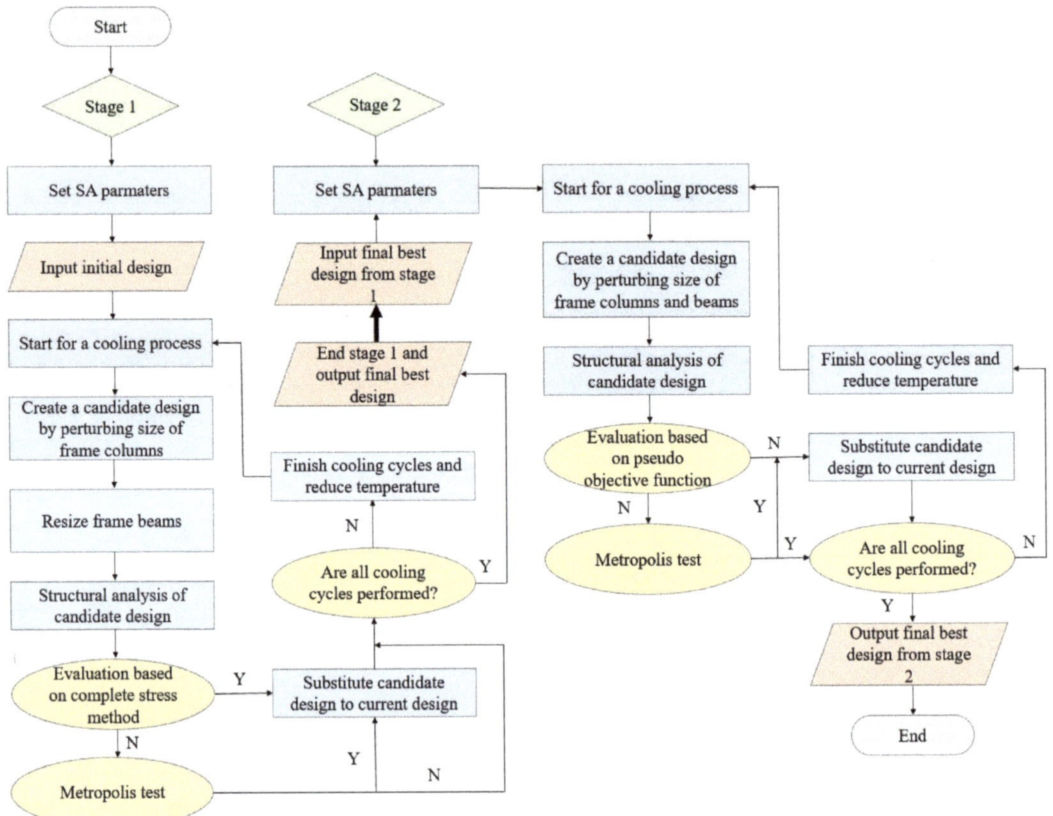

Figure 8. Flowchart of the two-stage simulated annealing algorithm.

Step 1. Initialization and setting of a cooling schedule:
The first step is initialization and setting of an appropriate cooling schedule, including the starting temperature T_s, the finial temperature T_f and the cooling factor η.

$$T_s = -\frac{1}{\ln P_s} \tag{13}$$

$$T_f = -\frac{1}{\ln P_f} \tag{14}$$

$$\eta = \left[\frac{\ln P_s}{\ln P_f}\right]^{1/N_c - 1} \tag{15}$$

In Equations (13)–(15), P_s represents the starting acceptance probability, P_f represents the final acceptance probability and N_c represents the number of cooling cycles.
Step 2. Generation of an initial design:
The structural model built by the engineer based on experience is set as the preliminary scheme. The structural mechanics analysis of the initial design is completed by the model transformation module, and the calculation results of the structure are extracted and sorted based on the Python language. The pseudo-objective function of the initial design is calculated using Equation (12).
Step 3. Creating and resizing of candidate designs:

The new structural arrangement is obtained by randomly perturbing the cross-sectional dimensions of one or more frame columns in the initial design. Under the new structural design scheme, the sectional dimensions of all frame beams are adjusted using the heuristic method based on a complete stress design to generate candidate designs.

(1) Set the size variable of all frame beams to 1. It should be noted that the dimension variables of each component are represented by the index numbers corresponding to the selected sections in the component library. That is to say, the section of all frame beams is set to the smallest section type in the component library.

(2) Analyze each candidate design.

(3) Only check the stress constraints of each component, including normal stress constraints and stable stress constraints. For components with stress exceeding the limit, by increasing the size variable of the component, a larger cross-section is used from the list of component library and other variables are kept unchanged.

(4) Repeat (2) and (3) until all components meet the stress constraints or the section size of all components is set to the maximum section in the component library.

Step 4. Evaluating the candidate design and Metropolis test:

Each time a candidate design is generated, it will compete with the pseudo-objective function of the current design. If the candidate design provides a better design, the current design is automatically accepted and replaced; otherwise, Metropolis tests are performed using the acceptance probability P of bad candidate designs determined by Equations (16)–(18). Metropolis generates a random number r between 0 and 1. If $r \leq p$, the candidate design is accepted and the current design is replaced. Otherwise, the candidate design will be rejected, and the current design will be maintained.

$$\varphi^{(k)} = \varphi^{(k-1)} \sqrt[3]{\overline{P}_t^{(k-1)} / \overline{P}_p^{(k-1)}}, 0.9 \leq \varphi \leq 1.1 \tag{16}$$

$$\Delta\varnothing_{tra} = \tanh(\frac{0.35\Delta\varnothing}{K}) \tag{17}$$

$$P = \varphi exp(-\frac{\Delta\varnothing_{tra}}{KT^{(k)}}) \tag{18}$$

In Equations (16)–(18), φ represents the correction factor introduced to ensure that the actual average acceptance probability follows the theoretical average acceptance probability, $\overline{P}_t^{(k-1)}$ and $\overline{P}_p^{(k-1)}$ represent the theoretical and practical average acceptance probability for the $k-1$-th cooling cycle, $\Delta\varnothing$ represents the pseudo-objective function difference, $\Delta\varnothing_{tra}$ represents the change value of $\Delta\varnothing$, $T^{(k)}$ represents the temperature at the k-th cooling cycle, K is a Boltzmann parameter and its value is the average of $\varnothing$. The basic principle and more details of Equations (16)–(18) are given in Tort et al. [40]. For the sake of simplicity, they are not repeated here.

Step 5. Iterations of a cooling cycle:

Cooling cycle iteration refers to the case where the cross-sectional dimensions of all frame columns are selected to be perturbed once and the corresponding candidate designs are generated. The cooling cycle is generally iterated a certain number of times in the same way, thereby ensuring that the pseudo-objective function is reduced to a reasonable value related to the cooling cycle temperature. The number of cooling cycle iterations i_c can be determined as follows:

$$i_c = \text{int}\left[i_f + (i_f - i_s)(\frac{T - T_f}{T_f - T_s})\right] \tag{19}$$

In Equation (19), i_s, i_f represent the starting and final cooling cycles, both of which were taken as 1 to reduce the computation time.

Step 6. Reducing temperature:

When the iteration of a cooling cycle is completed, the temperature is reduced by the cooling factor and the temperature of the next cooling cycle is set, as shown in Equation (20).

$$T^{(k+1)} = \eta T^{(k)} \tag{20}$$

Step 7. Termination criterion:
Steps 3–6 are repeated until the whole cooling cycles are finished.

Stage 2:

In the second stage, the simulated annealing algorithm is iteratively implemented for all component sizes, and the optimal design obtained in stage 1 is used as the initial design for stage 2. Therefore, the search in stage 2 starts with a more reasonable design, eliminating the need for a highly detailed cooling schedule. A milder cooling schedule is chosen that makes the algorithm under a reduced number of cooling cycles with a new set of annealing parameters. The results of the examples show that stage 2 produces a solution comparable to the simulated annealing algorithm. However, it employs a milder cooling schedule and requires much less computation time, thereby reducing design costs.

4.4. Model Transformation II Module

This module can be regarded as the reverse operation of model transformation I module, through the component of the component library as the basic unit to complete the model transformation. Each component in the SATWE two-dimensional model contains the standard layer, the type of components and the two-dimensional position of components. In the BIMBase three-dimensional model, the component is only determined by its component type (standardized component coding corresponding to the component library) and the three-dimensional position coordinates of the components. Through the mapping relationship shown in Figure 6 and leveraging the Python language, the extracted information of each component from the SATWE two-dimensional model is transformed and packaged into a BIMBase-compatible data format. This process involves matching the components in the component library to complete the modeling of the corresponding three-dimensional model in BIMBase.

5. Illustrative Examples

In this section, a complete description of the considered design examples was provided that utilized for performance evaluation of the proposed framework. Two kinds of design examples of rural buildings were considered in this paper with different plans in order to evaluate the efficiency of the proposed framework.

5.1. Example I

The first example is a two-story light steel frame with 14 column components and 17 beam components. The three-dimensional model of the structure built from the components in the component library selected by the designer based on experience is shown in Figure 5. The standardized component coding, the three-dimensional position coordinates of the components and the structural global parameters were extracted from the BIMBase three-dimensional model. The algorithm program corresponding to the Model transformation I of this framework was used to transform and package the data into a format compatible with the structural calculation software SATWE and generated the SATWE two-dimensional model corresponding to this example. It should be noted that the beam and column components in this example were smaller in section type compared to the primary scheme in the component library, as the conservative structural design scheme was chosen based on experience. This building was designed using both the SA and two-stage SA algorithms by performing three independent runs. Tables 1 and 2 display the results of the runs in terms of the optimized weight of the design scheme and the computation time in each run of the SA and two-stage SA algorithms, respectively. It can be seen from this example that the average performance of the two-stage SA algorithm is slightly better than that of the SA algorithm, even though the former located the optimum approximately two times faster. The comparison of structural design layout schemes of example I is shown in Figure 9. The comparison of key design indexes of example I is presented in Table 3. The results show that the steel consumption of the optimized structure is saved by 12.66%. The calculation results of the scheme are similar to those of the manual optimization scheme, which has an obvious optimization effect while meeting the calculation efficiency. The stress ratio of the structural components of example I is presented in Figure 10. It can be concluded that for the optimized design obtained by the proposed framework, the stress ratios of the structural components have high values and are close to the allowable value while retaining a certain safety reserve. This also verifies that the provided optimum design has less accessible cross-sections, thereby reflecting an economically favorable design perspective.

Table 1. Optimization results of simulated annealing algorithm of example I.

Run	Optimized Weight (kg)	Time (min)	Mean		Standard Deviation	
			Weight (kg)	Time (min)	Weight (kg)	Time (min)
1	2440.13	51				
2	2410.57	48	2426.17	52	21.00	6
3	2427.81	56				

Table 2. Optimization results of two-stage simulated annealing algorithm of example I.

Run	Weight (kg)		Time (min)		Mean		Standard Deviation	
	Stage 1	Stage 2	Stage 1	Stage 2	Weight (kg)	Time (min)	Weight (kg)	Time (min)
1	2590.43	2382.29	7	19				
2	2583.17	2371.95	6	21	2371.55	27	15.48	2
3	2575.84	2360.41	6	22				

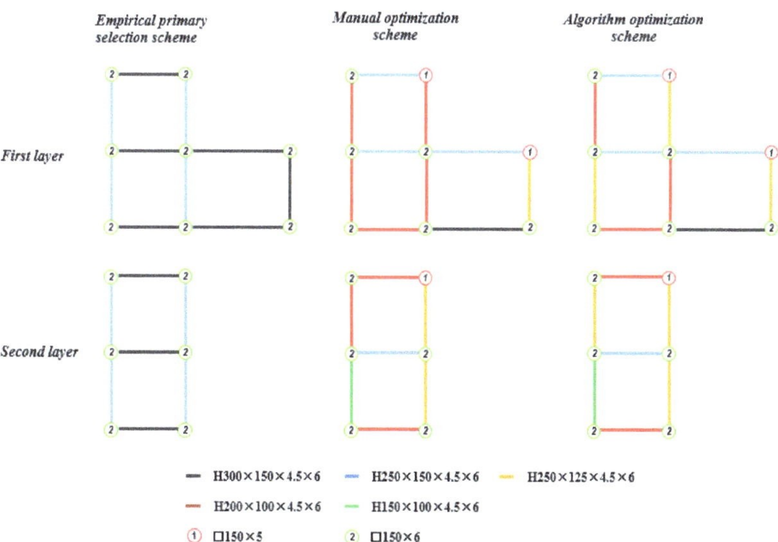

Figure 9. Comparison of structural design layout schemes of example I.

Table 3. Comparison of key design indexes of example I.

Key Design Indexes	Empirical Primary Selection Scheme	Manual Optimization Scheme	Algorithmic Intelligent Optimization Scheme
Total steel consumption (kg)	2702.70	2335.07	2360.41
Maximum interlayer displacement angle under X-directional earthquake	1/738	1/715	1/722
Maximum interlayer displacement angle under Y-directional earthquake	1/642	1/631	1/633
First mode and cycle	X-directional side vibration (0.5601 s)	X-directional side vibration (0.5849 s)	X-directional side vibration (0.5823 s)
Second mode and cycle	Y-directional side vibration (0.5372 s)	Y-directional side vibration (0.5586 s)	Y-directional side vibration (0.5547 s)
Third mode and cycle	Torsional vibration (0.4307 s)	Torsional vibration (0.4503 s)	Torsional vibration (0.4491 s)

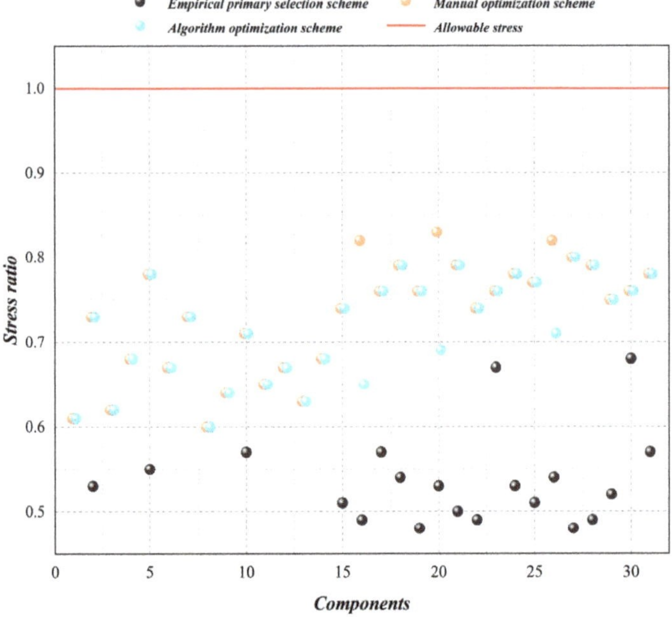

Figure 10. The stress ratio of the structural components of example I.

After the structural design scheme was optimized, the final module of the proposed framework was executed. The optimized SATWE two-dimensional model was returned to the BIM platform to generate a BIMBase three-dimensional model for subsequent detailed design. Both geometric and material information were correctly extracted and mapped, as demonstrated above. The process of implementing the proposed framework of example I is shown in Figure 11. The overall running time of the entire framework for this example is a little over half an hour. In contrast, the conventional structural layout scheme completed by experienced designers takes about 3 h, which is six times longer than the intelligent structural design enabled by the proposed framework. This shows that the proposed framework is an effective approach.

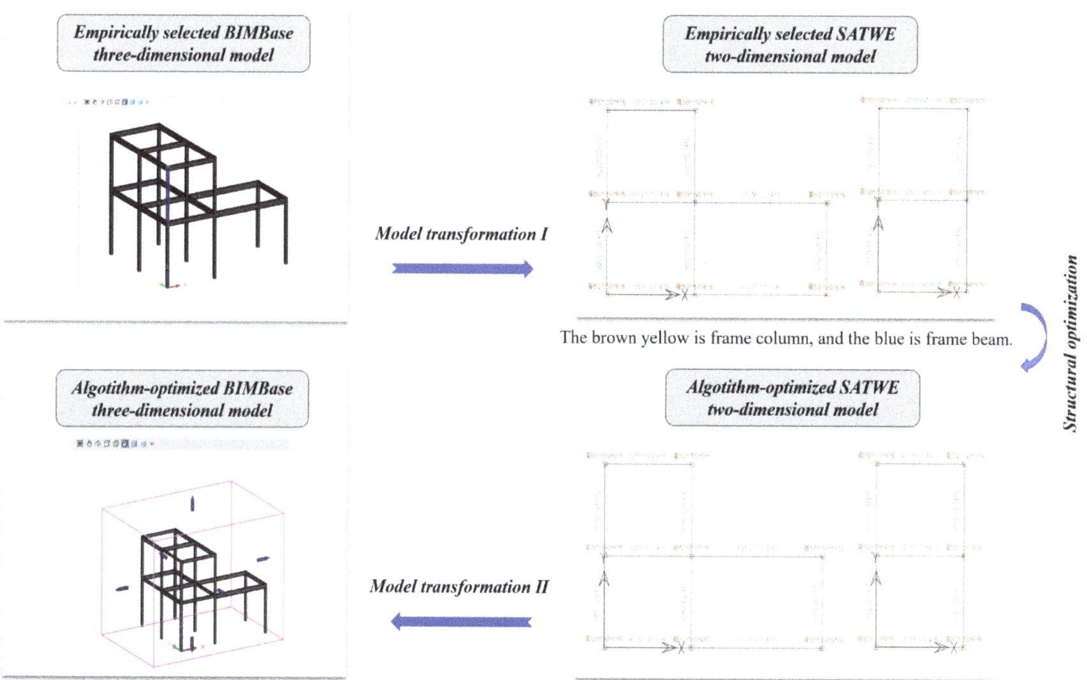

Figure 11. The process of implementing the proposed framework of example I.

5.2. Example II

The second example is a two-story light steel frame with 35 column components and 50 beam components. The three-dimensional model of the structure is shown in Figure 12. Again, this building was designed using both the SA and two-stage SA algorithms by performing three independent runs each. The results are reproduced in Tables 4 and 5 in terms of the optimized weight of the design scheme and the computation time in each run of the SA and two-stage SA algorithms, respectively. For this example, the two-stage SA algorithm shows slightly better performance than the SA algorithm on average, with a significant reduction in computation time. The comparison of structural design layout schemes of example II is shown in Figure 13. The comparison of the key design indexes of example II is presented in Table 6. After example II is intelligently optimized by the proposed framework, the steel consumption of the algorithm optimization scheme is significantly reduced by 11.67% compared with the empirical primary scheme. Compared with the design scheme given by manual optimization, the difference between the two results is 2.75%. The stress ratio of the structural components of example II is presented in Figure 14. It can be concluded from the comparison results of the three schemes that for example II, the proposed framework can better optimize the structure of the empirical primary scheme. After the cooling calculation, the steel consumption is significantly reduced, and the given structural design scheme can meet the design requirements of the ultimate state of bearing capacity and the ultimate state of normal use. Compared with the results of the manual optimization scheme, the key design indexes of the algorithm optimization scheme are close to the manual optimization scheme.

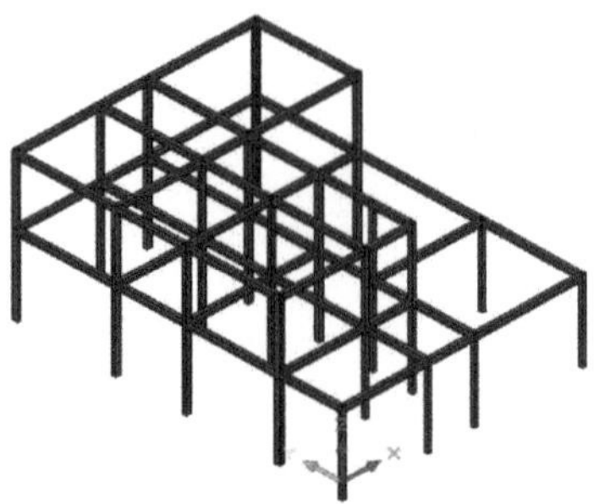

Figure 12. BIMBase three-dimensional structural model diagram of empirical primary scheme.

Table 4. Optimization results of simulated annealing algorithm of the example II.

Run	Optimized Weight (kg)	Time (min)	Mean		Standard Deviation	
			Weight (kg)	Time (min)	Weight (kg)	Time (min)
1	3079.82	127				
2	3074.36	133	3072.58	127	11.69	7
3	3063.57	124				

Table 5. Optimization results of two-stage simulated annealing algorithm of the example II.

Run	Weight (kg)		Time (min)		Mean		Standard Deviation	
	Stage 1	Stage 2	Stage 1	Stage 2	Weight (kg)	Time (min)	Weight (kg)	Time (min)
1	3220.38	3049.25	12	57				
2	3215.47	3054.49	14	55	3054.54	70	3.93	3
3	3198.72	3059.89	11	61				

Table 6. Comparison of key design indexes of the example II.

Structural Design Key Indicators	Empirical Primary Selection Scheme	Manual Optimization Scheme	Algorithmic Intelligent Optimization Scheme
Total steel consumption (kg)	3535.11	2959.91	3049.25
Maximum interlayer displacement angle under X-directional earthquake	1/323	1/300	1/323
Maximum interlayer displacement angle under Y-directional earthquake	1/385	1/367	1/385
First mode and cycle	X-directional side vibration (0.5183 s)	X-directional side vibration (0.5473 s)	X-directional side vibration (0.5392 s)
Second mode and cycle	Y-directional side vibration (0.5074 s)	Y-directional side vibration (0.5334 s)	Y-directional side vibration (0.5310 s)
Third mode and cycle	Torsional vibration (0.4342 s)	Torsional vibration (0.4535 s)	Torsional vibration (0.4480 s)

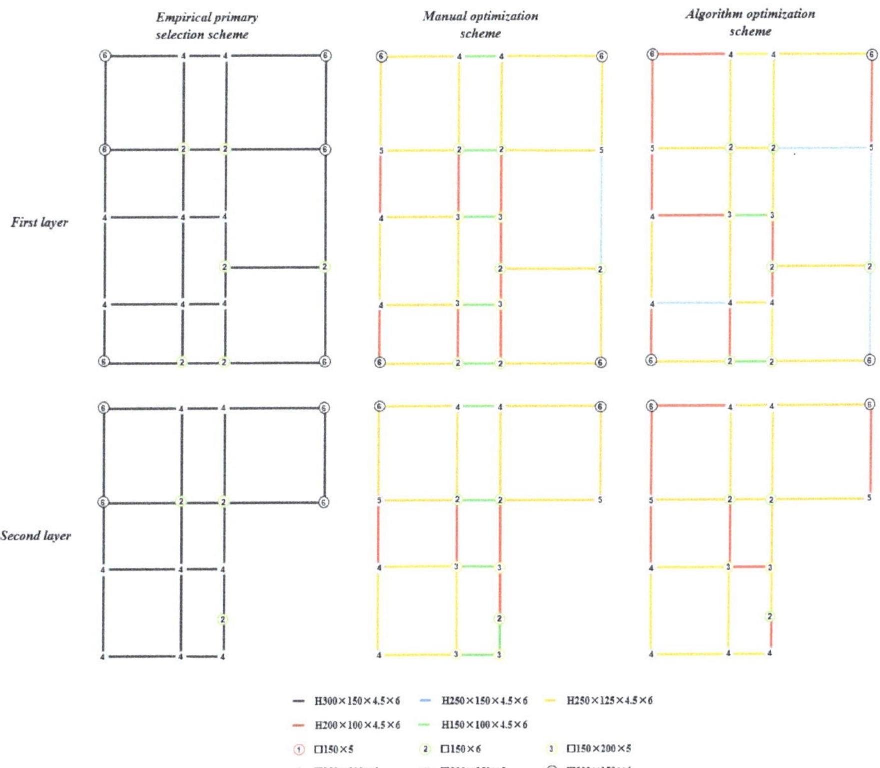

Figure 13. Comparison of structural design layout schemes of example II.

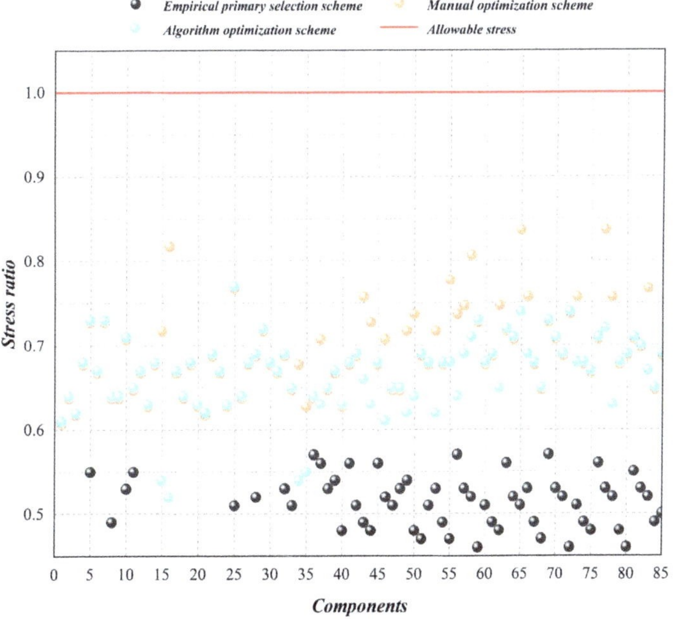

Figure 14. The stress ratio of the structural components of example II.

The optimized SATWE two-dimensional model was returned to the BIM platform to generate a BIMBase three-dimensional model. Other professional designers could conduct the detailed design based on the existing optimized structural design model to guide the detailed design and construction of rural buildings. The results showed a 100% precision rate in generating the BIMBase three-dimensional model, and both geometric and material information were correctly extracted and mapped, as demonstrated above. The process of implementing the proposed framework of example II is shown in Figure 15. The running time of the whole framework for this example is about one and a half hours, while the conventional structural layout scheme completed by experienced designers takes about 8 h, which is almost six times longer than the intelligent structural design facilitated by the proposed framework. This demonstrates the promising practical applications of the proposed framework.

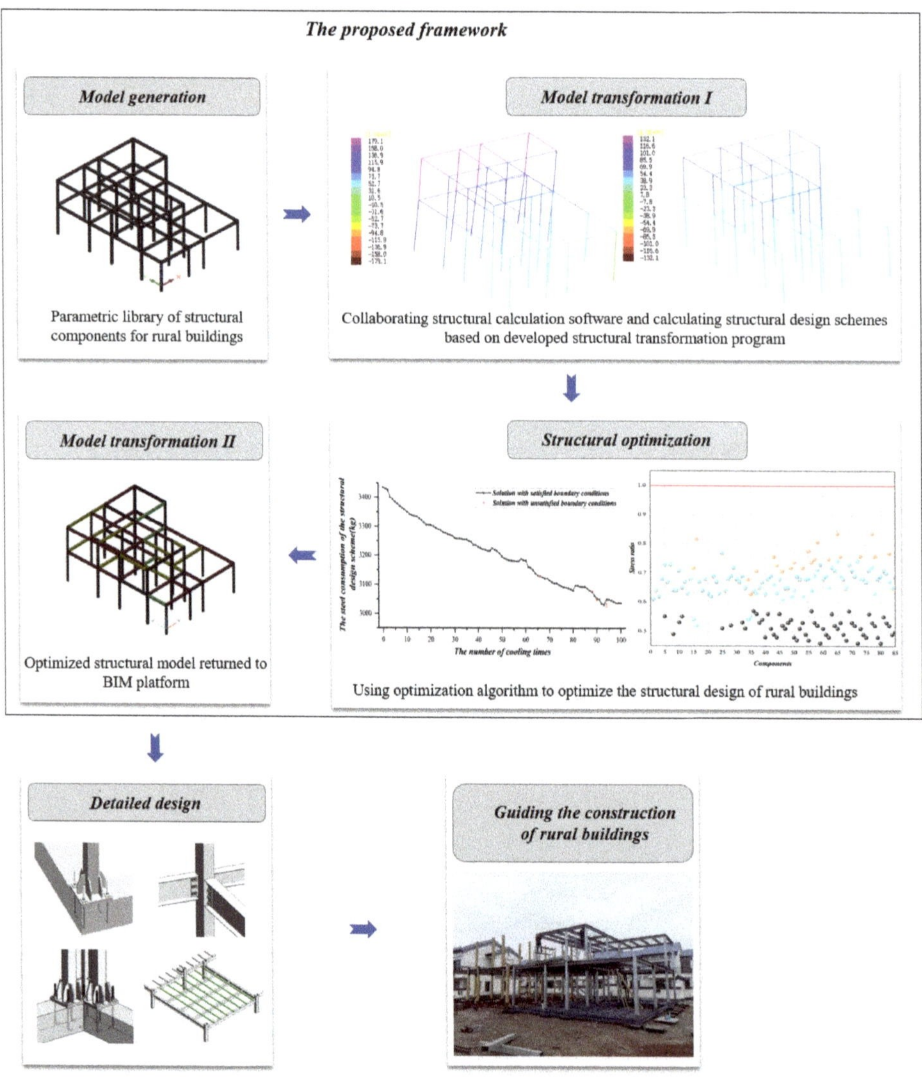

Figure 15. The process of implementing the proposed framework of example II.

6. Conclusions

The structural design of rural buildings in China is limited by economic conditions. It is not realistic to carry out accurate and detailed structural design for each rural building. While only in accordance with the experience of rural building design, its safety and economy cannot be fully verified. Therefore, this paper presents a new framework with BIM and an optimization algorithm for the automated optimum design of rural light steel frame structures. The developed framework integrates BIM technology with structural analysis applications and an improved two-stage SA algorithm to automate and simplify the structural design process of rural buildings. Two different rural light steel frame structures are used to verify the applicability and efficiency of the developed framework. Based on this study, the following conclusions are obtained:

(1) Under the guidance of the corresponding standardized component coding, the parametric library of structural components for rural buildings based on the state standard established on the BIM platform can realize the integrated management of BIM components, which is convenient for the subsequent retrieval of components and rapid modeling of structure.

(2) The model transformation method based on the component as the basic unit prevents data loss and conversion errors. It realizes the transformation of the BIMBase three-dimensional model and SATWE two-dimensional model and provides a data basis for subsequent structural intelligent optimization.

(3) An optimization method for rural light steel frame structures based on a two-stage SA algorithm is proposed. The example results show that the material consumption and key structural indexes of the algorithm optimization scheme are comparable to those of the manual optimization scheme. The proposed optimization algorithm reduces the iteration time of the simulated annealing algorithm to search for the optimal result and has good convergence and better optimization performance.

(4) The intelligent structural design reduces the computational time by six times compared to the conventional design for both example I and example II (i.e., 0.5 h versus 3 h/1.5 h versus 8 h). The proposed framework requires less adjustment and is shown to be automated, effective and efficient.

In addition, this study has some limitations for which recommendations are suggested for future work. The intelligent design method proposed in this paper still requires engineers to intervene in terms of data input and information transmission and is limited to light steel frame structures. In future research, joint connections and the seismic performance of the structure and multi-objective optimization should also be addressed.

Author Contributions: Conceptualization, T.Z. and Z.C.; methodology, T.Z.; software, H.L.; validation, Z.Y.; writing—original draft preparation, K.S.; writing—review and editing, K.S. All authors have read and agreed to the published version of the manuscript.

Funding: This research was sponsored by National Key R&D Program of China (No. 2019YFD1101005), National Construction Engineering Technology Research Center Open Foundation Project (No. BSBE2022-13) and Research and Practice Project of Higher Education Teaching Reform in Hebei Province (No. 2021GJJG244).

Institutional Review Board Statement: Not applicable.

Informed Consent Statement: Not applicable.

Data Availability Statement: Not applicable.

References

1. Wang, X.D.; Su, P.F.; Liu, J.D. Seismic performance of light steel-natural timber composite beam-column joint in low-rise buildings. *J. Struct. Eng.* **2022**, *256*, 113969. [CrossRef]
2. Christiana, W.; Erin, D. Active living environments mediate rural and non-rural differences in physical activity, active transportation, and screen time among adolescents. *Prev. Med.* **2021**, *23*, 101422. [CrossRef] [PubMed]
3. Li, Q.; Wang, Y.Q.; Ma, L.Y. Effect of sunspace and PCM louver combination on the energy saving of rural residences: Case study in a severe cold region of China, Sustain. *Energy Technol. Assess.* **2021**, *45*, 101126. [CrossRef]
4. Zhang, X.C.; Zhang, X.Q. Comparison and sensitivity analysis of embodied carbon emissions and costs associated with rural house construction in China to identify sustainable structural forms. *J. Clean. Prod.* **2021**, *293*, 126190. [CrossRef]
5. Zhou, T.; Wang, Z.X.; Wang, Y.H.; Miao, S.Y. Mechanical properties and automatic design method for rural residences with hinged frames and cold-formed thin-walled stiffened walls. *J. Build. Eng.* **2023**, *68*, 106156. [CrossRef]
6. Li, M.Y.; Zhu, E.; Wang, B. Study on the seismic performance of rural houses masonry walls with different geometries of open-hole area reduction. *Structures* **2022**, *41*, 525–540. [CrossRef]
7. Hu, X.Y.; Xiang, Y.M.; Zhang, H. Active–passive combined energy-efficient retrofit of rural residence with non-benchmarked construction: A case study in Shandong province, China. *Energy Rep.* **2021**, *7*, 1360–1373. [CrossRef]
8. Meng, Q.; Zhang, Y.; Li, Z. A review of integrated applications of BIM and related technologies in whole building life cycle. *Eng. Constr. Archit. Manag.* **2020**, *27*, 1647–1677. [CrossRef]
9. Wang, Y.; He, X.; He, J. Virtual trial assembly of steel structure based on BIM platform. *Autom. Constr.* **2022**, *141*, 104395. [CrossRef]
10. Bryde, D.; Broquetas, M.; Volm, J.M. The project benefits of building information modelling (BIM). *Int. J. Proj. Manag.* **2013**, *31*, 971–980. [CrossRef]
11. Barlish, K.; Sullivan, K. How to measure the benefits of BIM—A case study approach. *Autom. Constr.* **2012**, *24*, 149–159. [CrossRef]
12. Mangal, M.; Cheng, C.P. Automated optimization of steel reinforcement in RC building frames using building information modeling and hybrid genetic algorithm. *Autom. Constr.* **2018**, *90*, 39–57. [CrossRef]
13. Liu, C.; Zhang, F.; Zhang, H. Optimization of assembly sequence of building components based on simulated annealing genetic algorithm. *J. Alex. Eng.* **2023**, *62*, 257–268. [CrossRef]
14. Talatahari, S.; Azizi, M. Optimum design of building structures using Tribe-Interior Search Algorithm. *Structures* **2020**, *28*, 1616–1633. [CrossRef]
15. GB50017-2017; Code for Design of Steel Structures. China Architecture & Building Press: Beijing, China, 2017.
16. GB55006-2021; General Specification for Steel Structures. China Architecture & Building Press: Beijing, China, 2021.
17. Leite, F. *BIM for Design Coordination: A Virtual Design and Construction Guide for Designers, General Contractors, and MEP Subcontractors*; John Wiley & Sons: Hoboken, NJ, USA, 2019.
18. Hosseini, M.R.; Maghrebi, M.; Akbarnezhad, A. Analysis of citation networks in building information modeling research. *J. Constr. Eng. Manag.* **2018**, *144*, 04018064. [CrossRef]
19. Vilutiene, T.; Hosseini, M.R. Building information modeling (BIM) for structural engineering: A bibliometric analysis of the literature. *Adv. Civ. Eng.* **2019**, *19*, 5290690. [CrossRef]
20. Asl, M.; Zarrinmehr, S. BPOpt: A framework for BIM-based performance optimization. *Energy Build.* **2015**, *108*, 401–412.
21. Sacks, R.; Kaner, I.; Eastman, C.M.; Jeong, Y.S. The rosewood experiment—Building information modeling and interoperability for architectural precast facades. *Autom. Constr.* **2010**, *19*, 419–432. [CrossRef]
22. Ren, R.; Zhang, J. Model information checking to support interoperable BIM usage in structural analysis. In Proceedings of the International Conference on Computing in Civil Engineering 2019, Atlanta, GA, USA, 17–19 June 2019.
23. Aldegeily, M.; Zhang, J. From architectural design to structural analysis: A data-driven approach to study building information modeling (BIM) interoperability. In Proceedings of the 54th Annual ASC International Conference, Fort Collins, CO, USA, 18–21 April 2018; pp. 537–545.
24. Ramaji, I.J.; Memari, A.M. Interpretation of structural analytical models from the coordination view in building information models. *Autom. Constr.* **2018**, *90*, 117–133. [CrossRef]
25. Hamidavi, T.; Abrishami, S. Optimization of structural design by integrating genetic algorithms in the building information modelling environment. *Int. J. Civ. Eng.* **2018**, *12*, 877–882.
26. Sheikhkhoshkar, M.; Rahimian, F.P.; Kaveh, M.H. Automated planning of concrete joint layouts with 4D-BIM. *Autom. Constr.* **2019**, *107*, 102943. [CrossRef]
27. Hamidavi, T.; Abrishami, S. Towards intelligent structural design of buildings: A BIM-based solution. *J. Build. Eng.* **2020**, *32*, 101685. [CrossRef]
28. Oraee, M.; Hosseini, M.R.; Edwards, D.J. Collaboration barriers in BIM-based construction networks: A conceptual model. *Int. J. Proj. Manag.* **2019**, *37*, 839–854. [CrossRef]
29. Gan, V.J.; Wong, C.L.; Cheng, J.C. Parametric modelling and evolutionary optimization for cost-optimal and low-carbon design of high-rise reinforced concrete buildings. *Adv. Eng. Inform.* **2019**, *42*, 100962. [CrossRef]
30. Tafraout, S.; Bourahla, N.; Bourahla, Y. Automatic structural design of RC wall-slab buildings using a genetic algorithm with application in BIM environment. *Autom. Constr.* **2019**, *106*, 102901. [CrossRef]

31. Zhou, X.H.; Wang, L.F.; Liu, J.P. Automated structural design of shear wall structures based on modified genetic algorithm and prior knowledge. *Autom. Constr.* **2022**, *139*, 104318. [CrossRef]
32. Gholizadeh, S.; Shahrezaei, A.M. Optimal placement of steel plate shear walls for steel frames by bat algorithm. *Struct. Des. Tall Spec. Build.* **2015**, *24*, 1–18. [CrossRef]
33. Gholizadeh, S.; Poorhoseini, H. Seismic layout optimization of steel braced frames by an improved dolphin echolocation algorithm. *Struct. Multidiscip. Optim.* **2016**, *54*, 1011–1029. [CrossRef]
34. Kaveh, A.; Khodadadi, N. Optimal design of large-scale frames with an advanced charged system search algorithm using box-shaped sections. *Eng. Comput.* **2021**, *37*, 2521–2541. [CrossRef]
35. Xia, Y.; Langelaar, M.; Hendriks, M.A. Automated optimization-based generation and quantitative evaluation of strut-and-tie models. *Comput. Struct.* **2020**, *238*, 106297. [CrossRef]
36. Li, M.; Wong, B.C.; Liu, Y. DfMA-oriented design optimization for steel reinforcement using BIM and hybrid metaheuristic algorithms. *J. Build. Eng.* **2021**, *44*, 103310. [CrossRef]
37. Azizi, M.; Ghasemi, M.; Ejlali, R.G. Optimum design of fuzzy controller using hybrid ant lion optimizer and Jaya algorithm. *Artif. Intell. Rev.* **2020**, *53*, 1553–1584. [CrossRef]
38. Farshchin, M.; Maniat, M.; Pezeshk, S. School based optimization algorithm for design of steel frames. *Eng. Struct.* **2018**, *171*, 326–335. [CrossRef]
39. Azizi, M.; Ghasemi, M.; Talatahari, S. Optimal tuning of fuzzy parameters for structural motion control using multiverse optimizer. *Struct. Des. Tall. Spec. Build.* **2019**, *28*, e1652. [CrossRef]
40. Tort, C.; Sahin, S.; Hasançebi, O. Optimum design of steel lattice transmission line towers using simulated annealing and PLS-TOWER. *Comput. Struct.* **2017**, *179*, 75–94. [CrossRef]
41. Bettemit, O.; Sonmez, R. Hybrid Genetic Algorithm with Simulated Annealing for Resource-Constrained Project Scheduling. *J. Manag. Eng.* **2015**, *31*, 04014082. [CrossRef]
42. Liu, S.; Meng, T.; Jin, Z. Optimal Deployment of Heterogeneous Microsatellite Constellation Based on Kuhn-Munkres and Simulated Annealing Algorithms. *J. Aerosp. Eng.* **2022**, *35*, 04022090. [CrossRef]
43. Barkhordari, M.; Tehranizadeh, M. Response estimation of reinforced concrete shear walls using artificial neural network and simulated annealing algorithm. *Structures* **2021**, *34*, 1155–1168. [CrossRef]
44. Guo, J.; Yuan, W.; Dang, X. Cable force optimization of a curved cable-stayed bridge with combined simulated annealing method and cubic B-Spline interpolation curves. *Eng. Struct.* **2019**, *201*, 109813. [CrossRef]
45. Wu, K.; Soto, B.; Zhang, F. Spatio-temporal planning for tower cranes in construction projects with simulated annealing. *Autom. Constr.* **2020**, *111*, 103060. [CrossRef]

 *sustainability*

Article

Thermal Bridges Monitoring and Energy Optimization of Rural Residences in China's Cold Regions

Mingqian Guo, Yue Wu * and Xinran Miao

School of Civil Engineering, Harbin Institute of Technology, Harbin 150090, China;
18b933040@stu.hit.edu.cn (M.G.); 1183300916@stu.hit.edu.cn (X.M.)
* Correspondence: wuyue_2000@hit.edu.cn

Abstract: With the worldwide dissemination of the "green development" concept and the advancement of China's new rural construction, the sustainable development of rural residences has gained significant attention within the construction industry. This article focuses on large-scale prefabricated insulation block houses used in China's cold regions, specifically examining the case of Defa Village in Nenjiang City, Heilongjiang Province. By utilizing thermal imaging cameras, the thermal bridge parts of these houses are detected, and a finite element model is established to optimize the comprehensive heat transfer coefficient of these areas. This optimization is achieved by expanding the insulation layer and implementing low thermal bridge structures, ultimately enhancing the insulation and energy-saving efficiency of the houses. Simultaneously, an energy-saving analysis is conducted based on an optimized enclosure structure scheme, considering seven key design factors that influence building energy consumption: span, depth, clear height, and window-to-wall ratio in all four directions. Through a comprehensive experimental method, the building energy consumption is evaluated, and a scheme with optimal values is proposed. The results demonstrate that the insulation block walls and the main structures with expanded insulation layers and low thermal bridge structures are easier to construct. When compared to the original scheme, the comprehensive heat transfer coefficient of the walls is reduced by 54.82%, while that of the beams and columns is reduced by 97%. Implementing the optimal value scheme leads to a reduction of 66.83% in the building's overall energy consumption. This research provides valuable guidance for the design and construction of large-scale insulated block rural residences, revealing the substantial potential of rural residences in terms of energy-saving and emission reduction.

Keywords: sustainable architecture; rural prefabricated housing; climate analysis; structural thermal bridges; energy optimization

Citation: Guo, M.; Wu, Y.; Miao, X. Thermal Bridges Monitoring and Energy Optimization of Rural Residences in China's Cold Regions. *Sustainability* **2023**, *15*, 11015. https://doi.org/10.3390/su151411015

Academic Editor: Davide Settembre-Blundo

Received: 30 May 2023
Revised: 4 July 2023
Accepted: 10 July 2023
Published: 13 July 2023

1. Introduction

With the advancement of society and the growth of the economy, the principles of green and eco-friendly design, energy efficiency, and low-carbon construction have become crucial factors influencing architectural design. Prefabricated housing, characterized by its efficiency and high level of industrialization, has progressively assumed a significant role in the construction industry. In China, the rural population constitutes 40% of the total population, while rural residences account for 38% of the national construction area. Furthermore, rural housing contributes to 24% of the overall energy consumption in the country [1]. As the new rural development initiatives gain momentum, the demand for rural housing will continue to rise, necessitating more stringent energy-saving standards. Therefore, prefabricated rural housing faces unprecedented opportunities for development as well as new challenges.

This article delves into the thermal bridge optimization and energy optimization of large insulated block rural residences in the cold regions of China. Using the Northeast region as a case study is characterized by a climate with long and extremely cold winters, which poses significant challenges for architectural design. Analysis reveals that heat loss from the building envelope is substantial, accounting for over 50% of the total heat loss. Thermal bridges, in particular, tend to be concentrated areas of heat loss. Therefore, enhancing the insulation performance of the building envelope and mitigating the impact of thermal bridges are essential strategies to achieve energy efficiency in construction. Presently, many rural dwellings in the Northeast have adopted simplistic construction techniques. These measures encompass the utilization of higher-quality insulation materials, double-glazed and energy-efficient windows, sloping roofs, and wall air gaps, as well as the application of polyurethane materials for sealing. These initiatives have effectively improved insulation performance and elevated heating levels [1]. Nevertheless, despite meeting indoor comfort requirements, they still result in significant energy resource consumption. As a result, this article focuses on analyzing insulation design for the building envelope and optimizing the energy source for heating systems.

Currently, there have been significant advancements in both theoretical and practical research regarding insulation and energy-saving in rural residences. These achievements primarily concentrate on two key areas:

The first aspect pertains to the investigation of thermal performance within building envelope structures. This involves the identification of thermal bridge areas within the components of the building envelope based on their specific construction forms. Furthermore, it entails conducting measurements and analyses to assess the thermal performance, heat loss, and effects of thermal bridging under specific environmental conditions. Ultimately, the aim is to enhance the construction forms to mitigate the impact of thermal bridging and minimize heat loss. For instance, Laura et al. [2] utilized numerical simulation methods to analyze the two-dimensional heat transfer characteristics at the junction of walls and floors, as well as the L-shaped corners within a building. Ascione [3] carried out measurements to evaluate temperature distribution and heat flow fields within masonry structures. Through the analysis of the obtained results, they were able to identify areas influenced by thermal bridging based on changes in temperature and heat flow. They also investigated the variations in temperature distribution and heat flow fields in masonry structures under two-dimensional transient conditions while considering different insulation thicknesses and mortar joint thicknesses. Baldinelli et al. [4,5] employed an enhancement algorithm to optimize thermal images of three thermal bridges present in a concrete structure. These optimized images were then validated against measured temperature fields, leading to the identification of the thermal bridge areas. Zalewski [6–8] analyzed steel structure walls, performing tests to observe the dynamic changes in heat flow and temperature at various positions along the walls. By comparing the results obtained from measurements and simulations, they assessed the impact of heat loss.

The second aspect involves the analysis and optimization of energy consumption levels for the entire building over a specific period. This encompasses calculating the heating load, cooling load, electrical usage, domestic hot water, and other energy consumption levels under specific climatic conditions. The primary objective is typically to achieve energy efficiency while ensuring comfort levels. The optimization design of the building focuses on architectural geometric parameters and the thermal performance of the building envelope. For instance, Mithraratne et al. [9] conducted a comprehensive analysis of a group of individual buildings in New Zealand, with a specific emphasis on energy consumption during the building maintenance phase. They compared the energy consumption costs of buildings with different envelope structures. Sartori et al. [10] examined 60 case studies from the literature and determined that operational energy constitutes the majority of the total energy demand throughout the building lifecycle. Fang et al. [11–13] investigated the relationship between design parameters and performance in buildings, including variables such as geometric shape, window and skylight sizes, and positions. They employed mathematical methods to explore the impact of these variables on daylighting performance and energy efficiency, ultimately proposing optimization strategies. G. Verbeeck [14] assessed the economic feasibility of five types of insulation measures, glass selection, and renewable energy in Belgian residential buildings. They proposed a hierarchical approach to energy-saving measures. Payyanapotta et al. [15] integrated lightweight building systems with passive house concepts to achieve energy efficiency and thermal comfort. They identified limitations in lightweight steel systems, encompassing issues such as overheating, airtightness, thermal bridging, moisture transfer, and solar control.

The design of building envelope structures and the optimization of overall energy consumption levels play a crucial role in achieving energy-saving and emission-reduction goals in buildings. However, existing research has predominantly focused on concrete and steel structures, with limited investigations into large-scale thermal insulation blocks. This study addresses this gap by conducting thorough tests and analyses on the thermal performance and energy consumption levels of large-scale thermal insulation block systems in severe cold climate environments. Our research aims to optimize the building envelope structures and architectural forms to achieve a comprehensive wall heat transfer coefficient of 0.24 W/(m^2·K) and a comprehensive building energy consumption value of 43.8 kWh/(m^2·a). The outcomes of this research can serve as a valuable reference for the design of large-scale thermal insulation block buildings in severe cold regions, facilitating the achievement of ultra-low energy consumption standards.

2. Materials and Methods

2.1. Project Overview

2.1.1. Site Selection and Layout

This study chooses a single-story rural residence constructed with large-scale thermal insulation blocks, situated in Defa Village, Nenjiang City, Heilongjiang Province, China (Figure 1), as the target for on-site measurements and analysis of energy consumption. The enhancements made to the thermal performance of the building envelope structure and the energy-saving optimization measures discussed in the paper are all derived from this rural residence serving as the reference model.

 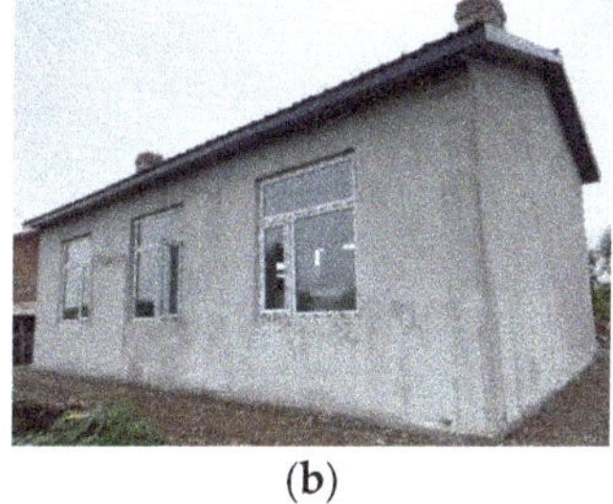

(a)   (b)

Figure 1. Benchmark building in real life. (**a**) South elevation of the building; (**b**) North elevation of the building.

For testing and analysis, the selected benchmark building in this study exhibits characteristics of a small-scale and uncomplicated spatial composition. The residential exterior is unobstructed by any other buildings, and the surrounding environment is open. The layout of the residence follows the traditional "one bright, two dark, three open spaces" arrangement commonly seen in the northeast region. It comprises two bedrooms and one living room, all interconnected (Figure 2). The house has a depth of 5 m, a span of 10 m, an interior clear height of 2.8 m, and a total floor area of 50 square meters. The south-facing windows come in two sizes: 2200 mm × 700 mm and 600 mm × 1700 mm, resulting in a window-to-wall ratio of 0.34. The north-facing windows measure 1500 mm × 1600 mm, with a window-to-wall ratio of 0.26.

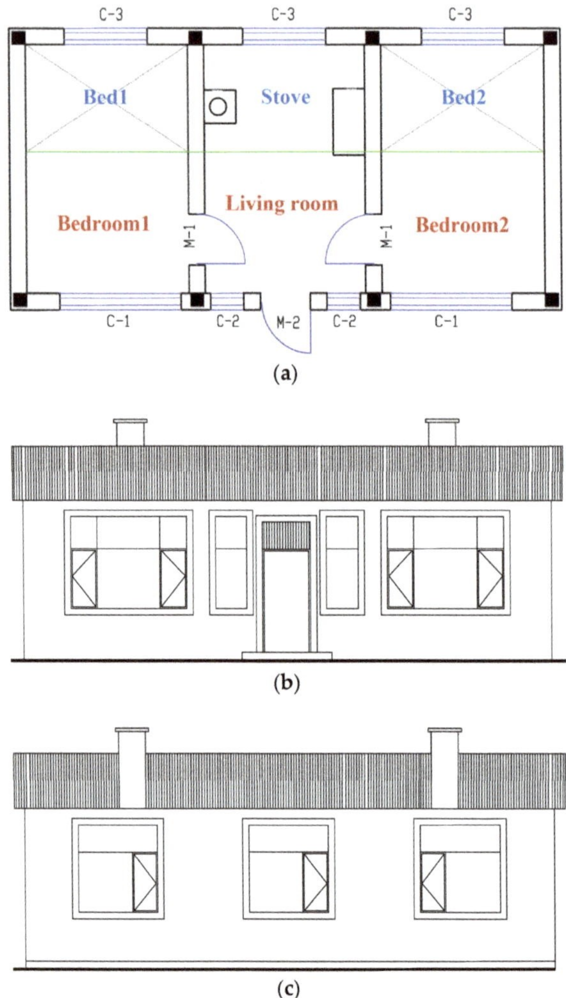

Figure 2. Design drawings of the benchmark building. (**a**) Architectural floor plan; (**b**) South elevation; (**c**) North elevation.

2.1.2. Materials of Walls

The building envelope structure of the benchmark building is composed of large-scale thermal insulation blocks. These blocks comprise expanded polystyrene foam boards (EPS boards), a protective layer of uniformly coated polymer insulation mortar, and a thermal bridge insulation layer made of extruded polystyrene foam boards (XPS boards). They are constructed as composite insulation modules with integrated construction grooves (Figure 3). The thermal insulation blocks have dimensions of 600 mm × 600 mm × 300 mm. The detailed construction sequence from the interior to the exterior includes a 10 mm layer of polymer insulation mortar, followed by a 230 mm layer of EPS, another 10 mm layer of polymer insulation mortar, and, finally, a 50 mm layer of XPS. The construction grooves surrounding the blocks have a depth of 30 mm. The exterior wall finish consists of a detailed construction sequence from the interior to the exterior: a 45 mm three-coat two-mesh system comprising 15 mm of phenolic board adhesive, mesh fabric, another 15 mm of phenolic board adhesive, additional mesh fabric, and a final 15 mm layer of cement mortar.

This is followed by a 2 mm sealer and a 2 mm layer of exterior wall latex paint. The thermal parameters of the materials can be found in Table 1.

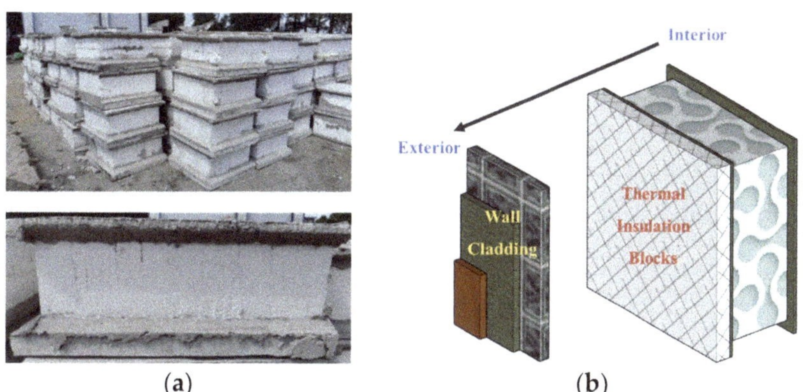

Figure 3. Thermal insulation block construction diagram. (**a**) Photograph. (**b**) Illustration.

Table 1. Thermal performance parameters of building materials.

Name	Density (kg/m^3)	Thermal Conductivity (W/m·K)	Specific Heat Capacity (J/Kg·K)	Heat Storage Coefficient W/(m^2·K)
XPS	30	0.03	4455	0.54
EPS	20	0.04	1380	0.29
Thermal insulation mortar	800	0.29	1050	4.44
Phenolic foam adhesive	350	0.07	1380	1.57
Cement mortar	1800	0.93	1050	11.37
C20	2300	1.51	920	15.24
C30	2420	1.62	970	15

2.1.3. Construction Methods

The thermal insulation blocks used in the benchmark building incorporate EPS and XPS as insulation core materials. These blocks are joined together using insulation mortar and feature 30 mm construction grooves. During block assembly, square holes measuring 60 mm by 230 mm are formed between the blocks. Horizontal and vertical HPB300 steel bars with a diameter of 6 mm are inserted within these square holes created via the module grooves. Subsequently, C20 concrete is poured to finalize the block assembly, resulting in a concrete mesh structure inside the wall that enhances its rigidity (Figure 4). Once the wall construction is complete, formwork is installed, and structural columns are cast at the L-shaped corners. Additionally, formwork is employed at the top of the wall to cast the ring beam (Figure 5). This construction method bears resemblance to traditional masonry techniques while incorporating rural construction practices and improved efficiency. It offers several advantages, including superior insulation performance, lightweight characteristics, and ease of construction.

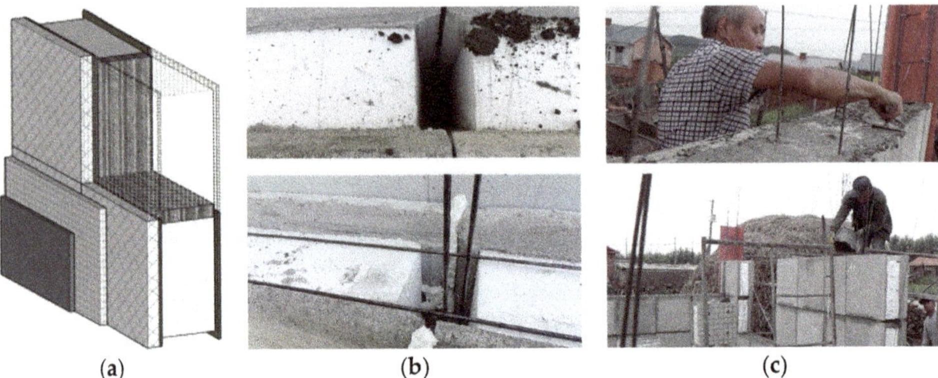

Figure 4. On-site construction of the wall. (**a**) Construction diagram of block assembly; (**b**) steel reinforcement placement within construction grooves; (**c**) pouring concrete into construction grooves.

Figure 5. On-site construction of beams and columns. (**a**) Pouring of columns. (**b**) Pouring of beams.

2.2. Methods for Environmental Analysis

Before conducting tests and analyses on building envelope structures and optimizing building energy consumption, it is essential to assess the climate and environmental conditions at the location of the reference building. This study focuses on Nenjiang County in Heihe City, Heilongjiang Province, situated in the northeastern region of China, which falls within the severe cold climate zone. The analysis was based on globally representative meteorological year data (specifically for the Nenjiang region) provided by the National Renewable Energy Laboratory of the U.S. Department of Energy. Using the meteorological data analysis model developed on the Grasshopper platform (Figure 6), statistical analysis methods were employed to examine the annual temperature, humidity, and wind direction, as well as the time range that satisfies the comfort requirements for annual temperature, humidity, and wind speed in Nenjiang City. This analysis allowed for the extraction of the fundamental climate characteristics of the region and the calculation of boundary conditions, facilitating a preliminary summary of sustainable measures that can be incorporated into the architectural design of the region.

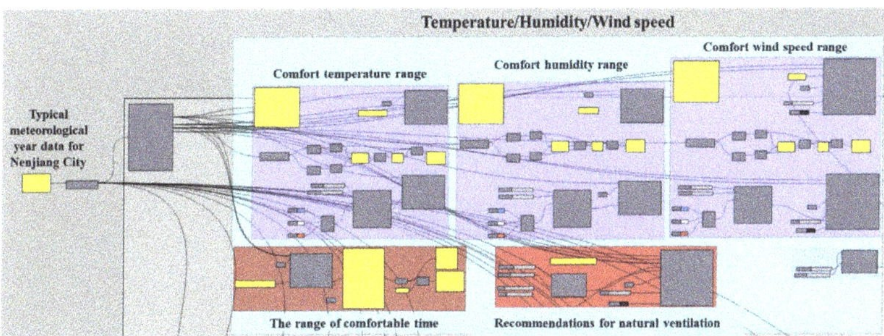

Figure 6. Meteorological data analysis model.

2.3. On-Site Methods for Identifying Thermal Bridges in Building Envelope Structures

Given the presence of residential occupants in the building being examined, this study utilizes the infrared thermography method to identify thermal bridges and capture the surface temperatures of the walls (Figure 7). These measurements are conducted in a manner that does not disturb the residents' daily routines.

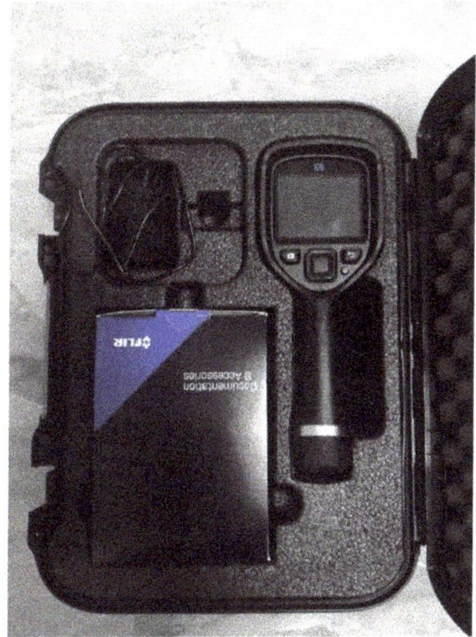

Figure 7. Infrared thermography camera.

A thermal bridge refers to an area within a building where thermal insulation performance is weakened, often caused by the uneven thermal conductivity of materials or insulation defects. Thermal bridges can lead to energy loss and waste, thereby diminishing the overall energy efficiency of the building.

The FLIR-E6-XT infrared thermal imager was utilized for data collection in this study. Data were gathered from the northwest wall of a reference building situated in Defa Village, Nenjiang City, Heilongjiang Province, China. The testing was conducted on 22 December 2020 at 11:35 a.m. At that time, the outdoor temperature was recorded as -13.0 °C, with a humidity level of 23.0% and wind speed ranging from 0.3 to 1.5 m/s. Inside the testing room, the temperature measured was 22.4 °C, with a humidity level of 49.3% and an indoor wind speed of 0.1 m/s. The captured thermal images were directly stored in the thermal imager and later analyzed and temperature-extracted using the accompanying FLIR-E6-XT-Tools software (version 6.0, FLIR Systems, Inc., Wilsonville, OR, USA).

2.4. Numerical Simulation Methods for Evaluating the Thermal Performance of Building Envelope Structures

Based on the results of climate analysis and thermal bridge identification in building envelope structures, the boundary conditions and analysis models for finite element analysis of building thermal bridges are determined. This study aims to utilize steady-state heat transfer calculation methods and utilize ABAQUS-6.14 finite element software to analyze and compare the heat flow fields of building thermal bridges at various locations and with different constructions. The finite element analysis was performed using ABAQUS/Standard (version 6.14, SIMULIA, Dassault Systèmes, Vélizy-Villacoublay, France).

The calculation of heat flux density follows the principles of steady-state heat transfer in solids, assuming a temperature difference between the two sides of the building wall, where heat transfers from higher to lower temperatures. This process encompasses heat conduction, convective heat transfer, and radiative heat transfer. According to the second law of thermodynamics, the heat flow passing through the wall is equal to the heat flow passing through the surface, which can be expressed as the following equation:

$$q = -\lambda \cdot [(\partial t)/(\partial n)] = (t_e - \theta_e) \cdot (\alpha_c + \alpha_r), \tag{1}$$

q is heat flux (W/m^2); λ is thermal conductivity (W/m·K); α_c is convective heat transfer coefficient (W/m^2·K); α_r is radiative heat transfer coefficient (W/m^2·K).

Given that the analyzed objects consist of complex structures with non-homogeneous and multi-layered configurations, it is essential to consider the impact of thermal bridge areas and conduct a quantitative evaluation of the overall thermal insulation performance of the components. To achieve this, the method outlined in the "Code for Thermal Design of Civil Buildings" (GB50176-2016) [16] will be employed to calculate the thermal resistance of the building envelope structures. By extracting the average heat flux density from the outer surface of the analysis model, the overall thermal resistance of the structure will be determined, enabling the subsequent calculation of the overall heat transfer coefficient. The thermal insulation performance of the analyzed objects will then be assessed and compared based on the overall heat transfer coefficient. The indoor calculation temperature is set at 18 °C, while the outdoor calculation temperature is set at −25 °C.

$$R = [(\Delta T)/q_w] - R_i - R_e \tag{2}$$

$$K = 1/R \tag{3}$$

R is overall thermal resistance of the structure (m^2·K/W); K is overall heat transfer coefficient of the structure (W/m^2·K); ΔT is indoor–outdoor temperature difference (K); q_w is average heat flux density (W/m^2); R_i is internal surface convective resistance [0.11 (m^2·K/W)]; R_e is external surface convective resistance [0.04 (m^2·K/W)].

2.5. Building Energy Analysis and Energy-Saving Optimization

2.5.1. Comprehensive Calculation of Building Energy Consumption

In this study, the comprehensive building energy consumption of the target building is simulated and calculated using the DesignBuilder-6.1 software (version 6.1, Designbuilder Software Limited, Gloucestershire, UK). The term "comprehensive building energy consumption" refers to the total energy consumed by the building over a specific period, usually one year, encompassing heating, cooling, electricity, and fuel consumption. Design-Builder is a specialized building energy simulation software that employs EnergyPlus as its computational engine and provides a wide range of energy simulation calculation modules. To facilitate the comparison of energy performance among different scenarios, the annual specific energy consumption (kwh/m^2·a) is adopted as the benchmark for comparison.

2.5.2. Single-Parameter Analysis of Building Energy Consumption

After conducting optimization of the building envelope structure, this study examines the geometric design factors that impact building energy consumption. These factors encompass building orientation, window-to-wall ratios in the four primary directions, building aspect ratio, depth, net height, and other relevant design considerations. By employing single-parameter analysis, this study investigates the correlation between these design factors and building energy consumption. Specifically, the research focuses on examining the influence of design parameters on the building's winter heating load, taking into account solar radiation gains. The goal is to explore the relationship between design parameters and energy consumption and establish an initial range of optimal variable values for each parameter, considering their individual effects. The initial value for each design parameter can be found in Table 2.

Table 2. Range of design parameter values.

Parameter Name	Range of Parameter Values
Building Orientation	0~3.1 rad
Window-to-Wall Ratio	0~0.9
Building Depth	6~20 m
Building Span	
Building Height	2.8~4.4 m

This study utilizes the integrated Grasshopper platform (Figure 8) for the parametric analysis process. Unlike the comprehensive calculation of building energy consumption, the primary focus of the parametric analysis is to investigate the relationship between design elements and design objectives. Consequently, the residential building is simplified into a box model, where the roof, walls, floor, and glass are identified as vital components of the building envelope system. The material properties of these components are defined based on real-world conditions. The windows are chosen with a heat transfer coefficient of 2 W/m^2·K, utilizing triple glazing. For insulation, the roof incorporates lightweight composite gray limestone panels with a thickness of 150 mm and a thermal conductivity of 0.022 W/m·K. The floor is composed of a 100 mm crushed stone concrete layer and a 100 mm sand bedding layer. The occupancy rate, calculation temperature, and equipment efficiency are set to ideal values, while the indoor temperature is constrained within the range of 18 °C to 25 °C. The equipment's coefficient of performance (COP) is set to 1, and the operation time is 24 h.

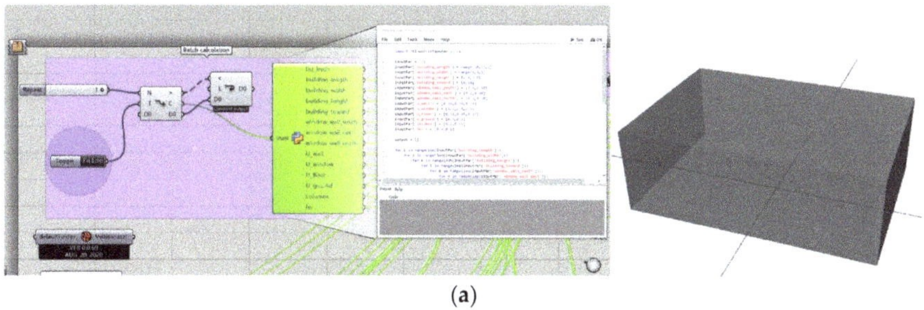

(a)

Figure 8. *Cont.*

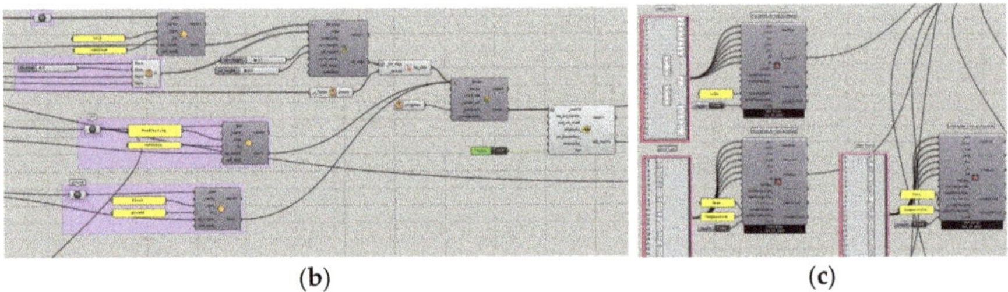

(b)   **(c)**

Figure 8. Grasshopper computational model. (**a**) Proprietary parametric modeling plugin and simplified analysis model; (**b**) parametric modeling of the building envelope structure; (**c**) configuration of additional boundary conditions.

2.5.3. Analysis of Building Energy Consumption Sensitivity and Optimization of Energy-Saving Measures

Based on the results of the single-parameter analysis, the value ranges for the key design parameters are determined, and the subset with lower energy consumption levels under each factor is chosen as the range of values. These key parameters include window-to-wall ratios in the four cardinal directions (N/W/S/E), as well as the length (X), width (Y), and height (Z) of the building envelope structure (Table 3).

Table 3. Design parameter value range.

Name	X/m	Y/m	Z/m	N	W	S	E
Range	7–10	6–9	3–4	0–0.2	0–0.3	0.3–0.6	0–0.3

A computational model is established using the Grasshopper analysis platform to calculate the cooling load, heating load, total load, and solar gains (Figure 9) for the target building. Comprehensive simulation calculations are conducted for all possible combinations of the seven variables, and the results are recorded in the Design-Explorer visualization platform.

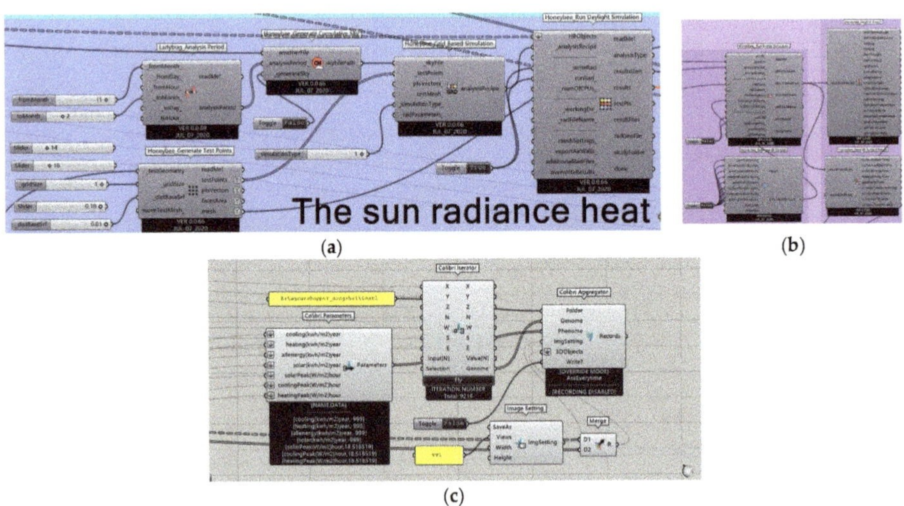

Figure 9. Grasshopper computational model. (**a**) Solar radiation heat gain calculation model; (**b**) Building load calculation model. (**c**) The Design-Explorer visualization platform for capturing and documenting models.

Once all simulation calculations are completed, this study will produce operational scenarios for all variables and various combinations of values. The Pearson correlation analysis method will be used to examine the correlation between each variable, including building aspect ratio, depth, net height, and window-to-wall ratios in the four cardinal directions, and each calculation objective, including cooling load, heating load, total load, and solar heat gain. This analysis aims to provide a comprehensive assessment of the strength and variability of the influence of different design elements on building energy consumption.

The Pearson correlation analysis method employed in this study is based on the Pearson correlation coefficient, which is a statistical measure that quantifies the strength and direction of the relationship between two variables. Correlation analysis utilizes covariance to evaluate the overall relationship between the variables and determine if they demonstrate synchronous changes, indicating whether they increase or decrease simultaneously. To mitigate the impact of units, the covariance is standardized, resulting in the correlation coefficient. This standardization enables comparisons and interpretations of the correlation independent of the specific units used.

The Pearson correlation coefficient is a widely used statistical measure for evaluating the strength and direction of the relationship between two variables. It falls within the range of −1 to +1, with the following interpretations:

- A coefficient close to +1 indicates a strong positive correlation, suggesting that the two variables change in the same direction, either increasing or decreasing simultaneously;
- A coefficient close to 0 signifies no correlation, indicating that there is no apparent relationship between the two variables;
- A coefficient close to −1 signifies a strong negative correlation, indicating that when one variable increases, the other variable decreases.

Subsequently, the correlation between different design elements and building energy consumption is assessed and ranked. Among all the operational scenarios, the optimal design solution is chosen based on the objective of achieving the lowest energy consumption.

3. Results

3.1. Climate and Environmental Characteristics

The Nenjiang region experiences a maximum annual temperature of 36 °C, with the lowest temperature dropping to −38.4 °C (Figure 10). In the driest period of winter, the relative humidity reaches a low of 6%. However, for the majority of the time, the relative humidity stays around 65% (Figure 11). Additionally, the maximum wind speed during winter can reach 15 m/s (Figure 12).

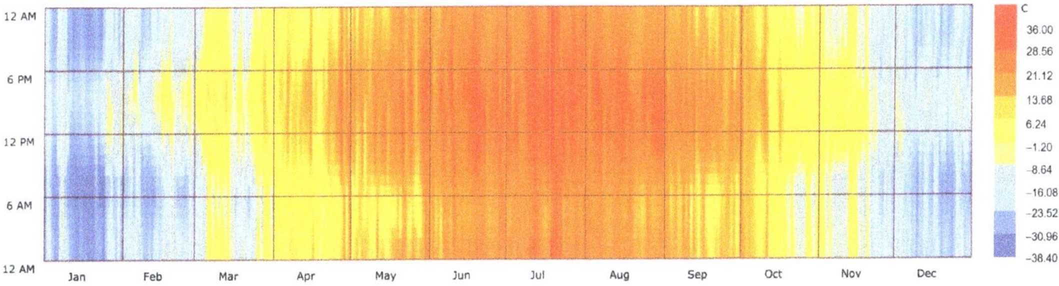

Figure 10. Annual temperature distribution map of Nenjiang City.

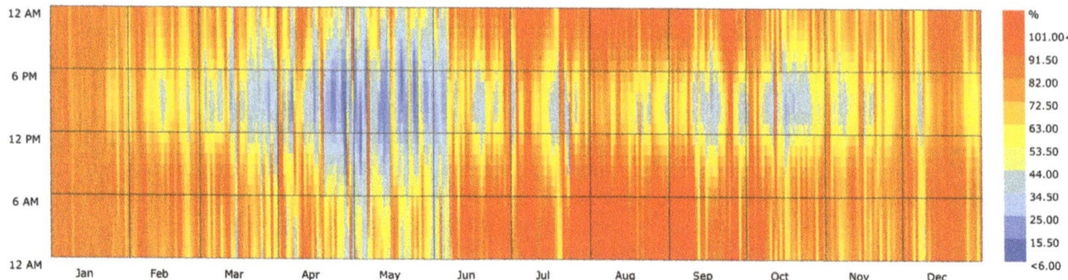

Figure 11. Annual humidity distribution map of Nenjiang City.

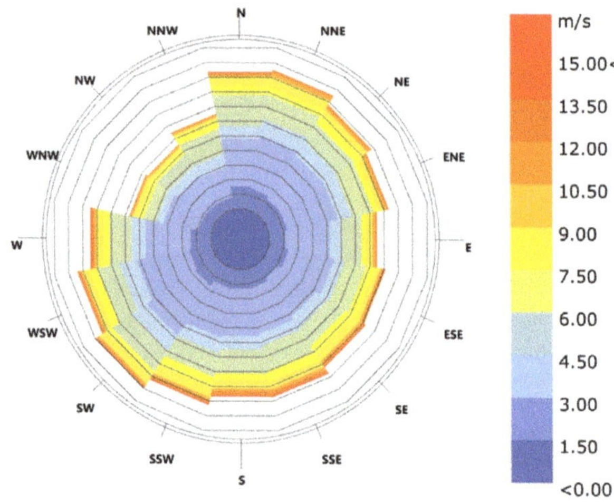

Figure 12. Annual wind speed distribution in Nenjiang City.

An analysis of the climate comfort in the Nenjiang region reveals significant findings. The recommended temperature range for human comfort is between 15 °C and 25 °C, while the optimal range for relative humidity falls between 45% and 65%. Wind speed is considered comfortable below 5 m/s. The results of this analysis have been presented visually and summarized in Table 4. It is noteworthy that the time spent in outdoor temperatures within the comfort range accounts for only 22.4% of the entire year, with no comfort time during the winter season (Figure 13). The duration within the comfortable range for relative humidity amounts to 26.2% of the year (Figure 14). Moreover, the period spent within the comfortable range for wind speed constitutes 59.9% of the year (Figure 15). Interestingly, the overlap of temperature, humidity, and wind speed falling within the comfortable ranges is merely 3.2% of the entire year, with no comfort time during the winter season.

Table 4. Outdoor environmental comfort time analysis table for Nenjiang City.

	Min-T	Max-T	Comfort	Min-H	Max-H	Comfort	Min-Wind	Max-Wind	Comfort	Overall Comfort
All year	−38.4 °C	36 °C	22.4%	6%	100%	26.2%	0 m/s	12 m/s	59.9%	3.2%
Winter	−38.4 °C	2.2 °C	0%	26%	96%	31.4%	0 m/s	12 m/s	28.9%	0%
Summer	4.3 °C	36 °C	76.3%	19%	100%	20.6%	0 m/s	12 m/s	29.8%	74.9%

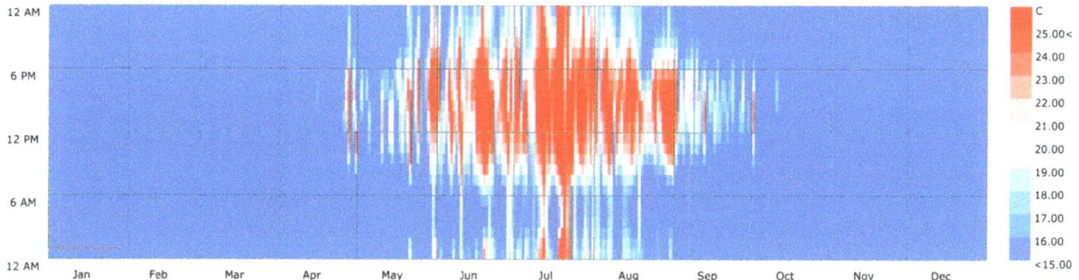

Figure 13. Annual distribution of comfortable temperatures in Nenjiang City.

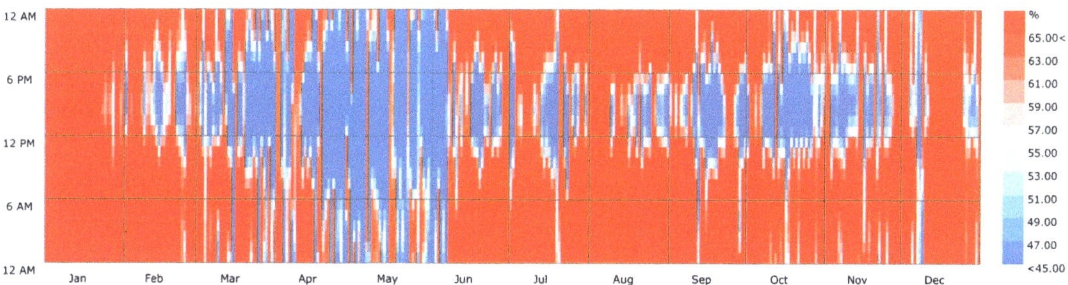

Figure 14. Annual distribution of comfortable humidity in Nenjiang City.

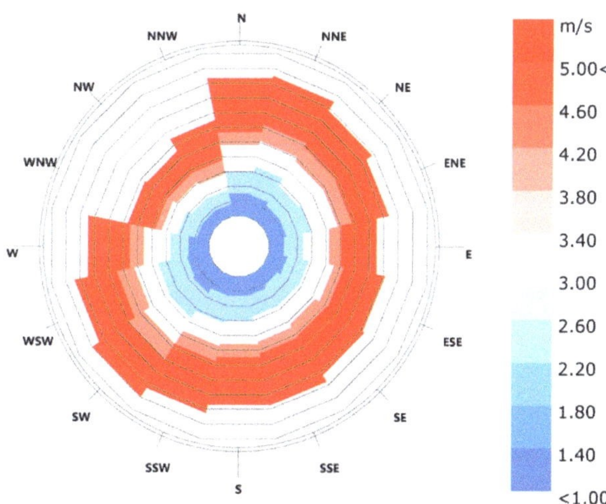

Figure 15. Annual distribution of comfortable wind speed in Nenjiang City.

The data indicate that the region's uncomfortable environment is primarily due to the combination of low temperatures and strong winds experienced during winter. Therefore, ensuring high-quality insulation in housing is a critical factor to be taken into account during architectural design to mitigate these challenges. Furthermore, considering the region's favorable climate conditions in summer, it is advisable to minimize the reliance on cooling equipment and instead promote the utilization of natural ventilation from the southwest and northeast directions as an effective means to achieve optimal comfort levels.

3.2. Thermal Bridge Identification in Buildings

During the test, the outdoor temperature was −13 °C, and the indoor temperature was 21 °C. To mitigate the impact of solar radiation, the test was conducted at the northwest corner of the building.

The thermal imaging conducted at the northwest corner provides a clear visualization of the temperature distribution within the interior wall of the building envelope (Figure 16), offering insights into the wall composition and its relationship with the main structure. The analysis of the results reveals that the insulation blocks exhibit a main body temperature of approximately 12 °C, while the temperature at the joints of the blocks is around 7.6 °C. Furthermore, the beam positions exhibit temperatures of approximately 6.8 °C, the column positions record temperatures of around 5.2 °C, and the intersection of beams and columns shows temperatures of approximately 3.7 °C. These findings highlight significant temperature variations among different structural positions, with lower temperatures indicating higher heat loss and inferior insulation performance. Moreover, larger temperature differentials indicate greater disparities in insulation effectiveness across various locations. Even in walls with an overall higher insulation performance, notable temperature differences persist across different positions (Figure 17). When considering the interior as a whole, the maximum temperature difference on the inner wall can reach up to 11.3 °C, underscoring the non-uniformity of the building envelope's thermal performance. This non-uniformity directs the flow of heat towards areas with lower insulation performance, exacerbating heat loss and leading to the formation of thermal bridges.

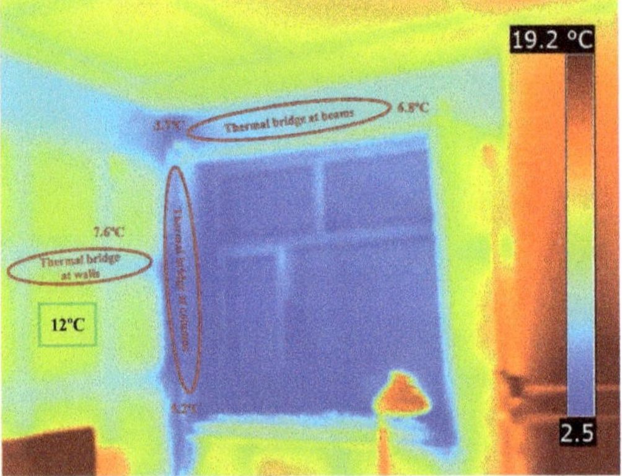

Figure 16. Thermal imaging of the benchmark building's northwest position.

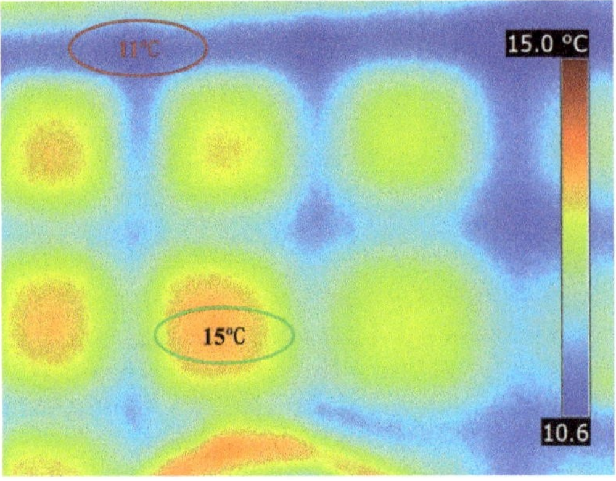

Figure 17. Thermal imaging of the west side wall of the benchmark building.

Based on the aforementioned analysis, it can be deduced that the benchmark building exhibits three primary types of thermal bridges: the joint thermal bridge generated via the concrete sections between blocks (Figure 18a), the thermal bridge caused by concrete columns (Figure 18b), and the thermal bridge resulting from concrete beams (Figure 18c). The temperature differential between the surfaces of these three types of thermal bridges and the building envelope ranges from 4 °C to 11.3 °C. By creating structural models at the locations of the thermal bridges, it can be concluded that the primary cause of their formation is the considerably lower thermal insulation performance of concrete in comparison to the insulation materials employed in the benchmark building, in addition to the substantial volume of concrete used.

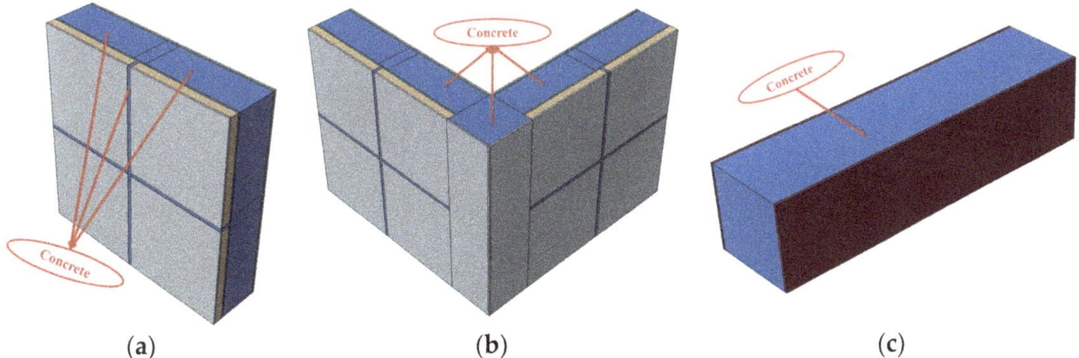

Figure 18. Construction of different thermal bridges. (**a**) Thermal bridge at block joints; (**b**) thermal bridge at column positions; (**c**) thermal bridge at beam positions.

3.3. Analysis and Optimization of Building Thermal Bridges

Calculation analysis was performed on the three thermal bridges of the benchmark building, and the comprehensive heat transfer coefficient was determined using the calculation method outlined in Section 2.4 (Table 5). The column position exhibits the poorest thermal insulation performance, whereas the beam position shows better insulation performance compared to the column position. The wall section demonstrates the highest insulation performance, aligning with the findings from the thermal imaging analysis. Notably, there is a significant disparity in heat loss between the joint of the blocks and the main body of the blocks, indicating a considerable loss of heat energy on a different scale (Figure 19a). At the column position, a substantial heat loss is observed compared to the surrounding building envelope structure (Figure 19b). Conversely, in the case of the beam position, the presence of a single concrete beam facilitates the smooth dissipation of relatively large heat flow through the beam (Figure 19c).

Table 5. The comprehensive heat transfer coefficients of various thermal bridge sections.

Name	Wall	Column	Beam
Value-W/(m^2·K)	0.54	10.24	7.25

To further enhance the insulation performance, structural modifications have been implemented for the three types of thermal bridges.

To mitigate the thermal bridge problem at the block joints, structural enhancements have been introduced during the block assembly phase. By incorporating XPS insulation strips and utilizing spray polyurethane foam for sealing, heat loss has been effectively reduced. For the horizontal joints of the walls, XPS insulation strips measuring 600 mm × 40 mm × 20 mm have been implemented (Figure 20a). Likewise, for the vertical joints of the walls, XPS insulation strips measuring 640 mm × 40 mm × 20 mm have been employed (Figure 20b).

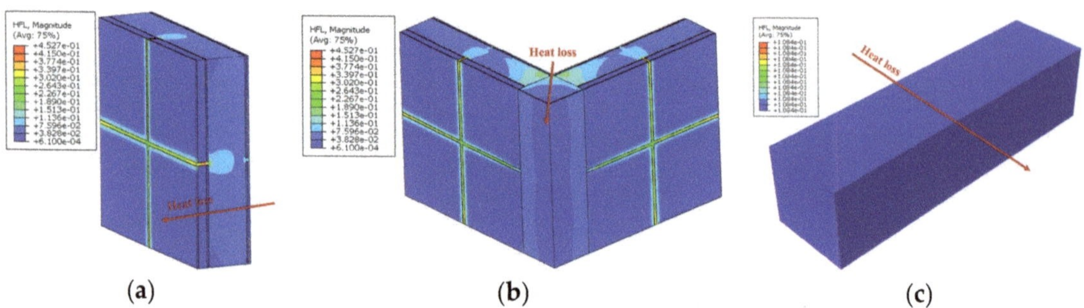

Figure 19. Thermal flow cloud maps of different thermal bridges. (**a**) Thermal flow cloud map at the joint of blocks; (**b**) thermal flow cloud map at the column; (**c**) thermal flow cloud map at the beam.

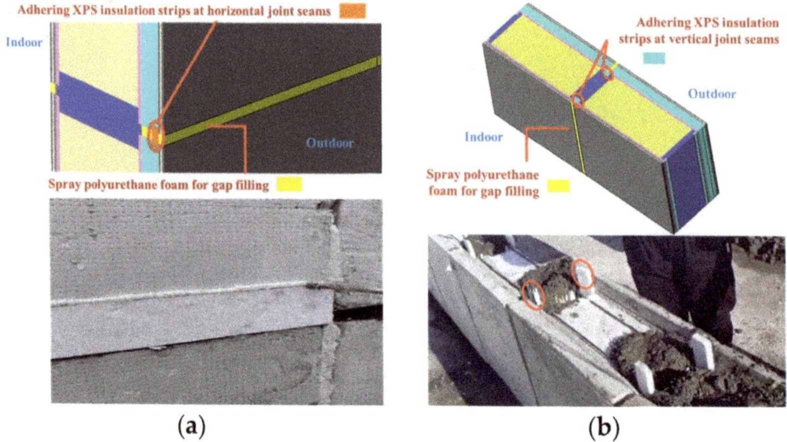

Figure 20. Mitigation techniques for thermal bridges at block joints. (**a**) Strategies for addressing horizontal joints in walls; (**b**) strategies for addressing vertical joints in walls.

To address the thermal bridge issue at the beam position, several methods have been implemented. Firstly, galvanized steel pipes with a thickness of 1.5 mm and a cross-section of 20 mm × 40 mm have been utilized to replace the longitudinal reinforcement of the beam (Figure 21a). This substitution helps minimize heat transfer. Secondly, XPS boards with a uniform application of 75 mm thick magnesium oxychloride cement on the outer surface have been employed as removable formwork (Figure 21b). These XPS boards act as insulation and provide support during the construction process. They are firmly secured on both sides of the beam reinforcement cage using self-tapping screws. To further enhance the insulation performance, XPS blocks measuring 40 mm × 40 mm × 20 mm have been strategically placed at the anchor points of the self-tapping screws (Figure 22). This additional insulation layer helps reduce heat loss. Finally, concrete is directly poured between the removable formwork, ensuring proper structural integrity and achieving the desired shape.

To mitigate the thermal bridge problem at the column position, we employ galvanized steel pipes with matching specifications as alternatives to the longitudinal reinforcement of the columns. Taking a corner column as an illustration, we utilize removable formwork and insulation blocks of identical specifications. These components are firmly attached to the outer side of the column reinforcement cage using self-tapping screws (Figure 23).

Ultimately, concrete is poured directly between the removable formwork to attain the desired structural configuration.

(**a**) (**b**)

Figure 21. Enhanced reinforcement cage and removable formwork. (**a**) Improved reinforcement cage for beams; (**b**) improved XPS formwork for easy removal.

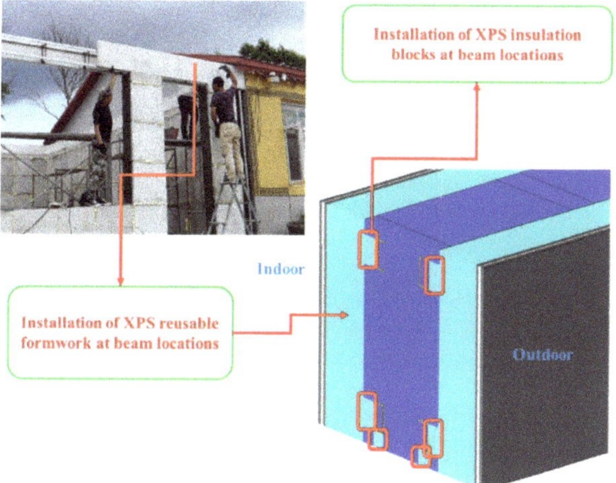

Figure 22. Thermal bridge treatment methods for beam positions.

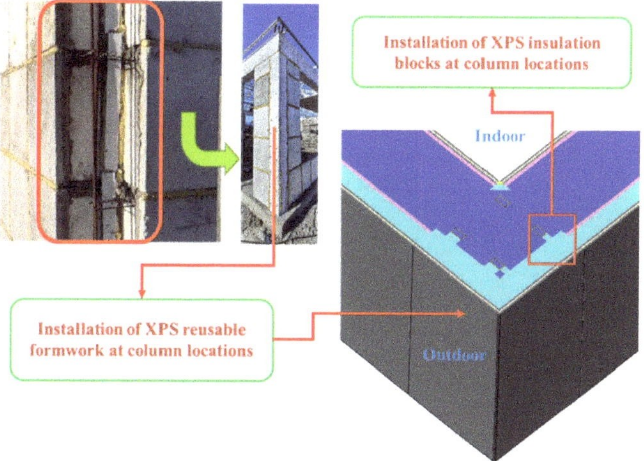

Figure 23. Thermal bridge treatment methods for column positions.

Following the aforementioned structural improvements, detailed construction drawings were generated for the block joints, column positions, and beam positions, incorporating the design principles of low thermal bridges (Figure 24).

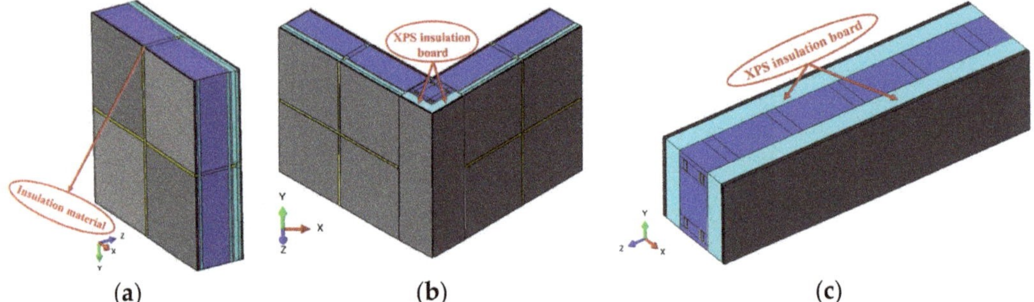

(a)   (b)   (c)

Figure 24. Design solutions for minimizing thermal bridges at various locations. (**a**) Measures to mitigate thermal bridges at block joints; (**b**) measures to mitigate thermal bridges at column positions; (**c**) measures to mitigate thermal bridges at beam positions.

By conducting a calculation analysis on the improved thermal bridges of the three types and utilizing the calculation method outlined in Section 2.4, the comprehensive heat transfer coefficients (Table 6) were determined. The results demonstrate a significant reduction in the overall heat transfer coefficients, indicating a substantial enhancement in insulation performance following the implementation of measures to mitigate thermal bridges. Specifically, the overall thermal resistance of the walls increased by a factor of 2.25 compared to the original design, while the thermal resistance at column positions increased by a factor of 38 and at beam positions by a factor of 36. In comparison to the original design, the overall heat transfer coefficient of the walls decreased by 54.82%, and the heat transfer coefficient at beam and column positions decreased by 97%. These improvements align with the requirements specified in the "General Specification for Building Energy Efficiency and Renewable Energy Utilization" (GB55015-2021) [17].

Table 6. The optimized overall heat transfer coefficients have been obtained.

Name	Wall	Column	Beam
Value-W/(m^2·K)	0.24	0.27	0.20

Based on the thermal flow map, it is apparent that there has been a substantial reduction in heat loss at the three types of thermal bridges (Figure 25). Thermal imaging tests were carried out on a residential building in Nenjiang, which was constructed with the implemented improvement measures (Figure 26). A comparison with the reference building showed that the wall surface temperatures in the original thermal bridge areas have become more uniform, resulting in a significant decrease in temperature differentials. These findings indicate excellent insulation performance.

3.4. Single-Parameter Analysis of Building Energy Consumption

A comparative analysis was conducted to evaluate the solar radiation received on the surfaces of residential blocks with improved insulation in the Nenjiang region, taking into account different orientations (Figure 27a). The vertical axis represents the amount of winter/summer radiation, while the horizontal axis denotes the angle between the normal of the south-facing facade and the north–south axis (with a clockwise direction as positive). It is hypothesized that positioning the building in a north–south orientation would be advantageous if it leads to higher radiation during winter and lower radiation during summer (Figure 27b).

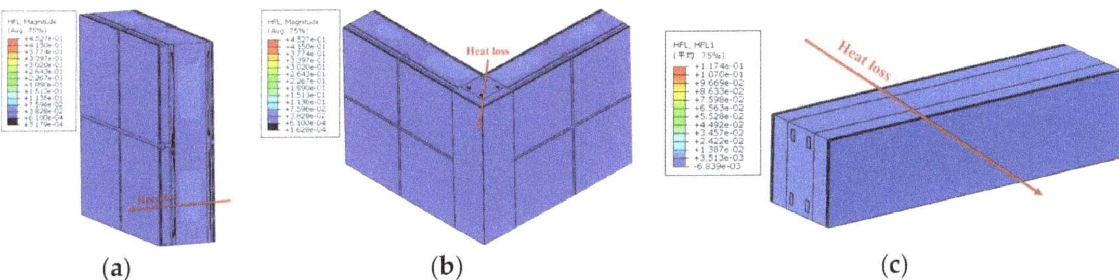

Figure 25. The thermal flow maps of the improved thermal bridges. (**a**) Thermal flow cloud map at the joint of blocks; (**b**) thermal flow cloud map at the column; (**c**) thermal flow cloud map at the beam.

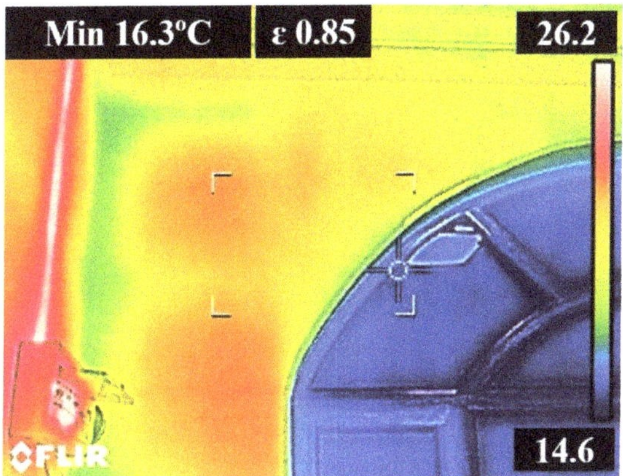

Figure 26. Thermal images showcasing the thermal bridges after the implementation of improvement measures.

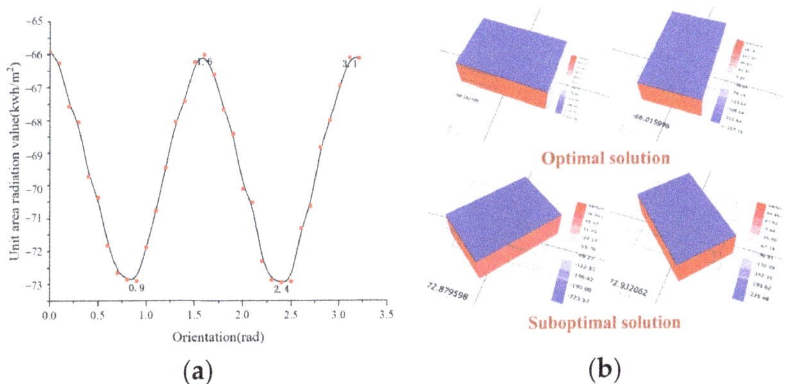

Figure 27. Analysis of solar radiation heat. (**a**) The correlation between building orientation and solar radiation heat per unit area; (**b**) good solutions versus bad solutions.

Based on the initial thermal imaging analysis, it is apparent that the transparent envelope structure often exhibits weaker thermal insulation characteristics. A comparison was conducted to assess the impact of different window-to-wall ratios (south, west, east, north) on winter heating loads (Figure 28). The findings indicate that the south-facing facade achieves the lowest specific heating energy consumption per unit area when the window-to-wall ratio is 0.3. In cases where the window-to-wall ratio is less than 0.5, the energy consumption increment is negative compared to having no windows. On the other hand, for the remaining three orientations, the winter heating energy consumption demonstrates a nearly linear increase as the window-to-wall ratio rises. Specifically, opening windows on the north-facing side exerts a greater influence on energy consumption compared to the east and west sides. It is noteworthy that when the window-to-wall ratio is below 0.25 for the west-facing facade, 0.2 for the east-facing facade, and 0.1 for the north-facing facade, the incremental energy consumption remains below 5% in comparison to having no windows.

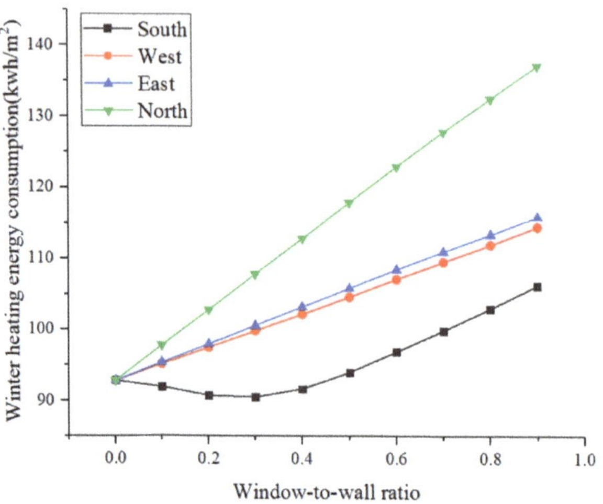

Figure 28. The impact of window-to-wall ratio on winter heating load in buildings.

Finally, a study was conducted to examine the impact of length, width, and height on heating load per unit volume by maintaining a fixed window-to-wall ratio of 0.5 on the south side and 0.1 on the north side (Figure 29). The results revealed a consistent decrease in heating load per unit volume as the dimensions of the building envelope increased. This reduction can be attributed to the exceptional thermal insulation performance of the structure, which effectively limits the increase in energy consumption relative to the expansion in volume. In the absence of windows, both the building's depth and span exhibited similar effects on energy consumption. However, in the presence of windows, an increase in the building's span resulted in a larger external window area and increased solar heat gain, thereby causing a slower decline in energy consumption compared to an increase in depth. Conversely, an increase in building height consistently led to a decrease in energy consumption, albeit at a slower rate than the scenario without windows. This can be attributed to the fact that an increase in height also enlarges the external window area, contributing to additional solar heat gain and moderating the rate of decrease in energy consumption. Overall, in terms of the influence on heating energy consumption per unit volume, the order of significance is height > depth > span.

3.5. Sensitivity Analysis of Building Energy Consumption and Optimization for Energy Savings

Based on the results, it is clear that $p < 0.01$ signifies a significant correlation between each variable and the analysis objectives. The absolute values presented in Table 7 reflect the strength of the correlation, where larger absolute values indicate a stronger correlation. The building's span and depth exhibit a negative correlation with the analysis objectives, whereas the remaining design parameters show positive correlations. When the objective is winter heating, the influence of building scale surpasses that of the window-to-wall ratio. Notably, building height exerts the most significant impact, followed by depth (which outweighs span). The effect of the window-to-wall ratio on the south-facing facade is relatively weak, as it is compensated via solar radiation for heating purposes. The impacts of other factors are relatively similar. In terms of solar radiation heat gain, the opening of windows on the south side yields the greatest influence, while building height also plays a substantial role due to its ability to widen the incidence angle. Regarding summer cooling, the presence of west-facing windows has a pronounced effect, indirectly highlighting the importance of addressing the impact of western sun exposure during this season.

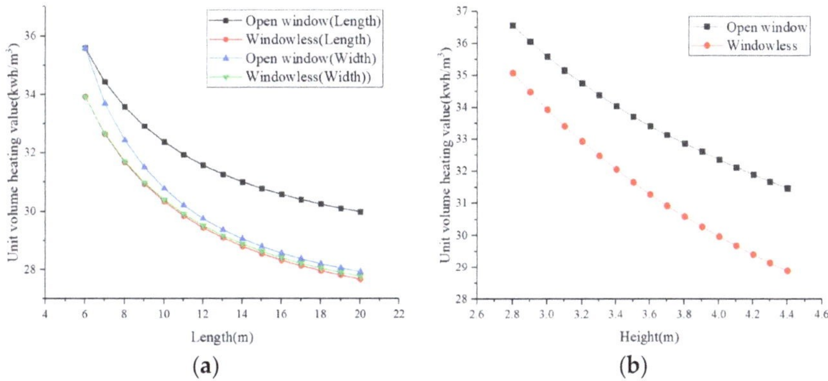

Figure 29. The correlation between building geometric variables and energy consumption. (**a**) The influence of building span and depth on winter energy consumption; (**b**) the influence of building height on winter energy consumption.

Table 7. Pearson correlation analysis (** $p < 0.01$).

	Cooling (kwh/m^2)	Heating (kwh/m^2)	All Energy (kwh/m^2)	Solar (kwh/m^2)
X	−0.203 **	−0.317 **	−0.303 **	−0.122 **
Y	−0.351 **	−0.441 **	−0.435 **	−0.398 **
Z	0.426 **	0.629 **	0.606 **	0.410 **
N	0.100 **	0.293 **	0.265 **	0.107 **
W	0.618 **	0.299 **	0.365 **	0.389 **
S	0.449 **	0.178 **	0.233 **	0.620 **
E	0.132 **	0.273 **	0.253 **	0.292 **

We analyzed the relationship between the window-to-wall ratio and both winter heating load and solar radiation heat gain for the four distinct orientations while maintaining a constant building scale (Table 8). The focused parameters employed in this analysis provide a more comprehensive understanding of the disparities in window placement compared to the previous round. Notably, there is a strong correlation between south-facing window openings and solar radiation heat gain, suggesting that variations in the window-to-wall ratio have minimal influence on winter heating. Conversely, the north-facing window openings, which receive minimal sunlight, exert the greatest impact on the heating load among the orientations, while the effects on the heating load for the east and west orientations are similar.

Table 8. Pearson correlation analysis (* $p < 0.05$, ** $p < 0.01$).

	Heating (kwh/m^2)	Solar (kwh/m^2)
N	0.796 **	0.061 **
W	0.404 **	0.174 **
S	0.025 *	0.965 **
E	0.425 **	0.163 **

Based on the aforementioned analysis, clear directions can be identified for selecting optimization strategies. In cold regions, the primary focus should be on winter heating and solar radiation heat gain. Firstly, it is crucial to control the building height to prevent excessive height while also considering an appropriate increase in the building span. Secondly, the window-to-wall ratio on the south side can be moderately increased, but strict control is necessary for window openings on the north side. Careful consideration should be given to the design of window openings on the west side to mitigate heat gain during summer.

Consequently, a selection of optimal solutions was made among the 9216 simulated cases (Figure 30) that exhibited lower energy loads. After careful comparison, it was determined that a design with a length of 10 m, width of 9 m, height of 3 m, a window-to-wall ratio of 0.4 on the south side, and a window-to-wall ratio of 0.1 on the north or east side is more favorable (Table 9). There are two groups of optimization schemes. (Figure 31) This optimized solution, with an energy load of 165 (kWh/m^2·a), achieved a significant 66.83% reduction compared to the original design's energy load of 497.38 (kWh/m^2·a).

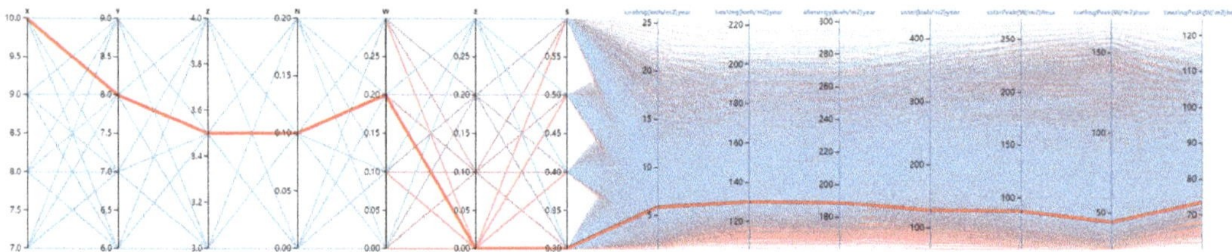

Figure 30. Visualization processing of cases using the Design-Explorer platform.

Table 9. Refined optimal solutions.

X	Y	Z	N	W	S	E	Energy (kwh/m^2)	Solar (kwh/m^2)
10	7	3	0	0	0.3	0	168	86
10	9	3	0	0	0.4	0	162	89
10	**9**	**3**	**0.1**	**0**	**0.4**	**0**	**166**	**94**
10	**8**	**3**	**0**	**0**	**0.3**	**0.1**	**167**	**86**
10	**9**	**3**	**0**	**0**	**0.4**	**0.1**	**165**	**100**

Note: The bold number is the recommended value for the optimal solution.

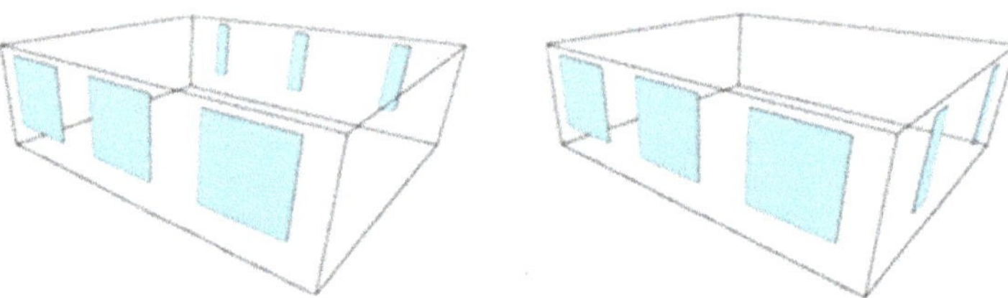

Figure 31. Optimal solutions.

It is important to note that the simulation calculations focused on optimizing building load and did not consider the presence of partition walls within the building. Additionally, the equipment was assumed to operate 24 h a day, which may have slightly overestimated the data. Nevertheless, the results still provide valuable insights into the impact of various design elements on building load.

4. Discussion

- This article offers a comprehensive analysis of the climatic conditions in the Nenjiang region and provides pertinent suggestions for integrating local climatic characteristics into architectural design. The proposed strategies, including harnessing winter solar radiation and leveraging summer natural ventilation, possess broad applicability.

- This article presents structural enhancements to the thermal bridging areas of large-scale insulated block buildings, resulting in a significant improvement in insulation performance and a reduction in thermal bridging effects. Furthermore, the refined construction methods have effectively increased construction efficiency. In practical applications, the proposed enhancements were implemented by the authors, and the primary structural elements were constructed using demountable formwork, achieving a remarkable completion time of just one day with a 100% formwork utilization rate. Additionally, the load-bearing capacity of the beams was enhanced by 3–6 times. The construction of the enclosure structure was efficiently accomplished in a mere three days, significantly expediting the overall construction process.

- Furthermore, there is still room for further improvement in this type of insulated block. In practical applications, the authors have replaced the EPS (expanded polystyrene) in the blocks with XPS (extruded polystyrene), resulting in a thermal resistance of $10 \, m^2 \cdot K/W$. Considering actual energy usage conditions (equipment not running all day), simulated calculations indicate a comprehensive energy consumption value of $43.8 \, kWh/m^2 \cdot a$, which is 73.45% lower than the ideal load value mentioned in this article (assuming equipment runs all day). This design achieves a 32.62% energy savings compared to the requirements outlined in the "Heilongjiang Province Ultra-Low Energy Residential Building Energy Efficiency Design Standard" (DB23/T 3337-2022) [18], thus fully meeting the standard's requirements.

- In comparison to the ultra-low energy consumption residential system utilizing porous bricks and EPS insulation boards [19], the rural housing system proposed in this study exhibits a thermal load of $29.3 \, kwh/m^2 \cdot a$, representing a 51.32% reduction compared to the thermal load of $60.19 \, kwh/m^2 \cdot a$, observed in such ultra-low energy consumption rural housing with porous bricks. Moreover, compared to ultra-low energy consumption rural housing equipped with a solar-air source heat pump coupled heating system, the annual cumulative thermal load index achieves a similar level [20]. This demonstrates that in practical applications, houses built with these improved insulated blocks will deliver even better performance and possess significant energy-saving potential.

5. Conclusions

This article provides an analysis of a particular type of insulated block residential housing in extremely cold rural areas of China, focusing on three aspects: the climate environment, thermal bridging in the enclosure, and energy consumption. It thoroughly examines the design requirements, construction characteristics, thermal performance, and energy consumption characteristics of the housing and proposes optimization strategies. Based on the analysis, the following conclusions are drawn:

- The cold climate conditions in the Nenjiang region play a crucial role in the absence of comfort during winter, resulting in a 0% comfort time. Conversely, summer offers a relatively higher comfort time, constituting 76.3% of the total annual comfort time. Hence, it is essential to prioritize the utilization of natural ventilation during the design phase, particularly in the summer. While ensuring effective insulation in the enclosure structure for winter, equal attention should be given to maximizing solar radiation heat gain to reduce the building's energy demand.

- Corresponding solutions have been proposed for three types of thermal bridging in large-scale insulated block residential buildings, leading to a remarkable 54.82% reduction in the overall heat transfer coefficient of the walls compared to the original design. Moreover, the overall heat transfer coefficient at the beam and column locations has been significantly reduced by 97%. These solutions have not only enhanced the construction efficiency of the main structure and enclosure but also yielded significant improvements in thermal performance.

- The impact of building scale on energy consumption load surpasses that of window size. In the case of residential buildings in extremely cold regions, ample solar radiation aids in reducing the building load. Hence, it is recommended to incorporate larger windows on the south side. Conversely, smaller windows with a window-to-wall ratio of approximately 0.1 are advisable on the north side. Windows on the west side are not recommended.
- To improve energy-saving effectiveness in building forms like insulated blocks, which already exhibit good insulation performance, it is essential to use more efficient insulation materials, implement measures to reduce thermal bridges, adopt a more suitable architectural exterior design, and establish a rational energy utilization plan. These collective efforts will contribute to further enhancing energy efficiency on the existing foundation.

Author Contributions: Conceptualization, M.G; methodology, M.G.; software, M.G.; validation, M.G., Y.W., and X.M.; formal analysis, M.G.; investigation, M.G.; resources, Y.W.; data curation, M.G.; writing—original draft preparation, M.G.; writing—review and editing, Y.W and X.M.; visualization, X.M.; supervision, Y.W.; project administration, Y.W.; funding acquisition, Y.W. All authors have read and agreed to the published version of the manuscript.

Funding: This research was funded by the National Key R&D Program of China, grant number 2019YFD1101004.

Institutional Review Board Statement: Not applicable.

Informed Consent Statement: Not applicable.

Data Availability Statement: The data presented in this study are available in the article itself.

Acknowledgments: The authors wish to express their gratitude for the financial support that has made this study possible.

Conflicts of Interest: The authors declare no conflict of interest.

References

1. Tsinghua University Research Center for Building Energy Efficiency. *Annual Development Research Report on Building Energy Efficiency in China*; China Architecture & Building Press: Beijing, China, 2022.
2. Dumitrescu, L.; Baran, I.; Pescaru, R.A. The Influence of Thermal Bridges in the Process of Buildings Thermal Rehabilitation. *Procedia Eng.* **2017**, *181*, 682–689. [CrossRef]
3. Ascione, F.; Bianco, N.; Rossi, F.D.; Turni, G.; Vanoli, G.P. Different methods for the modelling of thermal bridges into energy simulation programs: Comparisons of accuracy for flat heterogeneous roofs in Italian climates. *Appl. Energy* **2012**, *97*, 405–418. [CrossRef]
4. Baldinelli, G.; Bianchi, F.; Rotili, A.; Costarelli, D.; Seracini, M.; Vinti, G.; Asdrubali, F.; Evangelisti, L. A model for the improvement of thermal bridges quantitative assessment by infrared thermography. *Appl. Energy* **2018**, *211*, 854–864. [CrossRef]
5. Garay, R.; Uriarte, A.; Apraiz, I. Performance assessment of thermal bridge elements into a full scale experimental study of a building façade. *Energy Build.* **2014**, *85*, 579–591. [CrossRef]
6. Zalewski, L.; Lassue, S.; Rousse, D.; Boukhalfa, K. Experimental and numerical characterization of thermal bridges in prefabricated building walls. *Energy Convers. Manag.* **2010**, *51*, 2869–2877. [CrossRef]
7. Santos, P.; Martins, C.; Simoes, D.S.L. *Thermal Performance of Lightweight Steel-Framed Construction Systems*; Metallurgical Research & Technology I Cambridge Core: Cambridge, UK, 2014; pp. 329–338.
8. Soares, N.; Santos, P.; Gervasio, H.; Costa, J.J.; Da Silva, L. Energy efficiency and thermal performance of lightweight steel-framed (LSF) construction: A review. *Renew. Sustain. Energy Rev.* **2017**, *78*, 194–209. [CrossRef]
9. Mithraratne, N.; Vale, B. Life cycle analysis model for New Zealand houses. *Build. Environ.* **2004**, *39*, 483–492. [CrossRef]
10. Sartori, I.; Hestnes, A.G. Energy use in the life cycle of conventional and low-energy buildings: A review article. *Energy Build.* **2007**, *39*, 249–257. [CrossRef]
11. Fang, Y.; Cho, S. Design optimization of building geometry and fenestration for daylighting and energy performance. *Sol. Energy* **2019**, *191*, 7–18. [CrossRef]
12. Zhang, J.; Liu, N.; Wang, S. A parametric approach for performance optimization of residential building design in Beijing. *Build. Simul.* **2020**, *13*, 223–235. [CrossRef]

13. Chen, X.; Yang, H.; Lu, L. A comprehensive review on passive design approaches in green building rating tools. *Renew. Sustain. Energy Rev.* **2015**, *50*, 1425–1436. [CrossRef]
14. Verbeeck, G.; Hens, H. Energy savings in retrofitted dwellings: Economically viable? *Energy Build.* **2005**, *37*, 747–754. [CrossRef]
15. Payyanapotta, A.; Thomas, A. An analytical hierarchy based optimization framework to aid sustainable assessment of buildings. *J. Build. Eng.* **2021**, *35*, 102003. [CrossRef]
16. *GB50176-2016*; Code for Thermal Design of Civil Building. China Building Industry Press: Beijing, China, 2016.
17. *GB55015-2021*; General Specification for Building Energy Efficiency and Renewable Energy Utilization. China Building Industry Press: Beijing, China, 2021.
18. *DB23/T 3337-2022*; Energy-Saving Design Standard for Ultra-Low Energy Consumption Residential Buildings in Heilongjiang Province. Heilongjiang Provincial Department of Housing and Urban-Rural Development Press: Heilongjiang, China, 2022.
19. Deng, Q.Q.; Song, B.; Zhang, S.N. Study on Thermal Parameters of Enclosure Structure in Ultra-Low Energy Consumption Rural Housing. *Build. Energy Effic.* **2022**, *50*, 46–49.
20. Li, X.L.; Hao, M. Construction and Operation Research of Multi-Energy Coupled Heating System for Ultra-Low Energy Consumption Rural Housing. *Energy Sav.* **2023**, *42*, 22–25.

 *sustainability*

Article

Research on a New Plant Fiber Concrete-Light Steel Keel Wall Panel

Yuqi Wu [1], Yunqiang Wu [2,3,*] and Yue Wu [2,3]

[1] School of Mechanics and Civil Engineering, China University of Mining and Technology, Beijing 100083, China
[2] Key Lab of Structures Dynamic Behavior and Control of the Ministry of Education, Harbin Institute of Technology, Harbin 150090, China
[3] Key Lab of Smart Prevention and Mitigation of Civil Engineering Disasters of the Ministry of Industry and Information Technology, Harbin Institute of Technology, Harbin 150090, China
* Correspondence: qfmz_wyq@163.com

Abstract: With the growing worldwide attention towards environmental protection, the rational utilization of rice straw (RS) has gradually attracted the attention of scholars. This paper innovatively puts forward a solution for rational utilization of RS. A rice straw fiber concrete (RSFC) with good physical and mechanical properties and a rice straw concrete-light steel keel wall panel (RS-LSWP) with low comprehensive heat transfer coefficient and inconspicuous cold bridge phenomenon was designed. Firstly, the preparation method and process of RSFC is described in detail. Then, the physical and mechanical properties of RSFC, such as strength, apparent density, and thermal conductivity were tested. Finally, the thermal properties of the four new types of cold-formed thin-wall steel panels were analyzed using finite element simulation. The results show that the RSFC with a straw length of 5 mm, mass content of 12%, and modifier content of 1% is the most suitable for RS-LSWP. The standard compressive strength, tensile strength, and thermal conductivity of the RSFC are 2.2 MPa, 0.64 MPa, and 0.0862 W/(m·K), respectively. The wall panels with antitype C keel have a low comprehensive heat transfer coefficient and the best insulation effect. This study innovatively provides a technical method for the rational utilization of RS, promotes the application of RS and other agricultural wastes in building materials and the development of light steel housing.

Keywords: bio-based materials and structures; sustainable structures; plant fiber concrete; light steel keel wall panel; thermal performance

Citation: Wu, Y.; Wu, Y.; Wu, Y. Research on a New Plant Fiber Concrete-Light Steel Keel Wall Panel. *Sustainability* **2023**, *15*, 8109. https://doi.org/10.3390/su15108109

Academic Editor: Gianluca Mazzucco

Received: 27 March 2023
Revised: 10 May 2023
Accepted: 12 May 2023
Published: 16 May 2023

1. Introduction

Rice straw(RS) is one of the crop straw, whose annual output can reach billions of tons around the world [1]. Asia, one of the major rice-producing regions, produces about 620 million tons of straw every year [2]. However, only a small part of these straws are properly handled by livestock farming, construction materials, etc., and most of them is still burned directly, which not only wastes resources but also has a bad impact on the ecological environment [3].

Rice straw ash contains a lot of silica [4]. To reasonably solve the problem of RS, some scholars have tried to use straw ash as a building material. In 1983, Mehta used rice RS ash for cement concrete [5]. Josefa et al. characterized the ash from different parts of the RS and showed that the reuse of RS ash in the cement system is promising [6]. Blessen et al. have also confirmed the feasibility of using biomass ash from RS as cement concrete materials [6,7]. Rehman, Abou Sekkina, et al. studied how to increase the content of silica in straw ash [4,8]. Agwa et al. studied the use of rice husk ash in cement and showed that RHA-modified cement exhibits better mechanics and durability. Some scholars have studied the effects of RS ash applied in concrete on the setting time, workability, strength, water absorption, porosity and density of concrete [9–13].

As a straw fiber, RS has the advantages of low thermal conductivity and high sound insulation ability, there are many cases where it is directly used to make houses [14]. Some building materials, such as gypsum board and environmentally friendly brick, are also made by mixing straw with cement-based cementitious materials [15–17]. In addition, some scholars used rice husk, sugarcane bagasse ash (SBA), and other agricultural wastes to develop clay bricks [18]. Rice straw fiber reinforced concrete (RSFRC) developed by mixing RS with Portland cement is one of the ways of rational utilization of RS in recent years. Feraidon Ataie et al. studied the effects of adding rice straw fiber (RSF) on the compressive strength and bending strength of concrete, drying shrinkage rate, and heat of hydration of cement [19]. Chinh Van Nguyen et al. studied the properties of rice straw-reinforced alkali-activated cementitious composites (AACC) and found that RS had a very significant positive effect on the performance of AACC, and that alkali treatment was also an effective method for enhancing the bond between the RS and the matrix [20]. In the study of Liu et al., RS were pretreated with NaOH and grafted with Nano-SiO_2 [4]. The modified rice straw fiber reinforced concrete (RSFRC) prepared by them also promoted the application of plant straw fiber in civil engineering [6]. However, some scholars have proven that the plant fiber in Portland cement's strong alkaline environment (pH > 13) will be decomposed and mineralized, which will cause plant fibers to lose their strengthening effect [21–24]. In addition to Portland cement, Sorel et al. [25] developed a magnesium oxychloride cement (MOC) (pH: 8–9.5) with lower alkalinity suitable for mixing with RS. The cement is composed of light-burned magnesium oxide, magnesium chloride hexahydrate, water, and other proprietary materials [26]. MOC has outstanding advantages such as fire resistance, low cost, high strength, fast curing, high wear resistance, and low thermal conductivity [27]. However, low toughness and poor water resistance limit the application of MOC. To improve the performance of MOC, modified MOC have been prepared using fly ash, phosphoric acid, silica powder, silicone acrylic emulsion, and granite waste. All of these additives can improve the water resistance of MOC [28–31]. Xiao et al. invented the foam lightweight MOC material (MOCL) by adding an appropriate amount of H_2O_2 and $MgSO_4$ in the process of producing MOC. It is lighter, waterproof, and heat-resistant than traditional MOC materials, but the defects of poor toughness and easy cracking have not been completely solved [32]. Wang et al. combined MOC with straw treated with H_2O_2 foaming to develop straw/magnesium lightweight composite (SMLC). Compared with MOC, the production of SMLCS are more energy-saving and environmentally protected. The compressive strength and bending strength of SMLCS are up to 12.5 MPa and 4.8 Mpa, respectively, and the thermal conductivity is 0.06 W/(m·K). However, the additional amount of RS in SMLCS is not very large, amounting to only 0.9% [32].

Meanwhile, with the development of the social economy, houses in rural areas are changing from traditional brick and concrete structures to light and efficient thermal insulation structures. Therefore, light steel structure has gradually received the attention of researchers. The light steel structure is a low-rise and multi-story building structure with a large number of light insulation materials. It takes cold-formed thin-wall steel, light welding or high-frequency welding steel, and light hot rolled steel as the main stress components [33]. Since the 1950s, with the development of the steel structure industry, the characteristics of different types of light steel wallboards have been studied [34–40]. Light steel keel wall panels are mostly filled with glass fiber, rock wool, foamed concrete and other materials, which make the wall panels have a "hollow feeling", which seriously limits the popularization and application of cold-formed thin-wall steel structure buildings in China. Glass fiber cotton dust produced by rock wool will also directly enter the respiratory system of residents, causing respiratory tract related diseases. In addition, light steel keel have extremely poor thermal insulation performance, as such, scholars have proposed some measures to improve the thermal insulation and prevent the cold bridge phenomenon of wall panels at keel position [39,40].

Although previously scholars have proposed a better scheme of straw fiber combination with magnesium oxychloride cement (MOC), there are still some problems, such as the small amount of RS, the complicated processing process of straw, cracks appearing after combination; additionally, they do not consider the case of combining straw concrete with light steel wall panels. In this study, rice straw fiber was pulverized and combined with magnesium oxychloride cement (MOC) to produce a kind of RSFC with high content of straw and suitable for light steel keel wall panels. It is found that the mass content of RS in the RSFC used in the wall panel can reach about 12%. The rice straw concrete-light steel keel wall panels (RS-LSWP) proposed provides a technical means for rational utilization of RS and can promote the development of light steel residential buildings. The research results promote the application of straw and other agricultural wastes in building materials and also provide a way for the green, coordinated, and healthy development of rural areas.

2. Materials and Methods

2.1. Rice Straw Fiber Concrete

2.1.1. Raw Materials and Preparation Process

The rice straw fibers used are non-rotting, dried and crushed straw. To alleviate the problem of straw clumping, RS and husk were mixed in a mass ratio of 3:1 (Figure 1). Light-fired magnesium oxide powder (industrial magnesium oxide powder) with a content of 73.8% and magnesium sulfate crystal heptahydrate (chemical formula $MgSO_4 \bullet 7H_2O$) as the main compelling materials were used. The compound acid modifier, mainly composed of phosphoric acid, is provided by Shenyang Tianque Building Materials Co., LTD, Shenyang, China. The modifier can be used to improve the comprehensive performance of RSFC.

(a) (b)

Figure 1. Raw material: (**a**) rice straw; (**b**) husk.

The tool used to stir RSFC is the industrial electric mixer. Before preparing the specimen, $MgSO_4 \bullet 7H_2O$ was mixed with warm water at 40 °C at a mass ratio of 1:1 to form a magnesium sulfate solution. The preparation of RSFC begins by mixing rice straw with lightly burned magnesium oxide powder through the mixer. Next, the compound acid modifier was added to the magnesium sulfate solution and then added to the mixture of RS and magnesium oxide. The mixture was thoroughly stirred and then poured into the test mold to vibrate. Finally, it was cured under natural indoor conditions. Figure 2 shows the preparation process of RSFC.

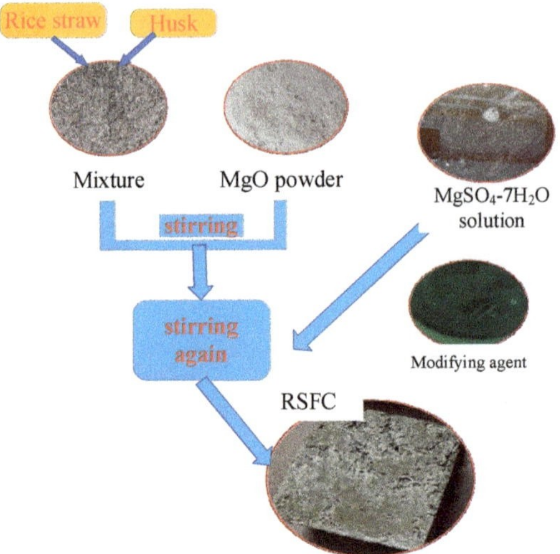

Figure 2. Flow chart.

2.1.2. Test Method

The apparent density, compressive strength test and tensile strength test of RSFC were measured according to the standard test method for mechanical Properties of ordinary Concrete GBT50081-2019 [41]. An MTS microcomputer-controlled electro-hydraulic servo universal testing machine with a range of 1000 kN was used to carry out the corresponding test, and the loading speed was 0.3 kN/s. Formulas (1)–(3) are the calculation formulas of apparent density, compressive strength and splitting strength of concrete, respectively:

$$\rho_a = \frac{m_d}{V_a}$$

(1)

ρ_a is the apparent density of concrete, m_d is the mass of the specimen under dry conditions, and V_a is the apparent volume of the specimen.

$$R = \frac{P}{A}$$

(2)

R is the standard compressive strength of RSFC(MPa), P is the ultimate load (N) at the time of failure, and A is the compression surface area (mm^2) of the RSFC.

$$R_t = \frac{2P}{\pi A}$$

(3)

R_t is the standard value of splitting tensile strength (MPa), P is the ultimate load (N) at the time of failure, and A is the area of the splitting surface of the cube test block of the RSFC (mm^2).

The hot-wire method (Figure 3) was used to measure the thermal conductivity of the test block. To improve the measurement accuracy, the average value of the test block was calculated after multiple measurements. The hot-wire method is an analytical method based on the proportional relationship between temperature rise ΔT and logarithmic heating time (ln(t)) within a long enough time (300 s) [42].

the volume expansion of the RSFC caused by the reaction of the active magnesium oxide with water which forms magnesium hydroxide during the condensation and hardening process of magnesium oxide and magnesium sulfate. However, when the straw content is 12%, the volume expansion generated by the magnesium hydroxide in the test block can be released in the gap due to the increase in RS content, and there are almost no cracks on the concrete surface. Thus, to avoid cracks, the content of RSFC should be more than 12%.

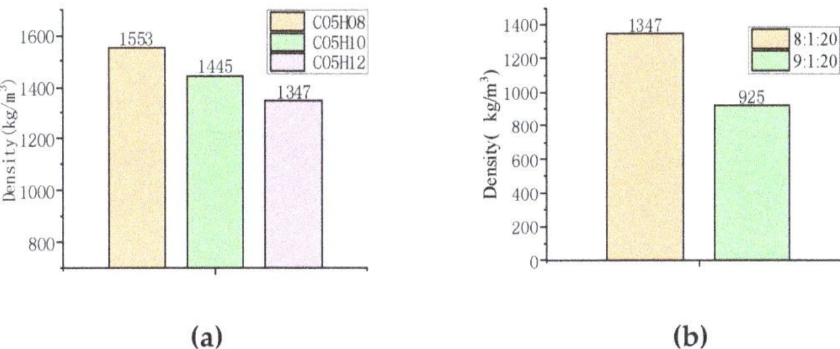

(a)   (b)

Figure 7. RSFC density: (**a**) Different RS content; (**b**) Different proportions of cementitious materials.

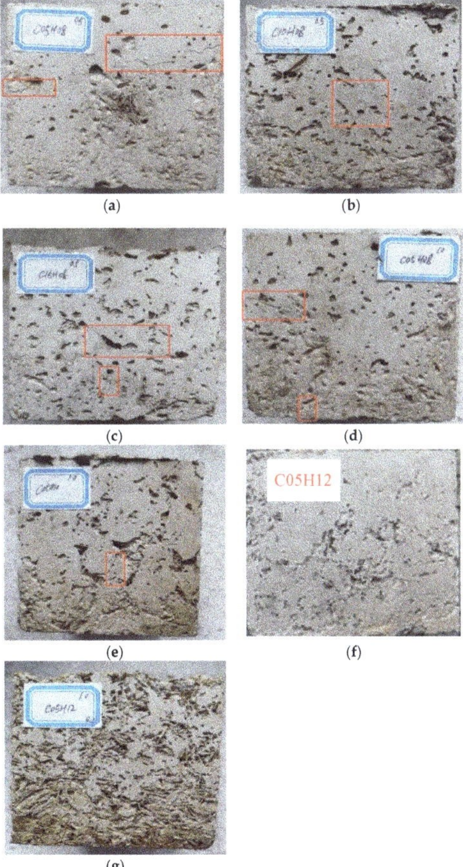

Figure 8. Surface crack(red box): (**a**) C05H08(0.5); (**b**) C10H08(0.5); (**c**) C15H08(0.5); (**d**) C05H08(1.0); (**e**) C05H10(1.0); (**f**) C05H12(1.0); (**g**) C05H12*(1.0).

Based on the above analysis, considering the RSFC cost, surface crack, density and insulation performance, it is recommended to adopt the RS ratio of 5 mm straw length, 12% mass content, 8:1:20 mass ratio of MgO:MgSO$_4$:H$_2$O, and 1% conditioner content for RS-LSWP. The standard compressive strength, tensile strength, and thermal conductivity of the mixture are 2.2 MPa, 0.64 MPa, and 0.0862 W/(m·K), respectively. The mass content of various materials per cubic meter of concrete under this mix ratio is shown in Table 5.

Table 5. Straw concrete ratio.

Composition	Content (kg/m^3)
Rice straw	81
Husk	27
Magnesium oxide	340
MgSO$_4$•7H$_2$O	235
Modifier	2.65

3.2. Thermal Performance of RS-LSWP

3.2.1. Type C

- Influence of different web widths on the thermal performance of RS-LSWP

The structure diagram of the RS-LSWP is shown in Figure 5a. Figure 9 shows the temperature field of the RS-LSWP when the web width of the steel keel is 100 mm, 150 mm, 200 mm, 250 mm and 300 mm, respectively. When the web width of the steel keel is below 150 mm, the influence range of the cold bridge effect of the steel keel gradually expands with the increase in web width. When the web width of the steel keel is greater than 200 mm, the cold bridge effect decreases with the increase in web width. Hence, the keel flange spacing should be more than 200 mm.

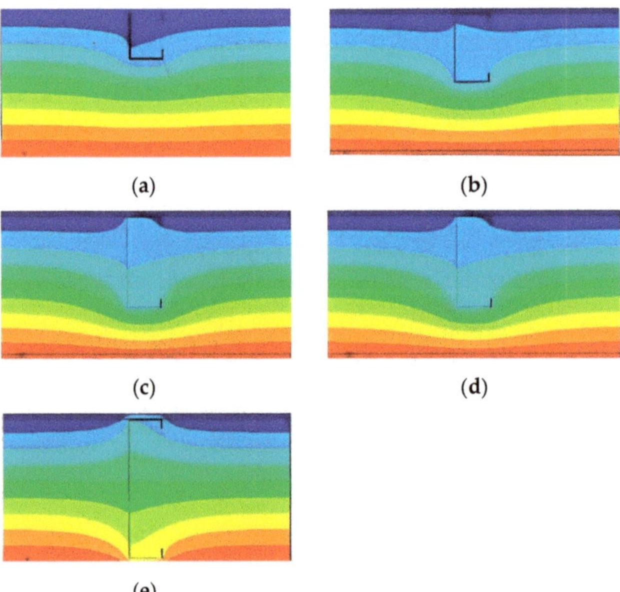

Figure 9. Temperature field of RS-LSWP: (**a**) 100 mm; (**b**)150 mm; (**c**) 200 mm; (**d**) 250 mm (**e**) 300 mm.

Figure 10 shows the variation rule of wall panel internal surface temperature with different web widths. It can be found that with the increase in the web width in the wall panel, the difference between the temperature of the inner surface of the wall and the surrounding temperature also gradually increases. When the height of the web width is 300 mm, there is a significant difference between the temperature at the cold bridge position and the surrounding temperature, which can reach 7~8 °C. Hence, the cold bridge effect can be significantly reduced by using a steel keel structure that does not penetrate the wall panel.

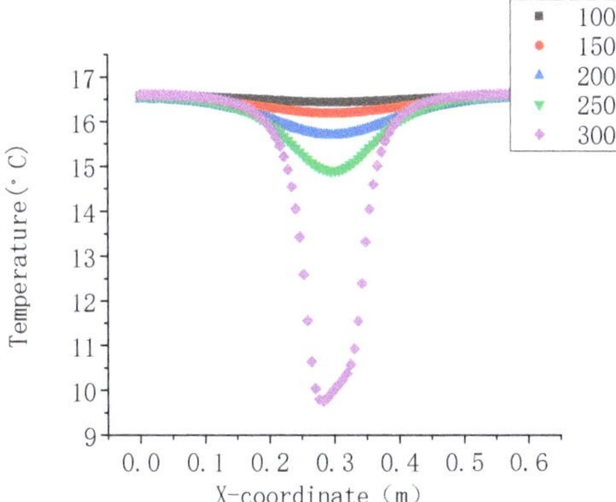

Figure 10. Internal surface temperature variation curve of type C wall panel in different insulation forms.

- Influence of different insulation forms on the thermal performance of RS-LSWP

The thermal performance of the RS-LSWP was analyzed considering the insulation forms of internal and external insulation and whether the steel keel and OSB board were padded with battens. It can be found that compared with external insulation, the internal surface temperature variation rule of the internal insulation wall panel is not affected by the keel cold bridge effect. The internal surface temperature variation trend of steel keel walls is almost the same as that of straw concrete internal insulation and batten internal insulation. This indicates that the keel cold bridge effect of steel keel does not influence the temperature change in the inner surface of the wall. In addition, the results show that the insulation effect of RS is better than that of batten (Figures 11 and 12).

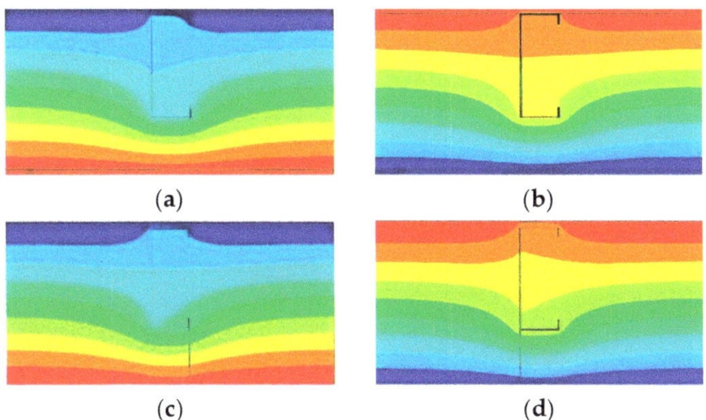

Figure 11. Temperature field of RS-LSWP in different insulation forms: (**a**) RS internal insulation; (**b**) RS external insulation; (**c**) Fill battens for internal insulation; (**d**) Fill battens for external insulation.

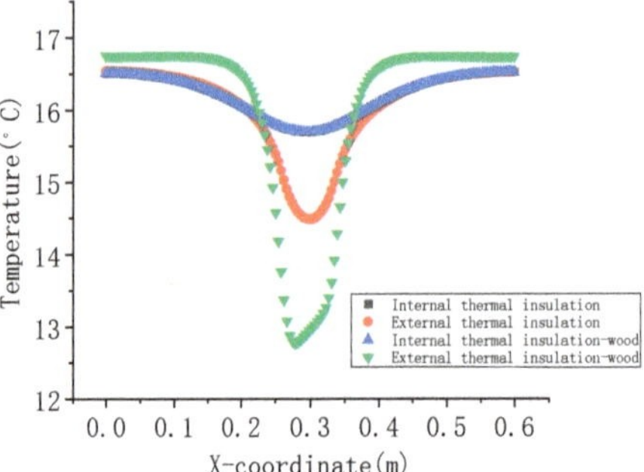

Figure 12. Internal surface temperature variation curve of type C in different insulation forms.

- Comprehensive heat transfer coefficient of RS-LSWP with different web widths and different insulation forms

GB50176-2016 requires that the comprehensive heat transfer coefficient of the external wall in different cold areas is not higher than a certain value. The influence of different web widths of steel keel on the comprehensive heat transfer coefficient is analyzed. Figure 13 shows that the comprehensive heat transfer coefficient of the wall panel increases with the increase in the web depth. When the steel keel does not penetrate the wall panel, the comprehensive heat transfer coefficient of the wall panel increases slowly, but when the keel penetrates the wall panel, the heat transfer coefficient increases. The effect of external insulation is better than that of internal insulation, but the distinction is not obvious. Battens will degrade the thermal performance of the wall panel. Hence, considering the indoor living comfort of the occupant and to prevent the condensation phenomenon, for this kind of wall panel, the internal insulation scheme should be preferred.

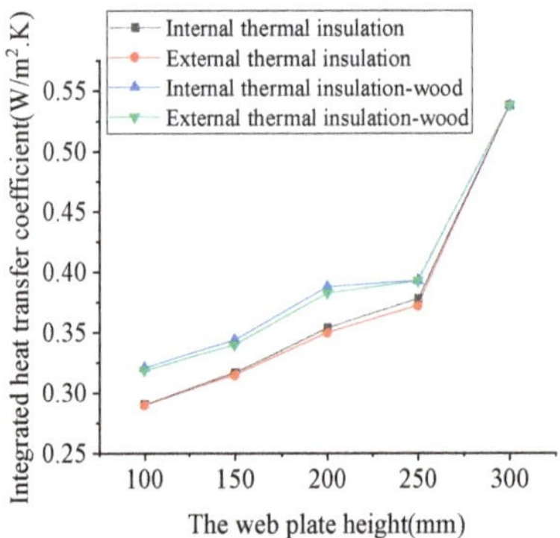

Figure 13. Comprehensive heat transfer coefficient curves of type C with different web depths and different insulation forms.

3.2.2. Antitype C

- Influence of different flange heights on the thermal performance of RS-LSWP

Considering the steel keel flange heights of 40 mm, 50 mm, 60 mm, and 70 mm, the thermal performance of the RS-LSWP was analyzed and parameterized. The structure diagram of the wall panel is shown in Figure 5b. The results of the temperature field calculation show that the influence range of the cold bridge effect of the steel keel expands gradually with the increase in flange height, but the influence range of the cold bridge effect of one side of the steel keel does not extend to the other side. The cross-section form of the antitype C steel keel can effectively reduce the transmission of the keel cold bridge effect in the thickness direction of the wall panel (Figure 14). Meanwhile, with the increase in the height of the steel keel flange, the minimum temperature of the inner surface of the wall panel decreases gradually. However, the phenomenon of the cold bridge is not obvious; the temperature of the inner surface of the wall panel is less than 2 °C, which is lower than the temperature around the wall. This means that an increase in the width of the steel keel flange does not change the range of temperature in the inner surface of the wall panel (Figure 15).

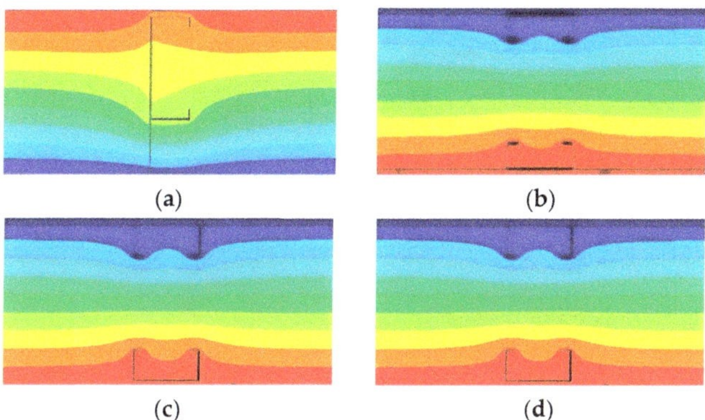

Figure 14. Temperature field of different flange heights in antitype C: (**a**) 40 mm; (**b**) 50 mm; (**c**) 60 mm; (**d**) 70 mm.

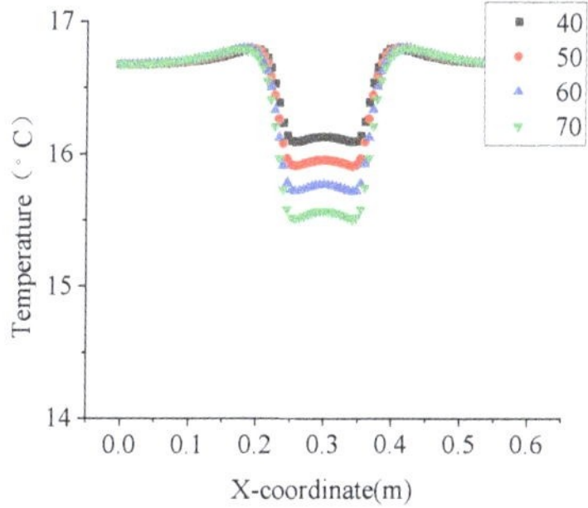

Figure 15. Internal surface temperature curves of different flange widths in antitype C.

- Influence of different offset distances on the thermal performance of RS-LSWP

The thermal performance of the RS-LSWP is analyzed considering that the opposite offset distance of the steel keel is 0 mm, 30 mm, 60 mm and 90 mm, respectively. The calculation results of the temperature field show that the cold bridge displacement of the steel keel is also offset by the increase in the opposite displacement distance of the steel keel. The offset of the opposite side of the steel keel has reduces the cold bridge effect of the keel (Figure 16). In addition, with the increase in the offset distance of the opposite side of the steel keel, the inner surface of the wall panel hardly changed. The phenomenon of the cold bridge effect of steel keel is not obvious, and the lowest temperature on the inner surface of the wall panel is only less than 2 °C lower than the temperature around (Figure 17).

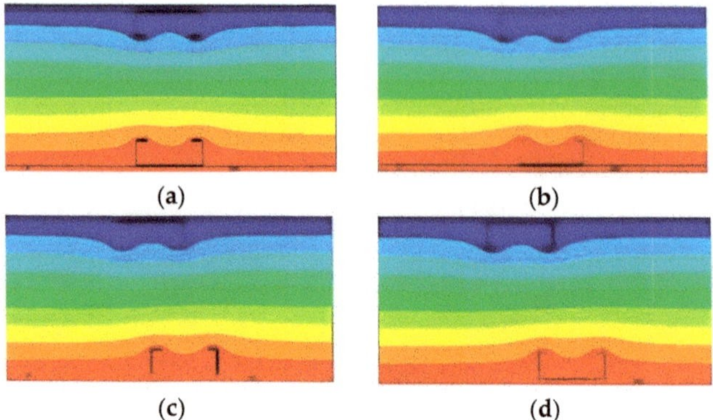

Figure 16. Temperature field with different offset distances in antitype C: (**a**) 0 mm; (**b**) 30 mm; (**c**) 60 mm; (**d**) 90 mm.

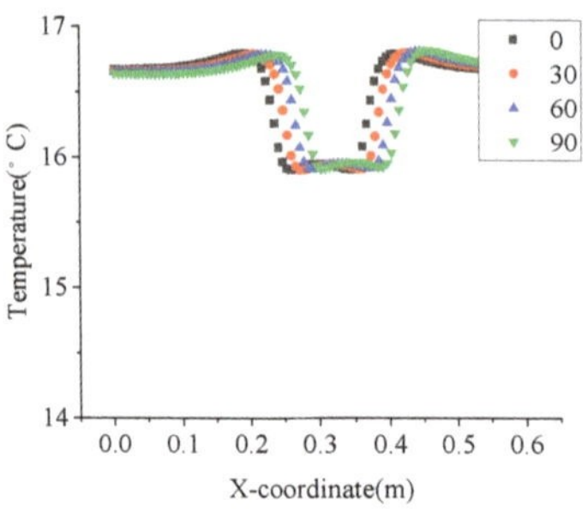

Figure 17. Internal surface temperature curves of different offset distances in antitype C.

- Comprehensive heat transfer coefficient analysis of wall panels with different flange heights and different offset distances

Figure 18 shows the influence of different flange heights and different offset distances on the comprehensive heat transfer coefficient of the steel keel. It can be found that the comprehensive heat transfer coefficient of the wall panel is not related to the opposite offset distance but to the flange height. The comprehensive heat transfer coefficient of the wall panel increases obviously with the increase in flange height. When the flange height of the steel keel is 40 mm, the comprehensive heat transfer coefficient of the wall panel is about 0.28 W/(m²·K). However, when the flange height of the steel keel increases to 70 mm, the comprehensive heat transfer coefficient of the wall panel is about 0.31 W/(m²·K).

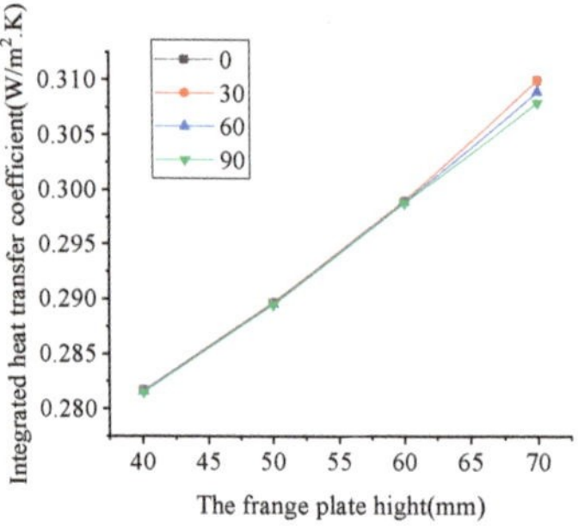

Figure 18. Comprehensive heat transfer coefficient curve in antitype C.

3.2.3. Type Z

- Influence of different web widths on the thermal performance of RS-LSWP

The thermal performance of the internal thermal insulation wall panel was analyzed considering that the web width is 100 mm, 150 mm, 200 mm, 250 mm, and 300 mm, respectively. The structure diagram of the wall panel is shown in Figure 5c. The results of temperature field calculation show that when the web width of the steel keel is less than 150 mm, the influence range of the cold bridge effect of the steel keel gradually expands with the increase in web widths. When the height of the steel keel web is greater than 200 mm, the influence range of the steel keel cold bridge effect decreases gradually with the increasing distance of the keel flange. This is mainly because with the increase in the flange distance of the steel keel, the cold bridge effect of the keel gradually can only be transmitted by the web. Therefore, the flange distance should be more than 200 mm. This is consistent with the temperature variation law of type C (Figure 19). The temperature curve shows that with the increase in steel keel web height, the minimum temperature of the inner surface of the wall panel decreases gradually, and the influence range of the cold bridge effect temperature of the steel keel expands gradually. When the web height of the steel keel is 300 mm, the inner surface temperature of the steel keel decreases, and the influence range of temperature expands. The use of a non-penetrating steel keel structure is conducive to reducing the cold bridge effect of the keel and increasing the inner surface temperature of the wall panel (Figure 20).

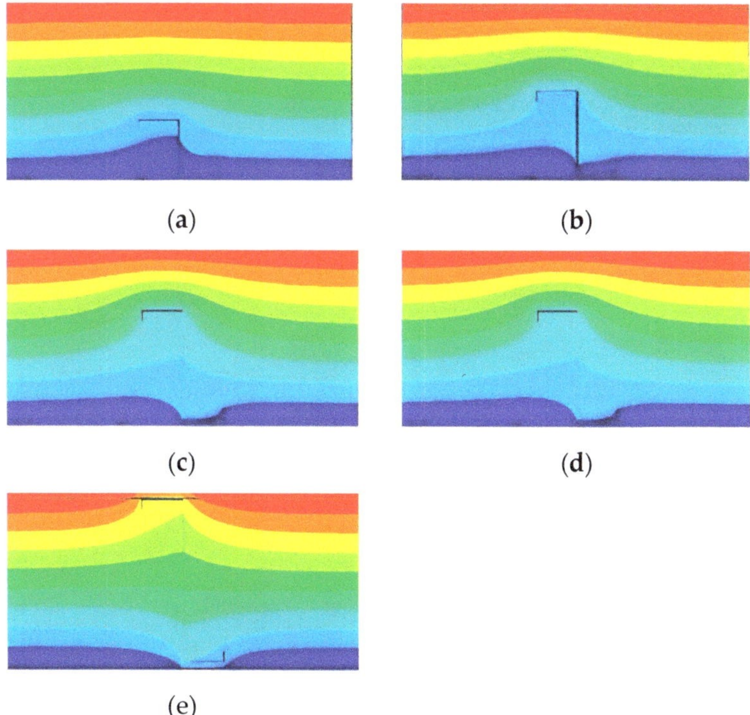

Figure 19. Temperature field with different web heights in type Z: (**a**) 100 mm; (**b**) 150 mm; (**c**) 200 mm; (**d**) 250 mm; (**e**) 300 mm.

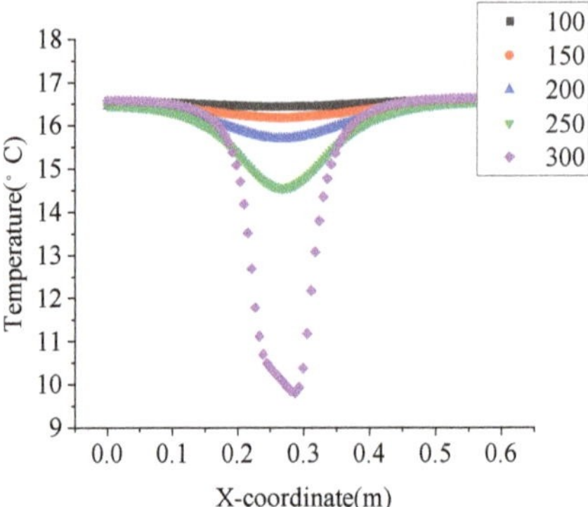

Figure 20. Internal surface temperature curves of different web heights in type Z.

- Comprehensive heat transfer coefficient analysis of wall panels with different flange heights and different offset distances

Figure 21 shows the influence curve of web widths of type Z steel keel section on the comprehensive heat transfer coefficient of the internal insulation wall panel. The results show that the comprehensive heat transfer coefficient of the wall panel increases with the increase in the height of the steel web, and the comprehensive heat transfer coefficient of the wall panel is 0.304 (W/m^2·K) when the height of the web is 100 mm. When the web height increases to 300 mm, the comprehensive heat transfer coefficient of the wall panel is 0.5381 (W/m^2·K). The width of the web has a great influence on the comprehensive heat transfer coefficient of the wall panel.

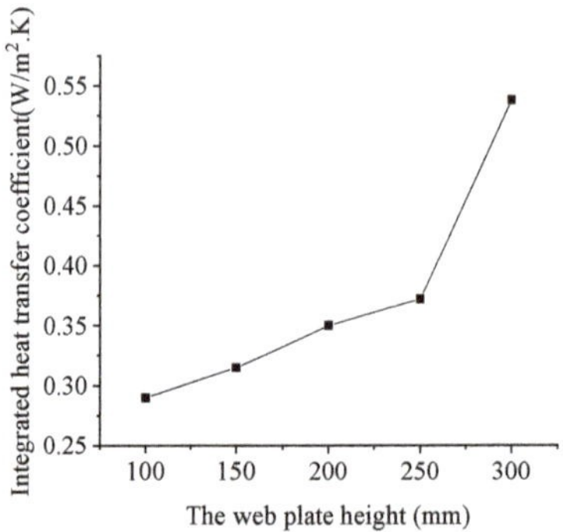

Figure 21. Comprehensive heat transfer coefficient curve in type Z.

3.2.4. Inverted Type Z

- Influence of different flange widths on the thermal performance of RS-LSWP

The thermal performance of the steel keel was analyzed with flange widths of inverted Z section of 75 mm, 100 mm, 125 mm and 150 mm, respectively. The construction diagram of the wall panel is shown in Figure 5d. With the increase in flange height of inverted type Z, the influence range of the cold bridge effect of the steel keel decreases gradually. The cold bridge effect has the largest influence range at 75 mm. With the increase in flange height, the influence of the web on the cold bridge effect decreases gradually. Thus, the flange height should be more than 100 mm (Figure 22). Meanwhile, with the increase in the height of the steel keel flange, the inner surface temperature of the steel keel wall panel gradually decreases, and the influence range of the keel flange on the inner surface temperature gradually expands. When the total flange height reaches 300 mm, the influence range of the temperature of the steel keel on the inner surface temperature of the wall panel increases. Therefore, for this form, non-penetrating steel keel sections should be selected (Figure 23).

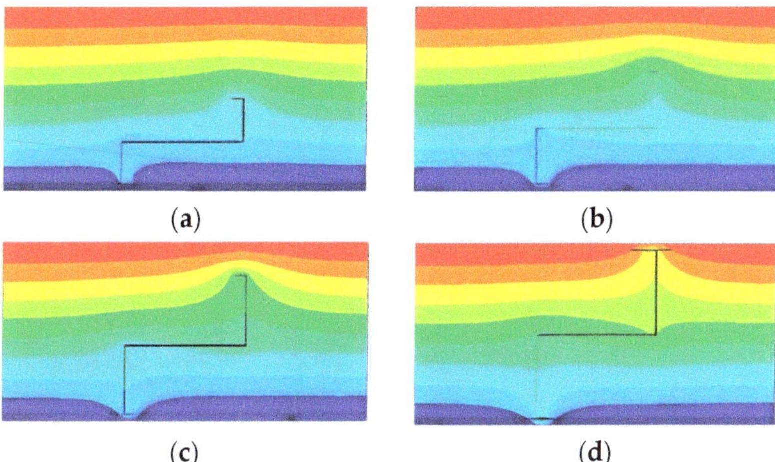

Figure 22. Temperature field with different flange widths in inverted type Z: (**a**) 75 mm; (**b**) 100 mm; (**c**) 125 mm; (**d**) 150 mm.

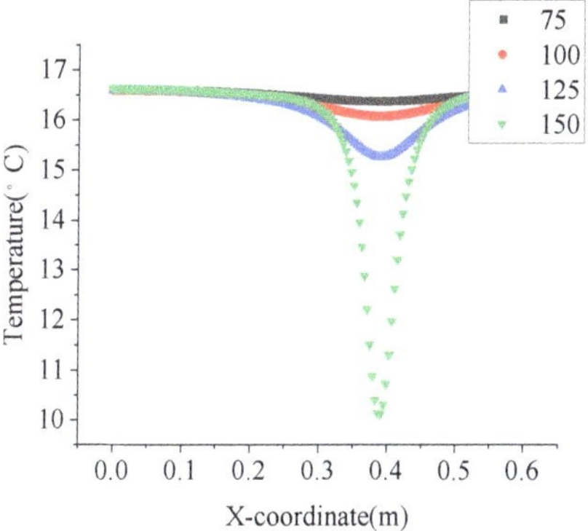

Figure 23. Internal surface temperature curves of different flange widths in inverted type Z.

- Influence of different web widths on the thermal performance of RS-LSWP

Considering that the height of the steel keel section web is 100 mm, 150 mm, 200 mm, 250 mm and 300 mm, respectively, the thermal performance of the wall panel is analyzed. The structure diagram of the wall panel is shown in Figure 5d. The results of temperature field calculation show that with the increase in web height of the steel keel, the influence range of the steel keel cold bridge effect increases gradually in the web height range, but decreases gradually in the flange width range. When the height of the web is above 250 mm, the cold bridge effect at the flange of the keel in the wall panel significantly decreases, and the influence range of the cold bridge effect of the keel also gradually decreases (Figure 24). The curve shows that the influence of the cold bridge effect on the inner surface temperature of the wall panel is unchanged because the steel keel is located outside of the wall panel. The increase in the web height of the steel keel does not affect the temperature variation rule of the inner surface of the wall panel (Figure 25).

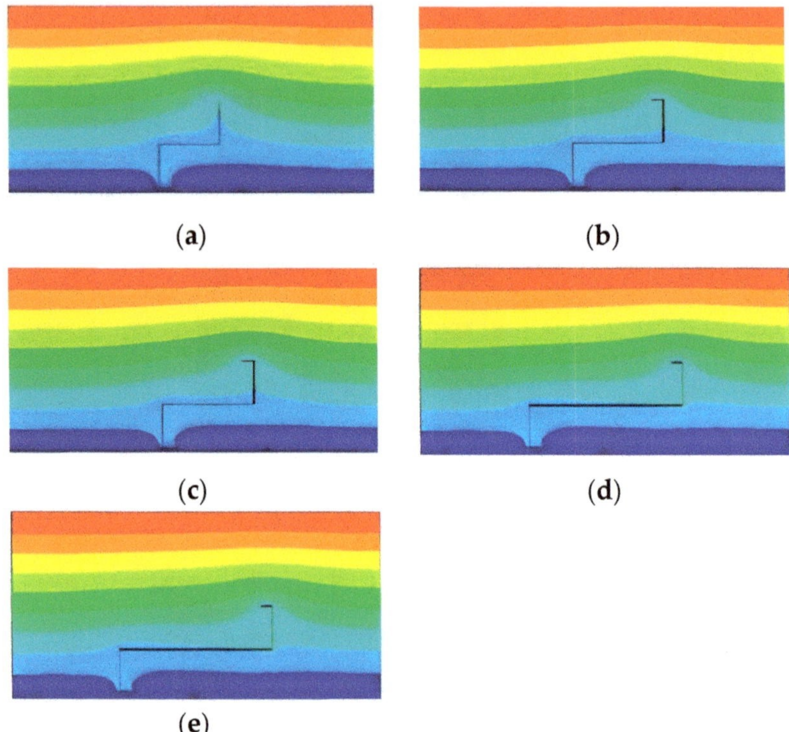

Figure 24. Temperature field with different web heights in inverted type Z: (**a**) 100 mm; (**b**) 150 mm; (**c**) 200 mm; (**d**) 250 mm; (**e**) 300 mm.

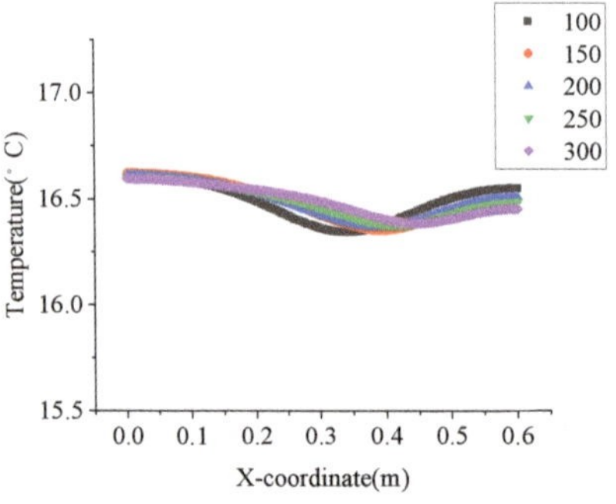

Figure 25. Internal surface temperature curves of different web heights in inverted type Z.

- Comprehensive heat transfer coefficient analysis of wall panels with different web widths and flange heights

Figure 26 shows the influence of different web heights and flange widths of inverted Z section steel keel on the comprehensive heat transfer coefficient of the wall panel. The results show that with the increase in the height of the steel keel web, the comprehensive heat transfer coefficient of the wall panel decreases gradually. The comprehensive heat transfer coefficient of wall panels with flange widths of 75 mm, 100 mm and 125 mm decreases slowly, while that with a flange width of 150 mm decreases rapidly. This is mainly because the keel with a width of 150 mm flange runs through the thickness of the wall panel, and the keel cold bridge effect is obvious. Therefore, for such a keel arrangement form, a non-penetrating steel keel arrangement is better.

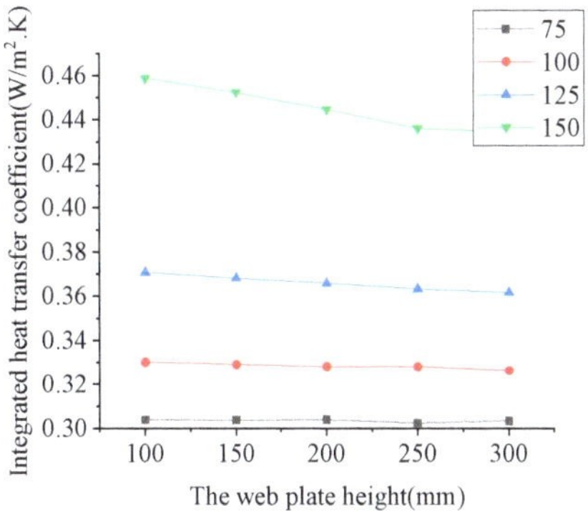

Figure 26. Comprehensive heat transfer coefficient curve of inverted type Z with different web heights and flange lengths.

3.2.5. Comparison of Thermal Performance of Different Steel Keel Structures

The comprehensive heat transfer coefficient and internal surface temperature of the four different keel arrangements were studied using quantitative comparative analysis. To ensure the principle of a single variable, the cross-section areas of different steel keel arrangements are calculated and analyzed. The results show that the inverted type Z and the antitype C keel structure can reduce the comprehensive heat transfer coefficient and improve the temperature variation law of the inner surface of the wallboard. Compared with the inverted type Z, the minimum temperature of the inner surface of the antitype C is higher, and the cold bridge phenomenon is the least obvious. Therefore, antitype C is the best keel arrangement scheme in the RS-LSWP (Figure 27).

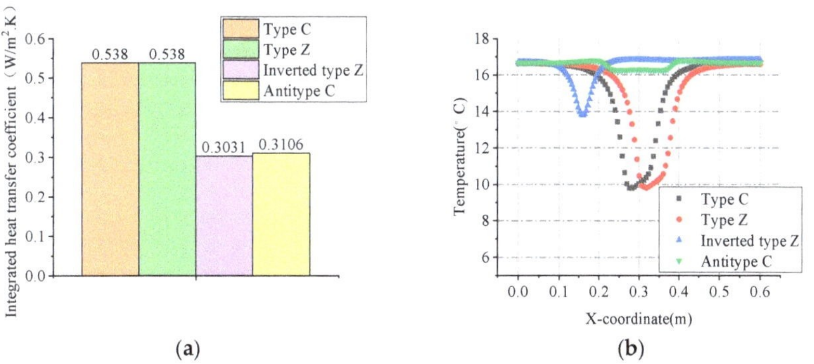

Figure 27. Comparison of thermal performance: **(a)** Comprehensive heat transfer coefficient; **(b)** Variation curve of inner surface temperature.

3.3. Discussion

To promote the rational utilization of crop fibers, such as rice straw, this study tries to combine rice straw and rice husk with magnesium cementitious material to develop an RSFC with high rice straw content and good physical and mechanical properties. The physical and mechanical properties, such as compressive strength, tensile strength and thermal conductivity of RSFC with different proportions were studied. In order to promote the practical application of RSFC, four different light steel keel structural forms are put forward in the light steel housing. The thermal performance of these wall panels is analyzed and some properties are discovered.

The physical and mechanical properties of RSFC are significantly affected by the change in RS content and length or the difference of cementing material ratio. When the length of RS increased from 5 mm to 15 mm, the tensile strength and compressive strength of RSFC first decreased and then increased. The higher the content of RS, the lower the compressive strength of RSFC. When the content is 12% of RS, the tensile strength and compressive strength of RSFC decrease. When the modifier increased from 0.5% to 1.0%, the tensile strength and compressive strength of RSFC increased. When the ratio of MgO: $MgSO_4$ changes from 8:1 to 9:1, the tensile strength and compressive strength of RSFC decrease. The 14-day strength of the RSFC reaches more than 85% of the final strength. Also, the higher the RS content, the lower the RSFC density. The ratio of $MgO:MgSO_4:H_2O$ changed from 8:1:20 to 9:1:20, and the density of RSFC decreased to a large extent. However, when the straw content is 12%, the volume expansion generated by magnesium hydroxide in the test block can be released in the gap due to the increase in RS content, and there are almost no cracks on the concrete surface.

It is recommended to adopt the RS ratio of length 5 mm, mass content 12%, the mass ratio of $MgO:MgSO_4:H_2O$ 8:1:20, and conditioner content 1% for RS-LSWP. The standard compressive strength, tensile strength, and thermal conductivity of the mixture are 2.2 MPa, 0.64 MPa, and 0.0862 W/(m·K), respectively. Compared with the results of previous studies [32], the straw content in RSFC was greatly improved using a simpler treatment method proposed in this study. Although the mechanical properties and thermal insulation properties of RSFC are lower than those of SMLCS to a certain extent, it does not affect the popularization and application of RSFC in certain scenarios. Rock wool and foamed concrete are commonly used as fillings in light steel housing wall panels [26–32,44]. The thermal conductivity of rock wool is 0.04 W/(m·K), but the strength is close to zero [43]. The thermal conductivity of foamed concrete is above 0.5 W/(m·K), and the compressive strength is about 0.65 MPa [43]. RSFC has good thermal conductivity and the highest strength. To sum up, RSFC has certain application prospects.

Further, to study and optimize the thermal performance of RS-LSWP, four different light steel keel structural forms, including type C, antitype C, type Z and inverted type Z were designed. The influences of the height of the web depth, the width of the flange, and the offset distance of the opposite keel on the internal temperature field of the wall panel, the variation law of the temperature on the inner and outer surfaces of the wall panel and the comprehensive heat transfer coefficient are discussed. Finally, to compare the advantages and disadvantages of the four keel construction methods, the comprehensive heat transfer coefficient and internal surface temperature of the four different keel arrangements were studied using a quantitative comparative analysis. The result shows that the inverted type Z and the antitype C keel structure can reduce the comprehensive heat transfer coefficient and improve the temperature variation law of the inner surface of the wallboard. Antitype C is the best keel arrangement scheme in the RS-LSWP.

4. Conclusions

To promote the rational application of RS, reduce environmental pollution, and improve the conditions of rural housing, the physical and mechanical properties of RSFC with different proportions were studied, and the suitable mix ratio of RSFC was selected in this paper. The thermal properties of four different types of RS-LSWPs with straw concrete were simulated using the finite element software. The main conclusions are as follows:

1. The mechanical properties of RSFC can be improved by increasing the length of RS and the content of the modifier. The mechanical properties of RSFC decreased with the increase in straw content and with the mix of MgO and MgSO4. When RS content is 12%, the compressive strength, tensile strength, apparent density, and thermal conductivity of straw concrete are reduced significantly, and the crack can be avoided effectively. The 14-day strength of RSFC reached more than 85% of the final strength.

2. The ratio of RSFC applied in RS-LSWP should be C05H12*(1.0). The standard compressive strength, tensile strength, and thermal conductivity of the mixture are 2.2 MPa, 0.64 MPa, and 0.0862 W/(m·K), respectively.
3. The inverted type Z and the antitype C keel structure can reduce the comprehensive heat transfer coefficient and improve the temperature curve of the inner surface of the wall panel. The antitype C keel structure is most suitable for RS-LSWP.

Author Contributions: Investigation, Y.W. (Yuqi Wu) and Y.W. (Yunqiang Wu); Data curation, Y.W. (Yunqiang Wu); Writing—original draft preparation, Y.W. (Yuqi Wu); Writing—review and editing, Y.W. (Yunqiang Wu) and Y.W. (Yue Wu); Visualization, Y.W. (Yuqi Wu) and Y.W. (Yunqiang Wu); Supervi-sion, Y.W. (Yue Wu); Project administration, Y.W. (Yue Wu); Funding acquisition, Y.W. (Yue Wu). All authors have read and agreed to the published version of the manuscript.

Funding: This research was funded by the National Key R&D Program of China, grant number 2019YFD1101004.

Institutional Review Board Statement: Not applicable.

Informed Consent Statement: Not applicable.

Data Availability Statement: The date presented in this study are available in the article itself.

Acknowledgments: The authors wish to express their gratitude for the financial support that has made this study possible.

Conflicts of Interest: The authors declare no conflict of interest.

References

1. Franz, J.; Morse, S.; Pearce, J.M. Low-Cost Pole and Wire Photovoltaic Racking. *Energy Sustain. Dev.* **2022**, *68*, 501–511. [CrossRef]
2. Syafiq, A.; Balakrishnan, V.; Ali, M.S.; Dhoble, S.J.; Rahim, N.A.; Omar, A.; Bakar, A.H.A. Application of Transparent Self-Cleaning Coating for Photovoltaic Panel: A Review. *Curr. Opin. Chem. Eng.* **2022**, *36*, 100801. [CrossRef]
3. Liu, J.; Xie, X.; Li, L. Experimental Study on Mechanical Properties and Durability of Grafted Nano-SiO$_2$ Modified Rice Straw Fiber Reinforced Concrete. *Constr. Build. Mater.* **2022**, *347*, 128575. [CrossRef]
4. Rehman, M.S.U.; Umer, M.A.; Rashid, N.; Kim, I.; Han, J.-I. Sono-Assisted Sulfuric Acid Process for Economical Recovery of Fermentable Sugars and Mesoporous Pure Silica from Rice Straw. *Ind. Crops Prod.* **2013**, *49*, 705–711. [CrossRef]
5. Mehta, P.K. *Pozzolanic and Cementitious Byproducts as Mineral Admixtures for Concrete—A Critical Review*; ACI Special Publication 79; American Concrete Institute: Farmington Hills, MI, USA, 1983; 46p.
6. Roselló, J.; Soriano, L.; Santamarina, M.P.; Akasaki, J.L.; Monzó, J.; Payá, J. Rice Straw Ash: A Potential Pozzolanic Supplementary Material for Cementing Systems. *Ind. Crops Prod.* **2017**, *103*, 39–50. [CrossRef]
7. Thomas, B.S.; Yang, J.; Mo, K.H.; Abdalla, J.A.; Hawileh, R.A.; Ariyachandra, E. Biomass Ashes from Agricultural Wastes as Supplementary Cementitious Materials or Aggregate Replacement in Cement/Geopolymer Concrete: A Comprehensive Review. *J. Build. Eng.* **2021**, *40*, 102332. [CrossRef]
8. Abou-Sekkina, M.M.; Issa, R.M.; Bastawisy, A.E.-D.M.; El-Helece, W.A. Characterization and Evaluation of Thermodynamic Parameters for Egyptian Heap Fired Rice Straw Ash (RSA). *Int. J. Chem.* **2010**, *2*, 81. [CrossRef]
9. Munshi, S.; Sharma, R.P. Utilization of Rice Straw Ash as A Mineral Admixture in Construction Work. *Mater. Today Proc.* **2019**, *11*, 637–644. [CrossRef]
10. Pandey, A.; Kumar, B. Effects of Rice Straw Ash and Micro Silica on Mechanical Properties of Pavement Quality Concrete. *J. Build. Eng.* **2019**, *26*, 100889. [CrossRef]
11. Agwa, I.S.; Omar, O.M.; Tayeh, B.A.; Abdelsalam, B.A. Effects of Using Rice Straw and Cotton Stalk Ashes on the Properties of Lightweight Self-Compacting Concrete. *Constr. Build. Mater.* **2020**, *235*, 117541. [CrossRef]
12. Pandey, A.; Kumar, B. Evaluation of Water Absorption and Chloride Ion Penetration of Rice Straw Ash and Microsilica Admixed Pavement Quality Concrete. *Heliyon* **2019**, *5*, e02256. [CrossRef] [PubMed]
13. Pandey, A.; Kumar, B. Investigation on the Effects of Acidic Environment and Accelerated Carbonation on Concrete Admixed with Rice Straw Ash and Microsilica. *J. Build. Eng.* **2020**, *29*, 101125. [CrossRef]
14. Bories, C.; Borredon, M.-E.; Vedrenne, E.; Vilarem, G. Development of Eco-Friendly Porous Fired Clay Bricks Using Pore-Forming Agents: A Review. *J. Environ. Manag.* **2014**, *143*, 186–196. [CrossRef]
15. Duppati, S.; Gopi, R. Strength and Durability Studies on Paver Blocks with Rice Straw Ash as Partial Replacement of Cement. *Mater. Today Proc.* **2022**, *52*, 710–715. [CrossRef]
16. Singh, S.; Maiti, S.; Bisht, R.S.; Balam, N.B.; Solanki, R.; Chourasia, A.; Panigrahi, S.K. Performance Behaviour of Agro-Waste Based Gypsum Hollow Blocks for Partition Walls. *Sci. Rep.* **2022**, *12*, 3204. [CrossRef]

17. Tachaudomdach, S.; Hempao, S. Investigation of Compression Strength and Heat Absorption of Native Rice Straw Bricks for Environmentally Friendly Construction. *Sustainability* **2022**, *14*, 12229. [CrossRef]
18. Rajput, A.; Gupta, S.; Bansal, A. A Review on Recent Eco-Friendly Strategies to Utilize Rice Straw in Construction Industry: Pathways from Bane to Boon. *Environ. Sci. Pollut. Res.* **2022**, *30*, 11272–11301. [CrossRef] [PubMed]
19. Ataie, F. Influence of Rice Straw Fibers on Concrete Strength and Drying Shrinkage. *Sustainability* **2018**, *10*, 2445. [CrossRef]
20. Nguyen, C.V.; Mangat, P.S. Properties of Rice Straw Reinforced Alkali Activated Cementitious Composites. *Constr. Build. Mater.* **2020**, *261*, 120536. [CrossRef]
21. Nazerian, M.; Sadeghiipanah, V. Cement-Bonded Particleboard with a Mixture of Wheat Straw and Poplar Wood. *J. For. Res.* **2013**, *24*, 381–390. [CrossRef]
22. Karade, S.R.; Irle, M.; Maher, K. Assessment of Wood-Cement Compatibility: A New Approach. *Holzforschung* **2003**, *57*, 672–680. [CrossRef]
23. Canovas, M.F.; Selva, N.H.; Kawiche, G.M. New economical solutions for improvement of durability of Portland cement mortars reinforced with sisal fibers. *Mater. Struct.* **1992**, *25*, 417–422. [CrossRef]
24. Ma, L.F.; Kuroki, Y.; Eusebio, D.A.; Nagadomi, W.; Kawai, S.; Sasaki, H. Manufacture of bamboo cement composites. 1. Hydration characteristics of bamboo-cement mixtures. *Mokuzai Gakkaishi* **1996**, *42*, 34–42.
25. Sorrel, S. On a new magnesium cement. *C. R. Acad. Sci.* **1867**, *65*, 102–104.
26. Zhou, X.; Li, Z. Light-Weight Wood–Magnesium Oxychloride Cement Composite Building Products Made by Extrusion. *Constr. Build. Mater.* **2012**, *27*, 382–389. [CrossRef]
27. Sglavo, V.M.; De Genua, F.; Conci, A.; Ceccato, R.; Cavallini, R. Influence of Curing Temperature on the Evolution of Magnesium Oxychloride Cement. *J. Mater. Sci.* **2011**, *46*, 6726–6733. [CrossRef]
28. Chau, C.K.; Chan, J.; Li, Z. Influences of Fly Ash on Magnesium Oxychloride Mortar. *Cem. Concr. Compos.* **2009**, *31*, 250–254. [CrossRef]
29. Deng, D. The Mechanism for Soluble Phosphates to Improve the Water Resistance of Magnesium Oxychloride Cement. *Cem. Concr. Res.* **2003**, *33*, 1311–1317. [CrossRef]
30. Xiao, J.; Zuo, Y.; Li, P.; Wang, J.; Wu, Y. Preparation and Characterization of Straw/Magnesium Cement Composites with High-Strength and Fire-Retardant. *J. Adhes. Sci. Technol.* **2018**, *32*, 1437–1451. [CrossRef]
31. Li, Y.; Yu, H.; Zheng, L.; Wen, J.; Wu, C.; Tan, Y. Compressive Strength of Fly Ash Magnesium Oxychloride Cement Containing Granite Wastes. *Constr. Build. Mater.* **2013**, *38*, 1–7. [CrossRef]
32. Wang, J.; Zuo, Y.; Xiao, J.; Li, P.; Wu, Y. Construction of Compatible Interface of Straw/Magnesia Lightweight Materials by Alkali Treatment. *Constr. Build. Mater.* **2019**, *228*, 116712. [CrossRef]
33. Dao, T.N.; van de Lindt, J.W. Seismic Performance of an Innovative Light-Gauge Cold-Formed Steel Mid-Rise Building. In Proceedings of the Structures Congress 2012, Chicago, IL, USA, 29–31 March 2012; American Society of Civil Engineers: Reston, VA, USA, 2012.
34. Li, Z.; Guan, K. Research on Housing Industrialization and Necessity of Its Development. *J. Harbin Univ. Civ. Eng. Archit.* **1999**, 98–102.
35. Girard, J.; Tarpy, T. Shear Resistance of Steel-Stud Wall Panels Shear Resistance of Steel-Stud Wall Panels. In Proceedings of the 6th International Specialty Conference on Cold-Formed Steel Structures, St. Louis, MO, USA, 16–17 November 2020.
36. He, B.-K.; Guo, P.; Wang, Y.-M. Experimental investigation on high strength cold-formed steel framing wall studs under axial compression loading. *J. Xi'an Univ. Archit. Technol.* **2008**, *40*, 567–573+579.
37. Bian, J.; Cao, W.; Xiong, C. Experiment on the compression performance of composite wall of cold-formed steel and tailing microcrystalline foam glass slab. *J. Harbin Inst. Technol.* **2019**, *51*, 86–93.
38. Qian, Z.; Pan, J.; Zhang, L. Research on Flexural Behavior of Light-Gauge Steel Stud Concrete Composite Wall Panels. *Ind. Constr.* **2020**, *50*, 32–37+4.
39. Wu, J.; Wang, Q.; Li, B. A study on compressive properties of cold-formed thin-walled steel frame fly ash ceramsite concrete wallboard. *New Build. Mater.* **2018**, *45*, 97–100. [CrossRef]
40. Lu, M.J. Thermal Optimization Strategy of External Wall in Low-Rise Light Steel Keel Structure Based on the Theory of Thermal Bridge. *AMR* **2013**, *671–674*, 2150–2153. [CrossRef]
41. GBT50081-2019; Standard Test Method for Physical and Mechanical Properties Of Concrete. China Building Industry Press: Beijing, China, 2019.
42. Wang, S.; Zhang, D.; Liu, G.; Wang, W.; Hu, M. Application of Hot-Wire Method for Measuring Thermal Conductivity of Fine Ceramics. *Mater. Sci.* **2016**, *22*, 560–564. [CrossRef]
43. GB50176-2016; Code for Thermal Design of Civil Building. China Building Industry Press: Beijing, China, 2016.
44. Li, F.; Li, J.; Xiao, M.; Ren, M. Impact Resistance of Lightweight Steel-Framed Foamed Concrete Composite Wall Panels. *Bull. Chin. Ceram. Soc.* **2022**, *41*, 68–75.

 *sustainability*

Article

An Improved Multi-Objective Optimization and Decision-Making Method on Construction Sites Layout of Prefabricated Buildings

Gang Yao, Rui Li and Yang Yang *

Key Laboratory of New Technology for Construction of Cities in Mountain Area, School of Civil Engineering, Chongqing University, Chongqing 400045, China
* Correspondence: 20121601009@cqu.edu.cn

Abstract: Construction site layout planning (CSLP) that considers multi-objective optimization problems is essential to achieving sustainable construction. Previous CSLP optimization methods have applied to traditional cast-in-place buildings, and they lack the application for sustainable prefabricated buildings. Furthermore, commonly used heuristic algorithms still have room for improvement regarding the search range and computational efficiency of optimal solution acquisition. Therefore, this study proposes an improved multi-objective optimization and decision-making method for layout planning on the construction sites of prefabricated buildings (CSPB). Firstly, the construction site and temporary facilities are expressed mathematically. Then, relevant constraints are determined according to the principles of CSLP. Ten factors affecting the layout planning on the CSPB are identified and incorporated into the method of layout planning on the CSPB in different ways. Based on the elitist non-dominated sorting genetic algorithm (NSGA-II), an improved multiple population constraint NSGA-II (MPC-NSGA-II) is proposed. This introduces the multi-population strategy and immigration operator to expand the search range of the algorithm and improve its computational efficiency. Combined with the entropy weight and technique for order preference by similarity to an ideal solution (TOPSIS), improved multi-objective optimization and decision for the CSLP model is developed on the CSPB. Practical cases verify the effectiveness and superiority of the algorithm and model. It is found that the proposed MPC-NSGA-II can solve the drawbacks of the premature and low computational efficiency of NSGA-II for multi-constrained and multi-objective optimization problems. In the layout planning on the CSPB, the MPC-NSGA-II algorithm can improve the quality of the optimal solution and reduce the solution time by 75%.

Keywords: sustainable construction; construction site of prefabricated buildings; multi-objective layout and optimization; MPC-NSGA-II algorithm

Citation: Yao, G.; Li, R.; Yang, Y. An Improved Multi-Objective Optimization and Decision-Making Method on Construction Sites Layout of Prefabricated Buildings. *Sustainability* **2023**, *15*, 6279. https://doi.org/10.3390/su15076279

Academic Editors: Zhihua Chen, Qingshan Yang, Yue Wu, Yansheng Du and Jurgita Antuchevičienė

Received: 24 February 2023
Revised: 22 March 2023
Accepted: 4 April 2023
Published: 6 April 2023

1. Introduction

With increasing attention to environmental protection and sustainability, the construction industry is becoming aware of its role in environmental protection and sustainability [1]. The new intelligent construction technology is a vital instrument for achieving energy management, environmental protection and safety, and improving the sustainability of buildings [2]. Prefabricated buildings are an important component of sustainable construction, and they are typically space-constrained projects. Prefabricated buildings require adequate resources for various materials, equipment, and temporary facilities for the extensive prefabrication operations on the CSPB [3,4]. Among these resources, hoisting equipment and prefabricated components occupy a large space on the CSPB [5]. Unreasonable layout planning of CSPB can cause conflict in the workspace, increase the distance of material transportation, and increase the workload of mechanical equipment. Furthermore, a cluttered layout of CSPB can increase the safety uncertainties on the construction site [6,7].

Relevant scholars have explored the CSPL problem of traditional cast-in-place buildings [8–11]. Currently, the CSPL optimization methods are mainly divided into mathematical programming [12,13] and heuristic algorithms [14–16]. The mathematical programming method mainly uses integer programming [17], linear programming [18], nonlinear programming [19], dynamic programming [20], and other mathematical methods [21] to produce a single-objective or multi-objective optimization function of the exact solution to provide a reasonable scheme of CSPL. However, its computational complexity is large, its solution time is long, and it is only suitable for solving small-scale problems, often making it impractical for engineering purposes [18,19]. Heuristic algorithms emphasize solving combinatorial optimization problems based on empirical rules. At this stage, heuristic algorithms usually focus on simulating natural selection and the natural evolution of organisms, such as genetic algorithms [22,23], particle swarm algorithms [24], and ant colony algorithms [25]. Heuristic algorithms are used to find the suboptimal solution of a problem or its optimal solution with a certain probability. Their good generality, stability, and fast convergence make them more commonly used in engineering [22,25]. Genetic algorithms have global search capabilities and can quickly solve complex non-linear problems. However, their programming implementation can be complex, and they have a slower search speed, which can tend to fall into prematureness [22]. The non-dominated sorting genetic algorithm (NSGA) proposes a non-dominated ranking criterion based on the classical genetic algorithm. The NSGA algorithm has shown significant advantages in solving multi-objective optimization problems. However, there are also complications, including great computational effort, lack of optimal individual retention schemes, and difficulty determining shared parameters [26]. The NSGA-II algorithm introduces improvements such as fast non-dominated sorting, crowding degree, and elite strategy based on the NSGA algorithm, which can be used without setting any parameters and reduces computational complexity. However, the NSGA-II algorithm tends to be premature in multi-constrained, multi-objective optimization problems, making it challenging to obtain the entire Pareto front surface and thus losing part of the optimal solution [27–29].

The heuristic algorithm commonly used in CSPL needs further improvement and optimization. For the characteristics of CSPB, a more reasonable and improved CSPL method is needed to achieve sustainable building construction [30,31]. The authors propose an improved multi-objective optimization and decision-making method in Section 2. Firstly, the construction site and temporary facilities are expressed mathematically, then three constraints are identified. Furthermore, ten factors affecting the layout planning on the CSPB are identified and incorporated into the method of layout planning on the CSPB in different ways. The MPC-NSGA-II algorithm applicable to layout planning on the CSPB is proposed. Furthermore, the solutions output from the MPC-NSGA-II algorithm is evaluated and selected by combining the entropy weight-TOPSIS method. In Section 3, a multi-objective optimization and decision model for layout planning of CSPB is proposed through practical engineering cases. The parameters required for calculating the improved algorithm are determined. The improved algorithm and the model for layout planning on the CSPB are validated using MATLAB 2020b software in Section 4. Meanwhile, the MPC-NSGA-II algorithm is compared with the NSGA-II algorithm in the layout planning on the CSPB. Finally, the conclusions of this study are drawn.

2. The Improved Multi-Objective Optimization and Decision-Making Method

2.1. Construction Site and Temporary Facilities Analysis

Many factors, including site area, structure type, duration, and transportation conditions, influence the CSPL. Necessary assumptions about the construction site and temporary facilities need to be made to facilitate the construction of a multi-objective optimization and decision model for the layout planning on the CSPB. Currently, the common assumption methods used on construction sites include location distribution, continuous space, and raster methods. The common assumptions for temporary facilities on construction sites include the ignoring facility size method, the actual shape method, and the approximate geometric shape method (AGSM). Different expressions of the construction site and temporary facilities can affect the process of searching for the optimal solution of the model.

After arranging and combining the construction site assumptions and construction site temporary facilities assumptions, all optional combinations are shown in Figure 1.

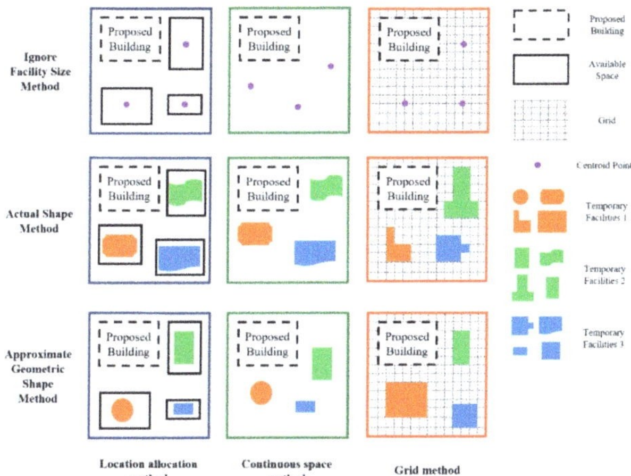

Figure 1. Construction site assumptions and construction site temporary facilities assumptions combination arrangement diagram.

The location assignment method has a simple calculation process. However, its way of determining the predetermined location in advance leads to the limitation of the model [32]. The continuous space method is closest to the actual site space, but its calculation process is complex and requires much time [33]. The raster method considers the advantages of the location assignment and continuous space methods. It balances the computational complexity and the accuracy of the optimization results [34]. Furthermore, the raster method is more flexible, thus, applicable to different construction sites. Representing temporary facilities by shape center point in the ignoring facility size method can simplify the calculation but is different from the actual situation [10]. In the actual shape method, the horizontal projection of the actual shape of the construction site facilities represents the temporary facilities closest to the actual site space [35]. However, it has more constraints on the construction function and is more difficult to calculate. Considering the influence of the facility shape on the search results and the complexity of the calculation, the basic geometry of the temporary facility represented by the AGSM can envelop the actual edge of the facility [31]. Therefore, the construction site of the model is assumed by the grid method, and the AGSM represents the temporary facilities.

Considering other influencing factors, the prerequisite assumptions of the improved multi-objective optimization and decision-making method include: (1) The construction site space is divided by the grid method, and the AGSM represents the construction site's temporary facilities; (2) It is flat and even inside the construction site; (3) The southwest corner of the construction site is considered the origin of the arrangement for later calculations; (4) The location of the fixtures is predetermined and will not be changed; (5) The model uses the centroid position of the field facilities to represent the real position of the temporary facilities; (6) The model assumes that the shape and dimensions of the site facilities remain unchanged throughout the project construction period.

2.2. Constraints Analysis

2.2.1. Site Boundary Constraints

The site boundary constraint means that temporary facilities must be placed within the red line boundary of the construction site regardless of any construction phase of the project. Furthermore, it should be ensured that the temporary facilities boundary of the construction site is kept at a sufficient safety distance from the construction fence.

Assume that the coordinates of the form center of the temporary facilities to be arranged at the construction site are (x_i, y_i). The length in the x-direction is l_i. The length in the y-direction is h_i. The red line horizontal coordinate range of the construction site is $a_1 \sim a_2$. The range of vertical coordinates is $b_1 \sim b_2$. Therefore, the coordinates of the temporary facility i should meet Equation (1).

$$\left\{ \begin{array}{l} y_i - \frac{h_i}{2} - b_1 - \varphi \geq 0 \\ y_i + \frac{h_i}{2} - b_2 + \varphi \leq 0 \\ x_i - \frac{l_i}{2} - a_1 - \varphi \geq 0 \\ x_i + \frac{l_i}{2} - a_2 + \varphi \leq 0 \end{array} \right\} \tag{1}$$

where φ is the safety distance between the consideration of temporary facilities and the construction fence. In the actual CSPB, the value of φ is usually taken as 3.0 m.

2.2.2. Facility Coverage Constraints

The facility coverage constraint means that there should be no coverage conflicts between individual construction facilities. Facility coverage constraints need to be satisfied to avoid spatial conflicts. With construction facility i and construction facility j, the required fire distance between the two facilities is W_{ij}. The facility coverage constraint stipulates that at least one of the criteria in Equation (2) should be met.

$$\left\{ \begin{array}{l} |x_i - x_j| - \frac{l_i + l_j}{2} \geq W_{ij} \\ |y_i - y_j| - \frac{h_i + h_j}{2} \geq W_{ij} \end{array} \right\} \tag{2}$$

2.2.3. Tower Crane Coverage Constraints

The tower crane boom needs to cover all temporary production facilities as far as possible. It can safety the fixed tower crane layout principles while avoiding the secondary handling of components and raw materials in the field as far as possible.

Assume that the site coordinates of the fixed tower crane are (x_t, y_t) and the boom length of the tower crane is R_t. The temporary construction site facilities and the tower crane should be met by Equation (3).

$$\sqrt{(x_t - x_c)^2 + (y_t - y_c)^2} \leq R_t \tag{3}$$

Temporary living spaces and office facilities should be as far from the tower crane's coverage as possible to protect staff safety.

2.3. Optimization Factor Ranking Analysis

A multi-objective optimization problem is one in which multiple objectives need to be achieved in each scenario. However, there is generally a conflict between objectives, and the optimization of one objective is at the cost of the deterioration of other objectives, so it is not easy to have a unique optimal solution. Therefore, the determination of optimization objectives must be based on the importance of the influencing factors.

Before determining the optimization objectives of the model, various factors affecting the layout of temporary facilities on CSPB need to be clarified. Through literature analysis, 16 influencing factors with high frequencies in relevant construction site temporary facilities arrangement studies were initially summarized.

An expert scoring method distributed questionnaires to industry experts and relevant practitioners. Based on their opinions, the importance of 16 influencing factors in determining the arrangement scheme of temporary facilities on CSPB was analyzed. The sources and composition of the experts are shown in Table 1. The contents of the questionnaire are in the Supplementary Materials.

Table 1. Number and source of experts.

Institution	Number of People	Percentage
Construction enterprise	16	48.5%
College and universities	8	24.2%
Design institute	9	27.3%
Total	33	100%

The reliability analysis of this questionnaire was performed using Cronbach's alpha coefficient method. The reliability coefficient value of 0.858 was obtained using SPSS 26 software analysis [36]. Therefore, the data obtained from this expert scoring is stable and reliable. The results of the questionnaire analysis are shown in Figure 2.

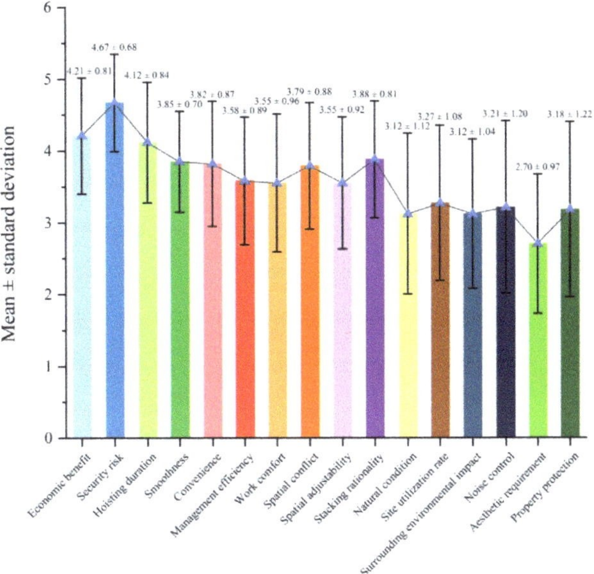

Figure 2. Statistical results of expert scores.

From Figure 2, we can see that the standard deviation of the five factors of "Natural condition", "Site utilization rate", "Surrounding environmental impact", "Noise control", and "Property protection" is greater than 1. It means that the expert's opinions are inconsistent, so they are excluded. The average score for "Aesthetic requirements" was only 2.70. According to the scoring rules, an average score of less than 3 is not a significant factor, so this factor was removed. A total of 10 influencing factors were finally included in the model consideration. In descending order of average score, they are: "Security risk", "Economic benefit", "Hoisting duration", "Stacking rationality", "Smoothness", "Convenience", "Spatial conflict", "Management efficiency", "Spatial adjustability", and "Work comfort". Among them, the three influencing factors, "Security risk", "Economic benefit", and "Hoisting duration", have an average score greater than 4. Therefore, these three influencing factors should be focused on the temporary facility layout of CSPB. The prefabricated component combination "stacking reasonableness" is taken as the input constraint of the model to ensure the proper stacking of prefabricated components. Two influencing factors, "space conflict" and "working comfort", are considered in the constraints. The rest of the factors are incorporated into the decision factors.

The top three influencing factors are taken as the optimization objectives. The objective functions are minimizing security risks, maximizing economic benefits, and minimizing hoisting durations.

2.4. Objective Functions Determination
2.4.1. Security Risk Function

Construction facilities can be divided into risk source facilities prone to security risks and vulnerable facilities that need protection. The construction security risk value can be quantified by analyzing the interaction process of the two types of facilities. Assuming that the hazard transfer value of the facility is H, and the vulnerability of the facility is V. Then the security risk interaction value R can be obtained from Equation (4).

$$R = H \times V \tag{4}$$

where H is the hazard transfer matrix and the individual element values represent the magnitude of the hazard transfer values.

The diagonal elements in H are the hazard levels between construction site facilities, which could be calculated from Equation (5).

$$H = \begin{bmatrix} h_{11} & & \\ & \ddots & \\ & & h_{nn} \end{bmatrix} \tag{5}$$

The hazard generated by the source decays with distance, and the remaining elements in H can be determined according to the law of risk decay from Figure 3 and Equation (6).

$$h_{ij} = \max \left\{ \begin{array}{c} h_{ii} + \frac{dH}{dd} \times d_{ij} \times \rho \\ 0 \end{array} \right. \tag{6}$$

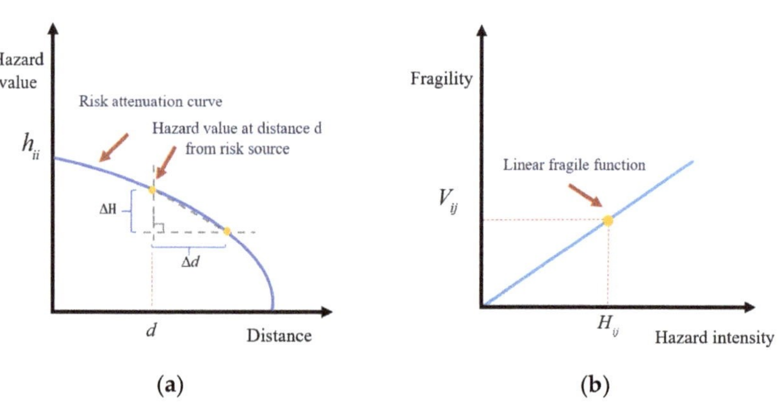

Figure 3. Risk attenuation curve and Linear fragile function (**a**) Risk attenuation curve; (**b**) Linear fragile function.

When $i = j$, $\rho = 0$, when $i \neq j$, $\rho = 1$, d_{ij} is the Euclidean distance. The relevant research suggests that the slope of the risk attenuation curve takes the value of 0.01 at the construction site [37].

The hazard transfer matrix is normalized in Equation (7).

$$h_{ij}{}^* = \frac{h_{ij}}{\max[h_{ii}]} \tag{7}$$

It is assumed that the hazards do not occur simultaneously. Therefore, the total risk of the construction site is the accumulation of the risks arising from each object. The objective function of security risk is set to minimize the potential security risk. The objective function of security risk can be deduced in Equation (8):

$$F_1 = \min \sum_{i=1}^{n} \sum_{j=1}^{n} R_{ij} = \min\{H^*V^*\} \tag{8}$$

H^* in the equation is the normalized hazard transfer matrix and V^* is the normalized fragility matrix.

2.4.2. Economic Benefit Function

The economy is one of the critical concerns of decision-makers in engineering project management. Some researchers have shown that the temporary facilities on the CSPL significantly impact the transportation costs of components and raw materials within the construction site. A reasonable scheme of CSPL can significantly reduce the related costs [38]. In addition, the location of temporary facilities may change during different stages of the project. This results in the cost of changes to the temporary facilities due to dismantling, relocation, and installation. The second objective function F_2 is to minimize the sum of the above two costs to maximize the economic benefits of the resulting construction site temporary facilities layout solution. The mathematical expression for F_2 is Equation (9).

$$F_2 = \min\left\{ \sum_{i=1}^{n} \sum_{j=1}^{i} C_{ij}d_{ij} + \sum_{k=1}^{n} \sum_{t=1}^{T} \left(C_{Dk} + C_k d_k^{(t-1,t)} + C_{Bk} \right) z_{kt} \right\} \tag{9}$$

Where C_{ij} is the transportation cost per unit distance between construction facility i and construction facility j; d_{ij} is the distance between construction facility i and construction facility j; C_{Dk} is the cost of dismantling temporary facilities k; C_k is the unit distance movement cost of temporary facility k; $d_k^{(t,t+1)}$ is the spatial distance between temporary facility k in stage 1 and stage 2; C_{Bk} is the cost of installation required for the rearrangement of temporary facilities k; z_{kt} is the value for judging the change of location of temporary facility k. When a change occurs, z_{kt} takes the value of 1, otherwise, it takes the value of 0. The value of C_{Dk}, C_k, and C_{Bk} need to be determined in accordance with the relevant standards, combined with the actual project works to determine the value of parameters.

2.4.3. Hoisting Duration Function

Prefabricated components are significant in number and individual weight, and installation machines are used frequently on the CSPB. The hoisting objective function is to make the shortest hoisting duration for prefabricated components through a reasonable layout of temporary facilities on the CSPB.

The hoisting process of a single prefabricated component contains six operations: tying, hoisting, alignment, temporary fixing, alignment, and final fixing [5]. The hoisting action can be divided into horizontal motion (horizontal tangential motion, horizontal radial running) and vertical motion. Therefore, the hoisting duration of a single prefabricated component can be divided into two parts: horizontal movement duration and vertical movement duration.

Assume that the fixed tower crane position is (x_t, y_t), supply point position is (x_s, y_s, z_s), demand point position is (x_d, y_d, z_d), and hoisting reserved safety operation height is h. The hoisting schematic diagram is shown in Figure 4.

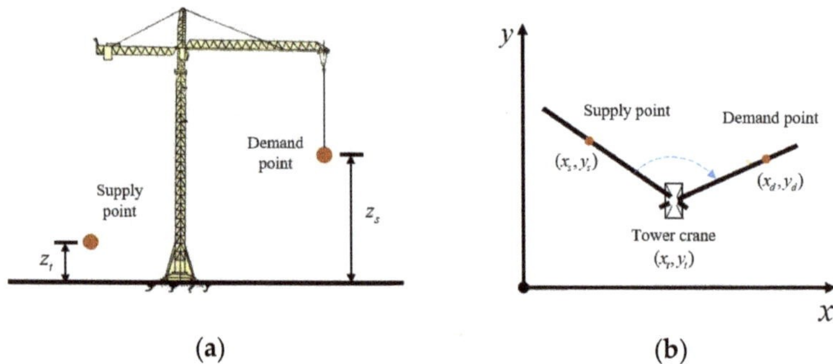

(a)   (b)

Figure 4. Schematic diagram of tower crane hoisting. (**a**) Tower crane hoisting front view; (**b**) Tower crane hoisting overhead view.

The hook horizontal movement duration can be divided into three types, including radial travel duration, tangential travel duration, and horizontal movement duration, and the calculation methods are shown in Equations (10)–(12), respectively.

$$T_r = \frac{\left| \sqrt{x_d^2 + y_d^2} - \sqrt{x_s^2 + y_s^2} \right|}{V_r} \tag{10}$$

$$T_a = \frac{\left| \tan^{-1}\left(\frac{y_d - y_t}{x_d - x_t} \right) - \tan^{-1}\left(\frac{y_s - y_t}{x_s - x_t} \right) \right|}{V_a} \tag{11}$$

$$T_h = \max\{T_r, T_a\} + \lambda \min\{T_r, T_a\} \tag{12}$$

The λ in Equation (12) considers the operator's ability to move the hook in the radial and tangential directions simultaneously. That is, considering the degree of overlap between the radial and tangential movements, λ takes a value between 0 and 1.

1. The hook vertical movement duration calculation method is given in Equation (13).

$$T_v = \frac{2(|z_d - z_s| + h)}{V_v} \tag{13}$$

2. Furthermore, the total hook travel duration can be calculated from Equation (14).

$$T_t = \mu(\max\{T_h, T_v\} + \eta \min\{T_h, T_v\}) \times Q \tag{14}$$

The μ parameter indicates the uncontrollable conditions of the construction site, such as extreme weather and obstructions to the view, and the value of μ is 0.1. The smaller the value of μ, the more favorable the site is for hoisting; the η parameter is the ability of the operator to move the hook in both horizontal and vertical directions, and the value of Q is the number of prefabricated components to be hoisted.

The hoisting duration objective function of F_3 can be expressed as Equation (15):

$$F_3 = \min[\mu(\max\{T_h, T_v\} + \eta \min\{T_h, T_v\})] \tag{15}$$

The objective function of the multi-objective optimization problem of temporary facilities arrangement on the CSPB can be expressed as F, and it is shown in Equation (16). The constraints of CSPB can be summarized in Equation (17).

$$F = \begin{cases} F_1 = \min\{H^*V^*\} \\ F_2 = \min\left\{\sum_{i=1}^{n}\sum_{j=1}^{i} C_{ij}d_{ij} + \sum_{k=1}^{n}\sum_{t=1}^{T}\left(C_{Dk} + C_k d_k^{(t-1,t)} + C_{Bk}\right)z_{kt}\right\} \\ F_3 = \min[\mu(\max\{T_h, T_v\} + \eta\min\{T_h, T_v\})] \end{cases} \quad (16)$$

$$s.t. = \begin{cases} y_i - \frac{h_i}{2} - b_1 - \varphi \geq 0 \\ y_i + \frac{h_i}{2} - b_2 + \varphi \leq 0 \\ x_i - \frac{l_i}{2} - a_1 - \varphi \geq 0 \\ x_i + \frac{l_i}{2} - a_2 + \varphi \leq 0 \\ |x_i - x_j| - \frac{l_i + l_j}{2} \geq W_{ij} \\ |y_i - y_j| - \frac{h_i + h_j}{2} \geq W_{ij} \\ \sqrt{(x_t - x_c)^2 + (y_t - y_c)^2} \leq R_t \end{cases} \quad (17)$$

2.5. MPC-NSGA-II Optimization Algorithm

Premature maturity is a highly likely phenomenon in multi-constraint and multi-objective optimization problems. In this case, the applicability of the NSGA-II algorithm is low [39]. Therefore, this study proposes a multi-objective optimization algorithm based on the NSGA-II algorithm with the constraint domination method to improve the initialization population and crowding distance in the NSGA-II algorithm. It proposes that the MPC-NSGA-II algorithm applies to the layout planning on the CSPB. The specific optimization elements of the improved MPC-NSGA-II algorithm are: (1) Adopting the Multi-population Strategy to expand the search range of the algorithm while achieving elite retention and thus avoiding premature maturity; (2) Reducing the interference of subjectively determined parameters through the constrained dominance method; (3) Introducing Harmonic distance to determine the congestion degree.

2.5.1. Multi-Population Strategy

Multi-population strategy is practiced by introducing three populations, the immigration operator and non-dominated sorting. The introduced populations are POP-a, POP-b, and POP-c. The POP-a population has a low variation probability and is responsible for searching for local optimal solutions. The POP-b population has a high variation probability and is responsible for searching for optimal global solutions. The POP-c population is an elite population responsible for recording the optimal solutions appearing in POP-a and POP-b populations. The formation process of the elite population POP-c is shown in Figure 5. The advantage of introducing this strategy is that by setting populations with different parameters, the global and local search capabilities of the algorithm are taken into account, thus expanding the search range. The elite population control algorithm process avoids the premature problem in the NSGA-II algorithm.

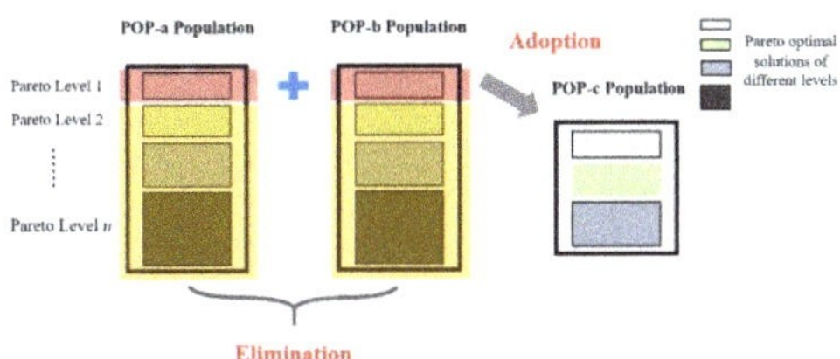

Figure 5. Elite population formation process.

The immigration operator is a procedural operator that periodically introduces the optimal solution in a population to other populations during the algorithm iteration, which works as shown in Figure 6. The optimal solutions in the two populations are replaced by the relatively inferior solutions of the other populations through the migration operator. On one hand, it realizes the synergistic exchange between the two populations of POP-a and POP-b to promote co-evolution. On the other hand, this exchange operation speeds up the elimination of inferior individual solutions and drives the convergence of the algorithm.

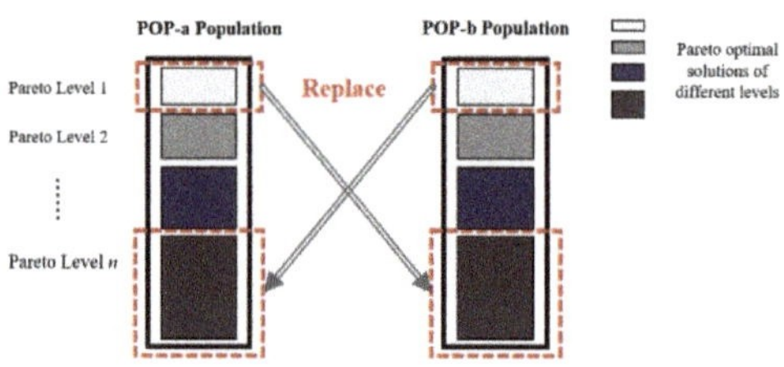

Figure 6. Working principle of the immigration operator.

The relatively inferior solutions within the set are removed by non-dominated sorting. It is possible to maintain optimal individuals without losing them and elite populations without crossover and mutation.

2.5.2. Constraint Domination Methods

Currently, the multi-constraint optimization problem is mainly solved by the following four methods [40]: (1) Considering feasible solution methods; (2) Penalty functions; (3) Random ordering methods; (4) Constraint domination methods. The constraint domination method avoids the artificial parameter interference present in the previous three methods while dealing with the constraints; hence, the method has been applied in this research.

Compared with the crowding distance in the NSGA-II algorithm, the Harmonic distance can better reflect the crowding level between individuals and is a more effective method. The crowding distance is introduced and given in Equation (18).

$$d_i = \frac{N-1}{\frac{1}{d_{i,1}} + \frac{1}{d_{i,2}} + \cdots + \frac{1}{d_{i,j}} + \cdots + \frac{1}{d_{i,N}}}, \quad i \neq j \tag{18}$$

where N is the population size and d_{ij} denotes the spatial Euclidean distance between individual X_i and individual X_j. The flow chart of the improved MPC-NSGA-II algorithm is shown in Figure 7.

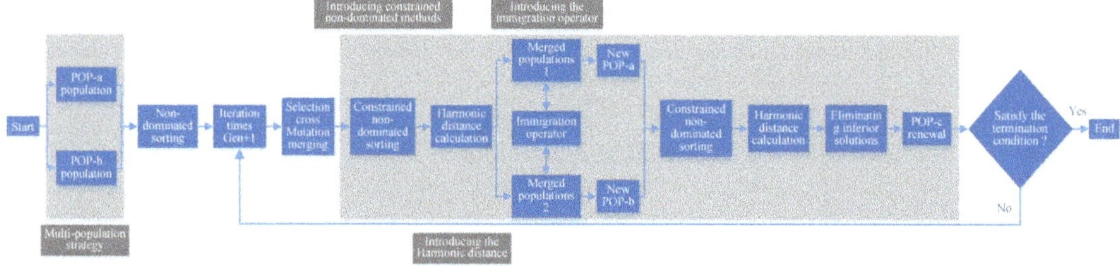

Figure 7. Algorithm flow chart of MPC-NSGA-II.

2.6. Entropy Weight-TOPSIS Decision-Making Method

After the MPC-NSGA-II optimization algorithm is used to output multiple construction site temporary facilities layout optimization schemes, the schemes are selected by a comprehensive evaluation and decision-making method by combining other influencing factors. To minimize the influence of human factors on the results, the entropy weight-TOPSIS integrated decision-making method is used to evaluate the output optimization solutions to obtain the best solution.

The entropy weight method determines the weight of indicators based on the amount of information reflected by the data of each indicator. Compared with subjective weighting methods (expert scoring, hierarchical analysis, etc.), the entropy weight method can reflect the importance of each index more objectively and accurately [41].

In an evaluation system with m options to be evaluated and n evaluation indicators, the weight ω_i of the i evaluation indicator is defined in Equation (19).

$$\omega_i = \frac{1 - \Phi_i}{n - \sum\limits_{i=1}^{n} \Phi_i} \tag{19}$$

where the entropy value of the i evaluation indicator Φ_i is defined in Equation (20).

$$\Phi_i = -\frac{1}{\ln m} \sum\limits_{j=1}^{m} Co_{ij} \ln Co_{ij} \tag{20}$$

where ω_i is the entropy weight coefficient, Φ_i is the entropy value of the first i evaluation index, and n represents that there are n evaluation indexes.

A larger ω_i means that the greater the amount of information represented by the indicator, the greater the effect on the comprehensive evaluation the greater the effect.

TOPSIS, also known as the "ideal solution method", is based on calculating the distance between the evaluation object and the optimal and inferior solutions. The basic principle of the TOPSIS method is to calculate the distance between the evaluation object and the optimal solution and the worst solution as the primary basis for evaluating the merits of the solution [42]. The ideal proximity C^* is calculated in this research from Equation (21).

$$C^* = \frac{S_i^-}{S_i^+ + S_i^-}, i = 1, 2, \ldots, m \tag{21}$$

where S^+ is the distance scale from the target to the ideal solution and S^- is the distance scale from the target to the anti-ideal solution. After calculating C^*, each solution is ranked according to the size of C^*. The larger C^* means the better scheme, and the best scheme is selected.

3. Engineering Analysis

3.1. Engineering Situations

The project has two teaching buildings with three floors. The length of the building is 40 m, the width is 20 m, the story height is 3.9 m, and the total height is 17.05 m. The building belongs to a Class A public building, with a total construction area of 4478 m^2. The BIM model of the building is shown in Figure 8.

The prefabricated components in the project are composite slabs and prefabricated stairway sections. The construction site layout size is 105 × 100 m. There is a 5.0 m wide proposed permanent circular road within the site. According to the principle of construction road layout, it will be used as a construction road. There are two entrances to the CSPB communication with the outside world. The main entrance is located on the south side of the site, and the secondary entrance is located on the east side of the site. According to the principle of fixed tower crane selection, the QTZ5010 was used on-site for hoisting work. The initial construction site of the project is shown in Figure 9.

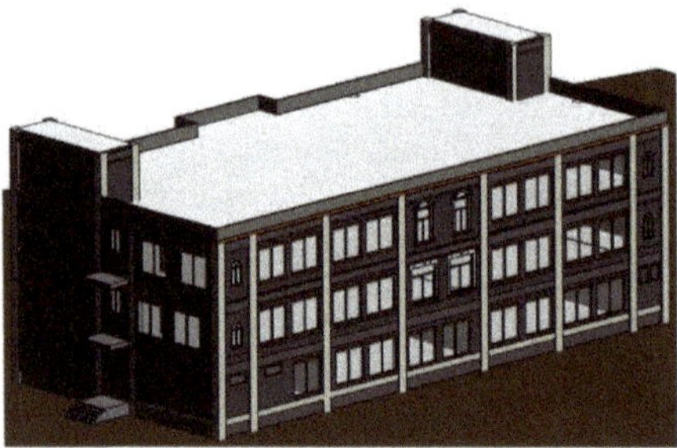

Figure 8. BIM model of the building.

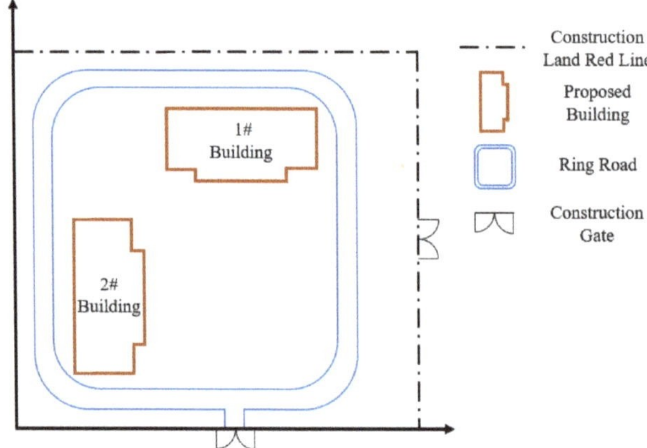

Figure 9. Initial construction site of the project.

3.2. Technical Analysis Route

The multi-objective optimization and decision-making model structure of temporary facilities arrangement on the CSPB is shown in Figure 10.

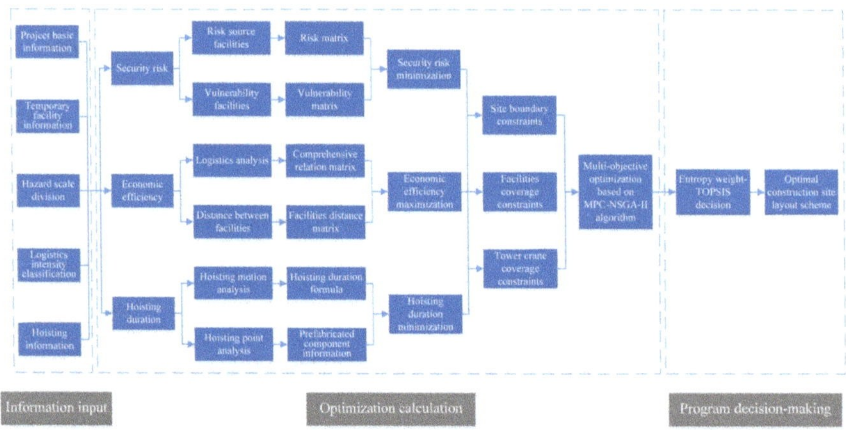

Figure 10. Multi-objective optimization and decision-making model structure of temporary facilities arrangement on the CSPB.

The MPC-NSGA-II multi-objective optimization algorithm consists of three main parts: initializing the populations, loop iteration, and loop termination.

Take the southwest corner of the site as the origin of the coordinate system of the whole construction site based on the basic information of the construction site. Divide the grid size (usually square grid) according to the actual situation and determine the coordinate information of fixed facilities. From Section 2.1, the model has too many constraints. If the initial population is randomly generated, it will increase the difficulty of searching for feasible solutions. Constraints must be checked and passed before a valid initial population is obtained. Therefore, the entire Pareto front surface can be obtained so that the solutions generated by the initialized population are all in the feasible domain.

The population evolves continuously in a loop iteration, so the iterative result gradually approximates the actual Pareto front surface. The loop iteration process of the model consists of three key components: population selection, crossover, mutation operations, immigration operator updates, and elite population updates.

The selection, crossover, and mutation of populations give the algorithm a powerful spatial search capability.

The selection is the operational process of transmitting good genes to the next generation by selecting high-quality individuals in the parent population. The binary tournament selection method places randomly selected individuals into the mating pool.

The crossover is performed by exchanging chromosomal information of two individuals in the mating pool, forming a new individual. Assume that the parents are X_i and X_j, respectively, and a single point of crossover is used between them to achieve the update of genetic information. The process is shown in Figure 11.

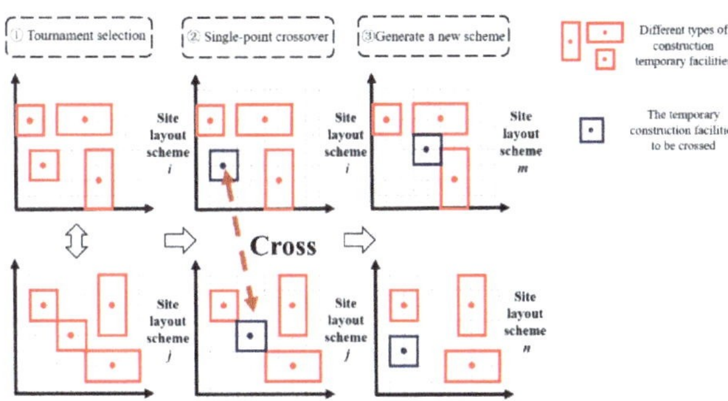

Figure 11. Schematic diagram of population crossing.

The mutation is a mutation of genetic information somewhere in a chromosome that results in the formation of a new individual. Due to the constraint relationship between individual facilities in a feasible temporary facility layout scheme, classical variation can easily lead to the generation of infeasible solutions. Therefore, the model does not use the classical variation approach. A new arrangement scheme is generated when the chromosome satisfies the mutation condition.

The model is terminated by the number of generations in which the optimal number of individuals remains constant with the maximum number of iterations. In other words, the model is stopped when the number of generations in which the optimal number of individuals in the elite population POP-c remains constant, reaches a preset value or when the maximum number of iterations is set.

3.3. Parameter Determination

The parameters to be determined are site and fixed facility parameters, temporary facilities parameters, hazard scale division parameters, logistics intensity classification situation, and hoisting parameters information parts.

3.3.1. Site and Fixed Facility Parameters

The southwest corner of the project construction site is taken as the origin of the coordinate axis, the AGSM is used to simplify the teaching building into a 40×20 m rectangular block, and the circular road within the site is regarded as a combination of four rectangular blocks to obtain the fixed facility coordinates. The length of the tower crane tail end of QTZ5010 is 12.72 m. According to the tower crane arrangement method, the proposed tower crane was arranged at $(50, 50)$. Table 2 provides the relevant location information of the fixed facilities.

Table 2. Dimensions and coordinates of fixed facilities.

Number	Facility Name	Coordinates	Size (Unit: m)
B1	1 # Teaching building	(60, 75)	40×20
B2	2 # Teaching building	(25, 35)	40×20
O1	Tower crane	(50, 50)	2×2
O2	Construction Road 1	(47.5, 7.5)	75×5
O3	Construction Road 2	(7.5, 45)	5×90
O4	Construction Road 3	(47.5, 92.5)	75×5
O5	Construction Road 4	(87.5, 45)	5×90
O6	South gate	(55, 0)	——
O7	East gate	(105, 50)	——

The simplified initial construction layout is shown in Figure 12.

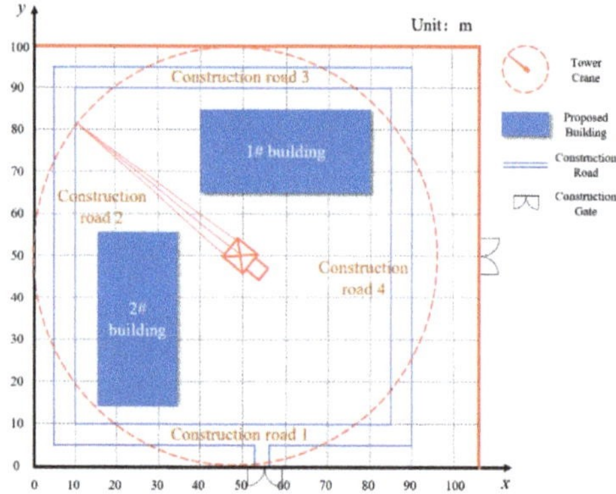

Figure 12. Simplified initial construction layout drawing.

According to the simplified preliminary construction layout plan, combined with the actual situation on-site, the coordinate range of the available sites for other temporary facilities is divided.

3.3.2. Temporary Facility Parameters

The prefabricated component yard is the key consideration in the temporary facilities layout on CSPB. Therefore, it is necessary to determine the area of the prefabricated components yard and its size first. The two buildings in the project have the same structure. A total of 76 prefabricated laminated panels are required for this floor, with a total of 2 types of sizes, of which 54 are required for DBS-67-3318 and 18 are required for DBS-67-4218. The prefabricated stairs are selected from SAT-39-25, and 2 stairs are required for each floor, with 4 stair sections.

The dimensions of all prefabricated components are summarized in Table 3.

Table 3. Summary of parameters of prefabricated components on the second floor.

Name	Size (mm)	Number per Layer
DBS-67-3318	3120 × 1800	54
DBS-67-4218	4020 × 1800	18
SAT-39-25	3660 × 1180	4

According to the construction plan and site conditions, a layer of prefabricated components needs to be reserved at the construction site. Therefore, 10 stacks of DBS-67-3318 precast laminated panels, 3 stacks of DBS-67-4218 precast laminated panels, and 1 stack of precast stairs were calculated. Considering the prefabricated components stacking requirements and the construction site space, the prefabricated laminated panels and stairs are placed in one yard.

The schematic diagram of the temporary storage of prefabricated components is based on the prefabricated component stacking requirements, as shown in Figure 13. The figure shows that the interval distance between prefabricated component stacks is 1 m, as reserved space for operation.

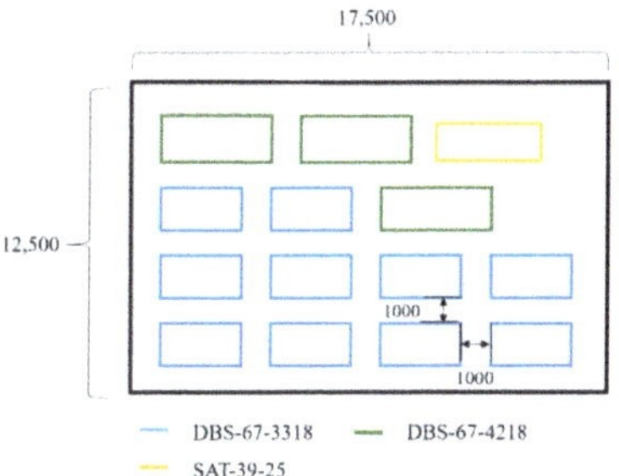

Figure 13. The layout of prefabricated components on the construction site.

In addition, seven temporary production facilities and two temporary living facilities were selected according to the actual situation on-site, and the information related to the temporary facilities is summarized in Table 4.

Table 4. Temporary facilities information.

Facilities Number	Facility Name	Size (Unit: m × m)	Facilities Properties
F1	Precast component yard	17.5 × 12.5	Non-fixation
F2	Steel processing shed	15 × 4	Non-fixation
F3	Steelyard	15 × 4	Non-fixation
F4	Woodworking processing shed	5 × 10	Non-fixation
F5	Woodworking yard	4 × 10	Non-fixation
F6	Construction waste yard	10 × 5	Non-fixation
F7	Small warehouse	8 × 5	Non-fixation
F8	Dormitories	5 × 30	Non-fixation
F9	Office building	4 × 30	Non-fixation

3.3.3. Hazard Scale Parameters

The hazard scales of each facility need to be divided in advance to calculate the objective function. Related research [37] generated the criteria for dividing the hazard scales of construction facilities in his research. The hazard scales of the facilities in this project are shown in Table 5.

Table 5. Hazard scale division of temporary facilities.

Temporary Facility	B1	B2	O1	F1	F2	F3	F4	F5	F6	F7	F8	F9
Hazard scale	2	2	4	3	3	3	3	3	2	4	1	1

3.3.4. Logistics Intensity Classification

We must grade the logistics intensity between the construction facilities before determining the objective function. According to the information on the engineering budget of the Ministry of Commerce for the project, the logistics intensity grading between construction facilities is shown in Figure 14.

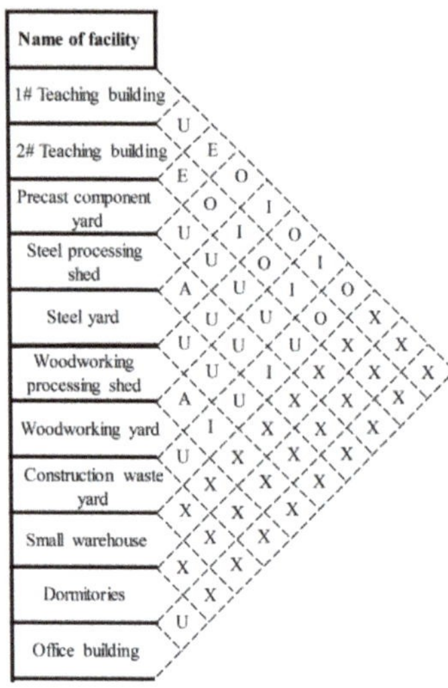

Figure 14. Logistics intensity classification.

Related research generated quantitative values of logistic intensity levels using the fuzzy set theory [28], proving the suggested values' validity by using practical projects [43]. Therefore, these suggested values are used in this case: A is 7776, E is 1296, I is 216, O is 36, U is 6, and X is 1. This project is the main structure construction phase, and the temporary facilities do not change their location during the construction period of 0.

3.3.5. Hoisting Information Parameters

The construction and installation machinery used in this project is the QTZ5010 tower crane. To ensure smooth installation work, the tower crane is used to hoist at four times the rate. Hook hoisting and radial and rotation speeds are 0.6 m/min, 20 m/min, and 0.5 rad/min, respectively. The prefabricated components of the proposed building are arranged symmetrically from left to right. Considering the flowing construction, the whole building is divided into two construction sections, and each section is considered as a whole, as shown in Figure 15. The center of the prefabricated staircase is used as the hoisting point, and the location of the prefabricated staircase demand point can be obtained. Because of many prefabricated laminated panels, the centers of the two construction sections are used as the demand point coordinates of the prefabricated laminated panels to consider the calculation volume and accuracy.

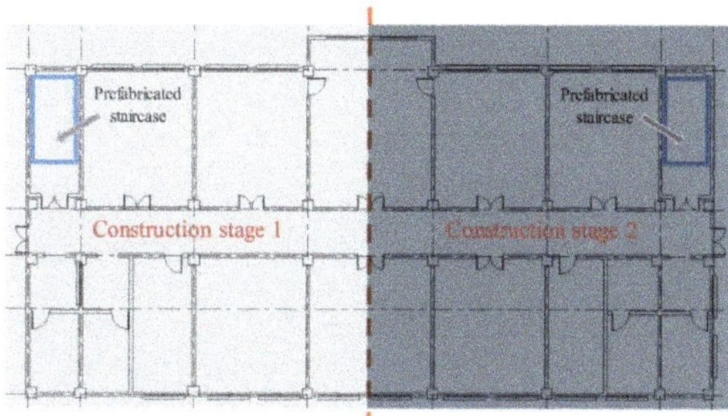

Figure 15. Division of construction section.

4. Results and Discussion

The computer hardware configuration for this validation simulation experiment is Intel (R) Core (TM) i5-13600KF CPU @ 5.10GHz, 32.0G of RAM, and a 64-bit operating system. The model was run in MATLAB 2020b. The input parameters were assigned to the MPC-NSGA-II optimization algorithm. The POP-a, POP-b, and POP-c population sizes were set to 150, the crossover rate was 0.9, the mutation rate was set to 0.05 and 0.7, respectively, and the maximum number of iterations was 200. A total of 23 optimal feasible solutions were generated from the model runs, and the results are shown in Figure 16 and Table 6.

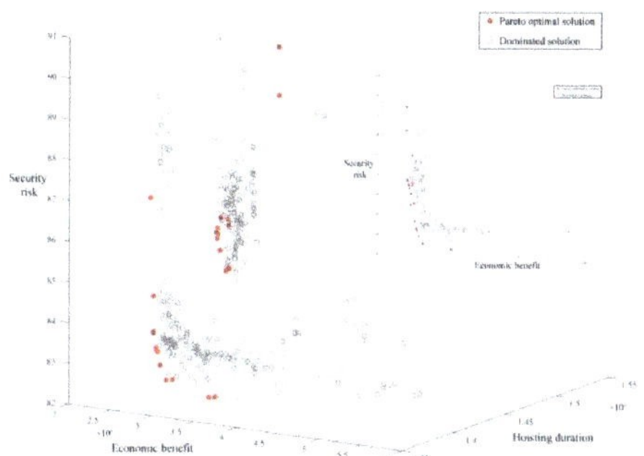

Figure 16. MPC-NSGA-II model iteration results.

Table 6. The fitness value of the Pareto optimal solution set.

Serial Number	Economic Benefit	Security Risk	Hoisting Duration	Serial Number	Economic Benefit	Security Risk	Hoisting Duration
1	269,245	82.85	13,824.48	13	261,066	84.52	13,824.48
2	266,083	83.18	13,824.48	14	276,176	82.51	13,824.48
3	253,839	84.46	14,574.44	15	276,145	82.47	13,884.65
4	251,443	85.29	14,509.67	16	246,257	89.40	15,152.66
5	250,740	85.44	14,509.67	17	246,332	88.21	15,152.66
6	248,977	85.50	14,639.03	18	259,524	84.56	14,556.61
7	254,804	85.02	14,509.67	19	259,534	84.54	14,556.61
8	247,711	85.74	14,574.44	20	264,183	83.24	13,824.48
9	256,963	86.94	13,824.48	21	264,178	83.26	13,824.48
10	255,990	85.59	14,477.65	22	261,232	83.63	13,824.48
11	248,274	85.65	14,639.03	23	261,231	83.66	13,824.48
12	256,693	85.45	14,477.65				

To provide a more precise illustration of the decision part of the model, four Pareto optimal solution schemes with significant layout differences were selected from the above optimization results for comparison. The simple arrangement of the four schemes is shown in Figure 17. The parameters are shown in Table 7.

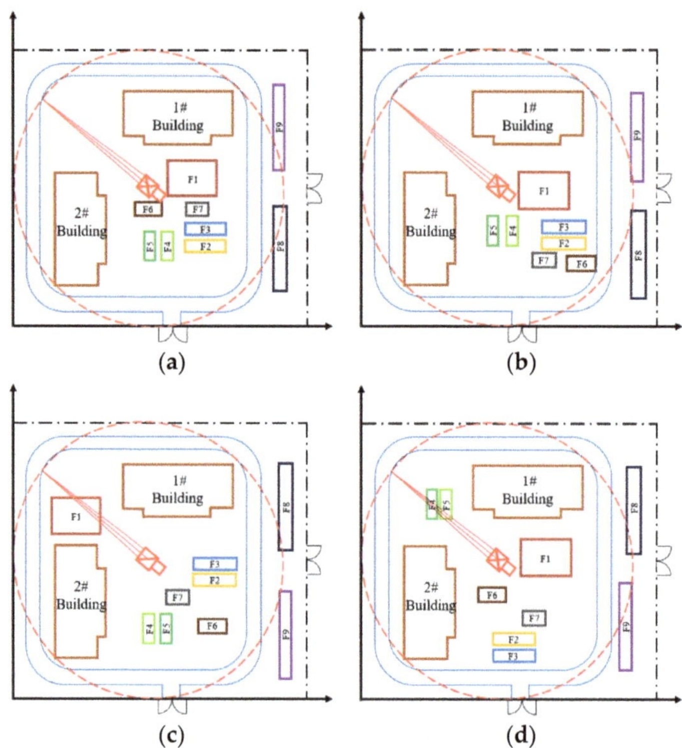

Figure 17. Schematic diagram of four Pareto optimal solution layout schemes. (**a**) Scheme 1; (**b**) Scheme 2; (**c**) Scheme 5; (**d**) Scheme 9.

Table 7. Four Pareto optimal solution layout schemes.

Scheme	Coordinate	F1	F2	F3	F4	F5	F6	F7	F8	F9
1	x	65	70	70	56	51	49	67	97	97
	y	52	28	34	28	28	41	41	27	70
2	x	65	72	72	54	47	78	65	98	98
	y	52	34	39	38	38	27	28	30	70
5	x	24	73	73	50	56	72	60	98	98
	y	68	46	51	29	29	30	40	71	27
9	x	65	54	54	30	25	46	61	99	97
	y	52	25	20	72	72	41	33	70	30

To choose the best solution, a comprehensive evaluation is proposed using the entropy weight-TOPSIS method. There are four evaluation attributes to be considered: "Smoothness", "Convenience", "Management efficiency", and "Spatial adjustability".

The project manager performs the fuzzy evaluation of the four judging factors of each program. First, the set of judging index factors U_f is established in Equation (22).

$$U_f = \{\text{Smoothness, Convenience, Management efficiency, Spatial adjustability}\} \quad (22)$$

The evaluation level V_e is established in Equation (23).

$$V_e = \{V_1, V_2, V_3, V_4, V_5\} = \{\text{Excellent}, \text{Good}, \text{Fair}, \text{Pass}, \text{Poor}\} \tag{23}$$

$V_1 \sim V_5$ correspond to 5, 4, 3, 2, and 1 scores, respectively. The initial decision matrix is obtained by combining the fuzzy evaluation results of the three project managers and then normalizing the decisionalization matrix.

The ideal and anti-ideal solutions are in Equations (24) and (25)

$$C^+ = \begin{bmatrix} 0.1344 & 0.3334 & 0.1090 & 0.0705 \end{bmatrix} \tag{24}$$

$$C^- = \begin{bmatrix} 0.0768 & 0.1192 & 0.0672 & 0.0434 \end{bmatrix} \tag{25}$$

Finally, the ideal proximity C^* is calculated, and the results are listed in Table 8.

Table 8. Entropy weight-TOPSIS evaluation calculation results.

Evaluation Object	Ideal Solution Distance	Anti-Ideal Solution Distance	Ideal Proximity C^*	Ranking
Scheme 1	0.1836	0.0471	0.2043	3
Scheme 2	0.2263	0.0107	0.0453	4
Scheme 5	0.0098	0.2250	0.9583	1
Scheme 9	0.0478	0.1829	0.7927	2

According to the evaluation results, scheme 5 is the best temporary facilities layout scheme on CSPB. To visually check whether the output best scheme is reasonable, a 3D temporary facilities layout model was established in BIMMAKE 2022, and the results are shown in Figure 18.

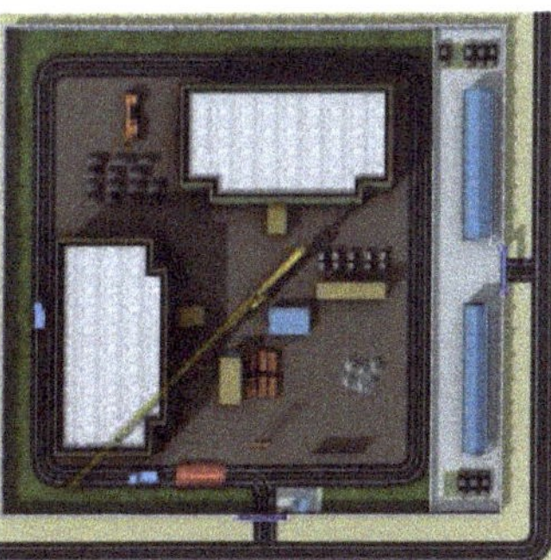

Figure 18. The layout of prefabricated components on the construction site.

To verify the superiority of the MPC-NSGA-II algorithm, the classical NSGA-II algorithm in the field of multi-objective optimization is used as a control experiment. The optimization results of the MPC-NSGA-II algorithm are compared with those of the NSGA-II algorithm in multiple dimensions. To eliminate the influence of chance factors as much as possible, the simulation tests were run 10 times, and the best optimization results were selected for comparison. The test parameters of the NSGA-II algorithm in the test were set as follows: population size N was 300, crossover probability was 0.8, variation probability was 0.1, and the maximum number of iterations was 200 generations.

The test results in Figure 19 show that for this case, the MPC-NSGA-II optimization algorithm is more computationally efficient, the number of Pareto optimal solutions obtained is higher, and the quality is higher. In terms of computational time, the NSGA-II algorithm takes 1702.0 s on average, while the MPC-NSGA-II algorithm takes 421.0 s on average, and its computational time is only 25% of that of the NSGA-II algorithm. In terms of computational results, the number of optimal solutions obtained by the NSGA-II algorithm is 9, and the number of optimal solutions obtained by the MPC-NSGA-II algorithm is 23. In terms of computational quality, the optimal solutions obtained by the MPC-NSGA-II algorithm dominate the optimal solutions obtained by the NSGA-II algorithm. Therefore, the MPC-NSGA-II optimization algorithm obtains more and better solutions with higher computational efficiency.

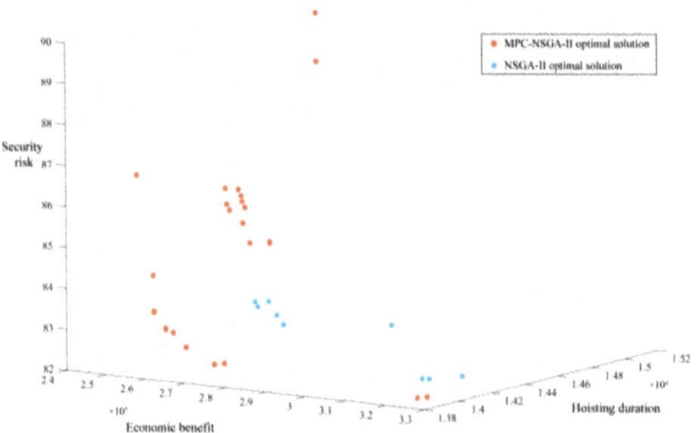

Figure 19. Comparison of results of two algorithms.

5. Conclusions

This study proposes that the MPC-NSGA-II algorithm applies to multi-constraint and multi-objective optimization problems based on the basis and general principles of layout on the CSPB. The main results of this study are as follows.

(1) The construction site and temporary facilities assumptions for the CSPB were determined by literature analysis. It is found that the grid method assumption of the construction site and the AGSM for temporary facilities are more suitable for CSPB analysis. With a large workload of hoisting work on the CSPB, the tower crane coverage constraint, site boundary constraint, and facility coverage constraint should be taken as constraints on the CSPB. Influencing factors of temporary facilities layout on the CSPB were screened out by literature analysis. The expert scoring method ranked the degree of importance of the influencing factors. Ten factors to be included in the model were finally identified and incorporated into the method of layout planning on the CSPB in different ways.

(2) The multi-objective optimization functions of CSPB were determined, and the quantitative formulas were proposed. Combined with the characteristics of CSPB, the Security risk function was quantified by the hazard interaction matrix and the vulnerability interaction matrix; the Economic benefits function was quantified by the systematic layout planning method; the Hoisting duration function was quantified by decomposing the hoisting process and calculating the horizontal and vertical running time respectively.

(3) The MPC-NSGA-II algorithm for multi-constraint and multi-objective optimization problems was proposed. It effectively improved the NSGA-II algorithm with disadvantages such as premature maturity and computational inefficiency. By introducing multiple swarm strategies, the global search and local search capabilities of the algorithm were taken into account, thus expanding the search range. Meanwhile, introducing elite populations avoided the loss of optimal solutions and improved the stability of the optimal solution set. The migration operator promoted collaborative communication among populations, sped up the elimination of inferior individual solutions, and improved the computational efficiency of the algorithm. Adopting the constraint domination method reduced the interference of the considered parameters. Harmonic distance improved the distribution of feasible solutions and sets and increased the efficiency of the algorithm.

(4) The multi-objective optimization and decision-making model of temporary facilities arrangement on the CSPB was established. The MPC-NSGA-II algorithm combined the entropy weight-TOPSIS decision-making method to output the best temporary facility arrangement scheme based on a practical case. The BIMMAKE 2022 established a visualized 3D construction site temporary facilities layout to verify the rationality of the best scheme and the theoretical model.

(5) The practical case verified the superiority of the MPC-NSGA-II algorithm and the theoretical model. The results of the MPC-NSGA-II and NSGA-II algorithms were compared in multiple dimensions. It was found that the MPC-NSGA-II algorithm has 3 times higher computational efficiency, 1.6 times higher number of optimal solutions, and higher quality of optimal solutions.

(6) There are also areas for improvement in this study. Only stationary tower cranes on the CSPB were considered. Mobile cranes also take on the critical role of transporting components in practical applications. Subsequent studies could consider the impact of each type of transport machinery working in concert with the temporary facility of the CSPL. Meanwhile, only one case was used in this study for validation analysis, and different cases should be used in future studies to demonstrate the model's generalizability.

Supplementary Materials: The following supporting information can be downloaded at: https://www.mdpi.com/article/10.3390/su15076279/s1, Expert Consultation Form.

Author Contributions: Conceptualization, G.Y. and Y.Y.; methodology, G.Y.; software, R.L.; validation, G.Y., Y.Y. and R.L.; formal analysis, Y.Y.; investigation, R.L.; resources, G.Y.; data curation, R.L.; writing—original draft preparation, R.L. and Y.Y.; writing—review and editing, G.Y., Y.Y. and R.L.; visualization, R.L.; supervision, G.Y.; project administration, G.Y. and R L.; funding acquisition, G.Y. and Y.Y. All authors have read and agreed to the published version of the manuscript.

Funding: This work was funded by the 111 project of the Ministry of Education and the Bureau of Foreign Experts of China (No. B18062) and the National Key R&D Program of the Ministry of Science and Technology (No. 2019YFD1101005-4).

Data Availability Statement: The data presented in this study are available on request from the corresponding author.

Conflicts of Interest: The authors declare no conflict of interest.

References

1. Ogunmakinde, O.E.; Egbelakin, T.; Sher, W. Contributions of the circular economy to the UN sustainable development goals through sustainable construction. *Resour. Conserv. Recycl.* **2022**, *178*, 106023. [CrossRef]
2. Guo, Z.; Li, L. A Conceptual Framework for Collaborative Development of Intelligent Construction and Building Industrialization. *Front. Environ. Sci.* **2022**, *10*, 802. [CrossRef]
3. Xie, L.; Chen, Y.; Chang, R. Scheduling Optimization of Prefabricated Construction Projects by Genetic Algorithm. *Appl. Sci.* **2021**, *11*, 5531. [CrossRef]
4. Yang, B.; Fang, T.; Luo, X.; Liu, B.; Dong, M. A BIM-Based Approach to Automated Prefabricated Building Construction Site Layout Planning. *Ksce J. Civ. Eng.* **2022**, *26*, 1535–1552. [CrossRef]
5. Lu, Y.; Zhu, Y. Integrating Hoisting Efficiency into Construction Site Layout Plan Model for Prefabricated Construction. *J. Constr. Eng. Manag.* **2021**, *147*, 04021130. [CrossRef]
6. Yi, W.; Wang, S.; Zhang, A. Optimal transportation planning for prefabricated products in construction. *Comput.-Aided Civ. Infrastruct. Eng.* **2020**, *35*, 342–353. [CrossRef]
7. Yuan, Z.; Zhang, Z.; Ni, G.; Chen, C.; Wang, W.; Hong, J. Cause Analysis of Hindering On-Site Lean Construction for Prefabricated Buildings and Corresponding Organizational Capability Evaluation. *Adv. Civ. Eng.* **2020**, *2020*, 8876102. [CrossRef]
8. Huang, C.; Wong, C.K. Optimisation of site layout planning for multiple construction stages with safety considerations and requirements. *Autom. Constr.* **2015**, *53*, 58–68. [CrossRef]
9. RazaviAlavi, S.; AbouRizk, S. Site Layout and Construction Plan Optimization Using an Integrated Genetic Algorithm Simulation Framework. *J. Comput. Civ. Eng.* **2017**, *31*, 04017011. [CrossRef]
10. Kaveh, A.; Vazirinia, Y. Construction Site Layout Planning Problem Using Metaheuristic Algorithms: A Comparative Study. *Iran. J. Sci. Technol. Trans. Civ. Eng.* **2019**, *43*, 105–115. [CrossRef]
11. RazaviAlavi, S.; AbouRizk, S. Construction Site Layout Planning Using a Simulation-Based Decision Support Tool. *Logistics* **2021**, *5*, 65. [CrossRef]

12. Duhbaci, T.B.; Ozel, S.; Bulkan, S. Water and energy minimization in industrial processes through mathematical programming: A literature review. *J. Clean. Prod.* **2021**, *284*, 124752. [CrossRef]
13. Slivkoff, S.; Gallant, J.L. Design of complex neuroscience experiments using mixed-integer linear programming. *Neuron* **2021**, *109*, 1433–1448. [CrossRef] [PubMed]
14. Hamza, M.F.; Yap, H.J.; Choudhury, I.A. Recent advances on the use of meta-heuristic optimization algorithms to optimize the type-2 fuzzy logic systems in intelligent control. *Neural Comput. Appl.* **2017**, *28*, 979–999. [CrossRef]
15. Halim, A.H.; Ismail, I. Combinatorial Optimization: Comparison of Heuristic Algorithms in Travelling Salesman Problem. *Arch. Comput. Methods Eng.* **2019**, *26*, 367–380. [CrossRef]
16. Ozdemir, G.; Karaboga, N. A review on the cosine modulated filter bank studies using meta-heuristic optimization algorithms. *Artif. Intell. Rev.* **2019**, *52*, 1629–1653. [CrossRef]
17. De Santis, M.; Grani, G.; Palagi, L. Branching with hyperplanes in the criterion space: The frontier partitioner algorithm for biobjective integer programming. *Eur. J. Oper. Res.* **2020**, *283*, 57–69. [CrossRef]
18. Aydemir, E.; Yilmaz, G.; Oruc, K.O. A grey production planning model on a ready-mixed concrete plant. *Eng. Optim.* **2020**, *52*, 817–831. [CrossRef]
19. Hammad, A.W.A.; Rey, D.; Akbarnezhad, A. A cutting plane algorithm for the site layout planning problem with travel barriers. *Comput. Oper. Res.* **2017**, *82*, 36–51. [CrossRef]
20. Hammad, A.W.A.; Akbarnezhad, A.; Rey, D. A multi-objective mixed integer nonlinear programming model for construction site layout planning to minimise noise pollution and transport costs. *Autom. Constr.* **2016**, *61*, 73–85. [CrossRef]
21. Aussel, D.; Svensson, A. Is Pessimistic Bilevel Programming a Special Case of a Mathematical Program with Complementarity Constraints? *J. Optim. Theory Appl.* **2019**, *181*, 504–520. [CrossRef]
22. Katoch, S.; Chauhan, S.S.; Kumar, V. A review on genetic algorithm: Past, present, and future. *Multimed. Tools Appl.* **2021**, *80*, 8091–8126. [CrossRef]
23. Farmakis, P.M.; Chassiakos, A.P. Genetic algorithm optimization for dynamic construction site layout planning. *Organ. Technol. Manag. Constr.* **2018**, *10*, 1655–1664. [CrossRef]
24. Zhu, H.; Hu, Y.; Zhu, W. A dynamic adaptive particle swarm optimization and genetic algorithm for different constrained engineering design optimization problems. *Adv. Mech. Eng.* **2019**, *11*, 1–27. [CrossRef]
25. Zouein, P.P.; Kattan, S. An improved construction approach using ant colony optimization for solving the dynamic facility layout problem. *J. Oper. Res. Soc.* **2022**, *73*, 1517–1531. [CrossRef]
26. Li, J.; Yang, B.; Zhang, D.; Zhou, Q.; Li, L. Development of a multi-objective scheduling system for offshore projects based on hybrid non-dominated sorting genetic algorithm. *Adv. Mech. Eng.* **2015**, *7*, 17. [CrossRef]
27. Zhao, Y.; Lu, J.; Yan, Q.; Lai, L.; Xu, L. Research on Cell Manufacturing Facility Layout Problem Based on Improved NSGA-II. *Cmc-Comput. Mater. Contin.* **2020**, *62*, 355–364. [CrossRef]
28. Elarbi, M.; Bechikh, S.; Gupta, A.; Ben Said, L.; Ong, Y.-S. A New Decomposition-Based NSGA-II for Many-Objective Optimization. *IEEE Trans. Syst. Man Cybern.-Syst.* **2018**, *48*, 1191–1210. [CrossRef]
29. Chen, G.; Qiu, S.; Zhang, Z.; Sun, Z.; Liao, H. Optimal Power Flow Using Gbest-Guided Cuckoo Search Algorithm with Feedback Control Strategy and Constraint Domination Rule. *Math. Probl. Eng.* **2017**, *2017*, 14. [CrossRef]
30. Xu, M.; Mei, Z.; Luo, S.; Tan, Y. Optimization algorithms for construction site layout planning: A systematic literature review. *Eng. Constr. Archit. Manag.* **2020**, *27*, 1913–1938. [CrossRef]
31. Hawarneh, A.A.; Bendak, S.; Ghanim, F. Construction site layout planning problem: Past, present and future. *Expert Syst. Appl.* **2021**, *168*, 114247. [CrossRef]
32. Li, Z.; Shen, W.; Xu, J.; Lev, B. Bilevel and multi-objective dynamic construction site layout and security planning. *Autom. Constr.* **2015**, *57*, 1–16. [CrossRef]
33. Al Hawarneh, A.; Bendak, S.; Ghanim, F. Dynamic facilities planning model for large scale construction projects. *Autom. Constr.* **2019**, *98*, 72–89. [CrossRef]
34. Ning, X.; Qi, J.; Wu, C.; Wang, W. A tri-objective ant colony optimization based model for planning safe construction site layout. *Autom. Constr.* **2018**, *89*, 1–12. [CrossRef]
35. Sanad, H.M.; Ammar, M.A.; Ibrahim, M.E. Optimal construction site layout considering safety and environmental aspects. *J. Constr. Eng. Manag. Asce* **2008**, *134*, 536–544. [CrossRef]
36. Tanner-Smith, E.E.; Tipton, E. Robust variance estimation with dependent effect sizes: Practical considerations including a software tutorial in Stata and SPSS. *Res. Synth. Methods* **2014**, *5*, 13–30. [CrossRef]
37. Abune'meh, M.; El Meouche, R.; Hijaze, I.; Mebarki, A.; Shahrour, I. Optimal construction site layout based on risk spatial variability. *Autom. Constr.* **2016**, *70*, 167–177. [CrossRef]
38. Ning, X.; Qi, J.; Wu, C.; Wang, W. Reducing noise pollution by planning construction site layout via a multi-objective optimization model. *J. Clean. Prod.* **2019**, *222*, 218–230. [CrossRef]
39. Deb, K.; Pratap, A.; Agarwal, S.; Meyarivan, T. A fast and elitist multiobjective genetic algorithm: NSGA-II. *IEEE Trans. Evol. Comput.* **2002**, *6*, 182–197. [CrossRef]
40. Amaran, S.; Sahinidis, N.V.; Sharda, B.; Bury, S.J. Simulation optimization: A review of algorithms and applications. *4OR Q. J. Oper. Res.* **2014**, *12*, 301–333. [CrossRef]
41. Markechova, D.; Riecan, B. Entropy of Fuzzy Partitions and Entropy of Fuzzy Dynamical Systems. *Entropy* **2016**, *18*, 19. [CrossRef]

42. Mirzanejad, M.; Ebrahimi, M.; Vamplew, P.; Veisi, H. An online scalarization multi-objective reinforcement learning algorithm: TOPSIS Q-learning. *Knowl. Eng. Rev.* **2022**, *37*, e7. [CrossRef]
43. Oral, M.; Bazaati, S.; Aydinli, S.; Oral, E. Construction Site Layout Planning: Application of Multi-Objective Particle Swarm Optimization. *Tek. Dergi* **2018**, *29*, 8691–8713. [CrossRef]

 *sustainability*

Article

Experimental Investigation on Thermal Conductivity of Straw Boards Based on the Temperature Control Box—Heat Flux Meter Method

Kuo Sun [1], Chaorong Zheng [1,2,*], Yue Wu [1,2] and Wenyuan Zhang [1,2]

[1] School of Civil Engineering, Harbin Institute of Technology, Harbin 150090, China; sunkuo586@163.com (K.S.)
[2] Key Lab of Structures Dynamic Behavior and Control of the Ministry of Education, Harbin Institute of Technology, Harbin 150090, China
[*] Correspondence: flyfluid@163.com

Abstract: Straw boards are the environmentally-friendly and sustainable materials used for building envelopes. To validate a novel alternative test method of thermal conductivity (λ), temperature control box–heat flux meter method (TCB-HFM), and better understand the thermal properties of straw boards, experimental investigation on two types of straw boards using such a method was carried out. Moreover, the control test via the conventional guarded hot plate method (GHP) was conducted to provide a benchmark. Results show that the fluctuation amplitudes of the temperature difference and heat flux density for the TCB-HFM during the steady state are much smaller than a generally acceptable limitation of 5%, and a 5.9% deviation of λ between the two test methods is in a reasonable range. It is indicated that the TCB-HFM can be regarded as an alternative test method to conduct investigations on the thermal properties of materials. Furthermore, the correlation between density and λ is explored and expressed by a linear fitting formula with the determination coefficient (R^2) of 0.9193, and the formula is verified to have the feasibility to predict the λ of different types of straw bio-based materials.

Keywords: straw bio-based material; wheat straw strand board; paper straw board; temperature control box—heat flux meter method; thermal conductivity

Citation: Sun, K.; Zheng, C.; Wu, Y.; Zhang, W. Experimental Investigation on Thermal Conductivity of Straw Boards Based on the Temperature Control Box—Heat Flux Meter Method. *Sustainability* **2023**, *15*, 10960. https://doi.org/10.3390/su151410960

Academic Editor: Syed Minhaj Saleem Kazmi

Received: 27 May 2023
Revised: 30 June 2023
Accepted: 11 July 2023
Published: 13 July 2023

1. Introduction

As one of the most high-yield agricultural wastes, straw has many advantages, such as low thermal conductivity, sustainability, low cost, waste reusing, and environmental protection, which contributes to the potential use for envelopes in green buildings. The relevant research about straw bio-based material shows that such material used for envelopes can benefit indoor comfort by adjusting indoor temperature and relative humidity [1,2], in which the straw bio-based material mainly includes straw bales and straw composites. Consequently, thermal properties, as well as some other key physical properties relating to building envelope performances of straw bio-based materials, have been widely concerned in recent years [3].

Thermal conductivity is the parameter that intuitively describes the insulation capacity of materials, which attracts most of the research attention. For the straw bales produced from the straws in the field by compression and baling, previous studies have reported that a range of thermal conductivity varies from 0.033 W/(m·K) to 0.19 W/(m·K), and it is found that the variability can be caused by the types of straw, fiber orientation, ambient temperature, relative humidity, and especially the density [4–13]. Such studies involve straw types of wheat, rice, barley, and rice, fiber orientations including perpendicular and parallel to the heat flow, ambient temperature ranging from 10 °C to 40 °C, relative humidity varying from 10% to 80%, and density within the range of 35 kg/m^3~350 kg/m^3. As for the straw composites manufactured from short straw fibers and binders that are generally molded as boards by pressing, the influence factors on thermal conductivity are similar to those of the straw bales but with an extra factor of binder type. The densities of straw composites cover a wider range compared with those of the straw bales, with a maximum value of 700 kg/m^3 [14]. Higher density usually results in greater thermal conductivity. For instance, the thermal conductivity of a pozzolanic straw composite with a density of 661 kg/m^3 is 0.146 W/(m·K) [15]. On the whole, the densities of straw bio-based materials involved in current studies are still in a small range. However, as the straw bio-based material is gradually applied in building envelopes, the requirement for load-bearing capacity presents, which promotes the development of such material towards higher density. Thus, investigations on the thermal properties of high-density straw bio-based materials are in demand to complement the gaps of the existing studies and serve for engineering applications.

At present, the steady-state experimental investigations on the thermal properties of insulation materials are mainly performed based on three test methods, the heat flux meter method, guarded hot plate method, and hot box method, respectively [16]. The corresponding test methods and principles have been specified in international and Chinese standards including ISO 8301: 1991 and GB/T 10295-2008 [17,18], ISO 8302: 1991 and GB/T 10294-2008 [19,20], and ISO 8990: 1994 and GB/T 13475-2008 [21,22]. Among the above three methods, the guarded hot plate method (GHP) is relatively straightforward and most commonly used for the measurement of thermal conductivity of small plate specimens, as used in [5,23,24]. The 0.3 m and 0.5 m are the most common dimensions of the specimen diameter or side length. For larger dimensions, it is hard to keep the specimens attached closely to the surfaces of heating and cooling units as the initial unevenness of the specimen surfaces is easier to happen on large-dimension specimens, and it is difficult to match the proper apparatus for such large-dimension specimens. In addition, the specimens with a large thickness are also not applicable to the test apparatus of GHP. Thus, the straw bales to be tested in previous studies are usually resized to match the requirement of the corresponding apparatus [23]. The heat flux meter method used in [4,25] is conventionally used for thermal property measurements of boards or walls in situ since the required temperature difference can be formed by warmer indoor environments and cooler outdoor environments. The bigger temperature difference between the indoor and outdoor environment leads to a more significant heat flow through the specimen and a smaller error effect, which means such a method relies closely on the test environment. As for the hot box method used in [8], it is usually implemented for thermal property investigation of boards or walls in the lab. Not only the guard hot box method but also the calibrated hot box method, the total heat flux that needs to be measured is determined according to the heating power, so the heat flux loss error is inevitable. For the guard hot box method, the error is reduced by setting a guard box and a metering box, but the error cannot be eliminated completely. While for the calibrated hot box method, an error calibration has to be performed before tests [26]. To sum up, the heat flux meter method or the hot box method is not the first choice for carrying out the thermal property tests of straw bio-based materials efficiently.

This study proposes an alternative test method of thermal conductivity and focuses on the thermal properties of straw boards, including a paper straw board (PSB) and five wheat straw strand boards (WSSBs) that can be regarded as an improved plate-shaped straw bale and a type of plate-shaped straw composite. Compared with the straw bio-based materials studied in previous works, the densities of the PSB and WSSBs in this study are greater, which contributes to a larger load-bearing capacity and more possibilities in engineering applications. The alternative test method refers to the temperature control box—heat flux meter method (TCB-HFM), which comes from a combination of the hot box method and the heat flux meter method with complementary advantages. The feasibility of such a test method is validated by comparing the monitored parameters, such as temperature difference and heat flux density, as well as the results of thermal conductivities with those of the guarded hot plate method (GHP), which is used to provide a control test. This study contributes to providing an alternative test method of thermal conductivity for relevant researchers and revealing the correlation between straw boards with different specifications and their thermal properties, which benefits the application of straw bio-based materials.

2. Experimental Program

2.1. Test Specimens

Figure 1 shows the PSBs of a specification and WSSBs of five specifications used for test specimen sampling. For the PSB, an improved plate-shaped straw bale, the inner straws are stacked and compacted along the length of the board, and the slurry overflows from the straw with the process of compaction and becomes the natural binder distributing among the straws. To provide restraint for inner straws and enhance the board integrality, surface strength, and smoothness, as well as protect the inner straws from the external environment, the paper made of high-density pulp is used to implement complete cladding, which makes the PSB more convenient to the application compared with the traditional straw bales. As for the WSSB, it is a type of plate-shaped straw composite manufactured from mixed smashed wheat straw fibers and the binder of isocyanate resin by compacting along the thickness of the board [27]. The isocyanate resin is a kind of healthy binder with no harmful gases such as formaldehyde released. The specification details of the straw boards are listed in Table 1, and the label of the board with different specifications is determined by the type and thickness d of it. As shown in the table, the densities of the PSB58 and WSSB12~WSSB25 are greater than the maximum densities of the straw bale and straw composite in previous studies [9,14], as stated in the Introduction.

(a)

(b)

Figure 1. Straw boards for test specimen sampling: (**a**) PSBs; (**b**) WSSBs.

Table 1. The specification details of a PSB and five WSSBs.

Label	Dimensions: Length l × Width w × Thickness d (mm × mm × mm)	Nominal Density ρ_n (kg/m³)	Actual Density ρ (kg/m³)
PSB58	3000 × 1200 × 58	400	379
WSSB12	2440 × 1220 × 12	1200	1209
WSSB15	2440 × 1220 × 15	1100	1092
WSSB18	2440 × 1220 × 18	1100	1175
WSSB25	2440 × 1190 × 25	1100	1094
WSSB30	2440 × 1220 × 30	620	621

The dimensions of a test specimen are 300 mm × 300 mm × d, and the test specimens are cut from the discontinuous portions away from the edges of the boards. For the control test conducted via GHP, only the WSSB25 specimens are tested as an example, and the two same specimens are tested simultaneously with no additional repeated tests. As for the tests performed via TCB-HFM, specimens covering all board specifications are in Table 1. The number of specimens for each board specification is determined by the error magnitude, which is introduced in detail as follows. Two specimens with the same specification are tested first, and the error between the result of each specimen and the average result is compared with 10%. If the errors are within that limitation, the number of specimens is no longer increased. Otherwise, the third same specimen is tested, and the error between the result of each specimen and the average result for three specimens is compared with 15%. Analogously, if the errors are within the limitation, the number of specimens is no longer increased. Otherwise, the fourth same specimen is being tested, and so on. The WSSB specimens are labeled, for instance, "WT−t12−2", which refers to WSSB, thermal properties, 12 mm thickness, and the second of the same specimens. For the PSB specimens, the label is shown as "PT−t58−1". The label "WT−t25*" is for the specimen of the control test.

2.2. Test Method and Apparatus

As shown in Figure 2a, the GHP tests with two specimens were conducted using the CD−DR3030 thermal conductivity tester (Shenyang Ziwei Mechanical and electrical equipment Co., LTD., Shenyang, China), and the test principle is presented in Figure 2b. For the actual apparatus, the cooling units, specimens, metering and guard surface plates, and metering and guard section heaters shown in Figure 2b are closely next to each other from the sides to the middle. Moreover, the dimensions of the metering section are 150 mm × 150 mm in this study. The standards [19,20] that the test method is based on have been stated in the Introduction, and the theory of the two-specimen test is expressed by Equation (1). The total heat flux Φ is determined by the heating power of the metering section heater, and the temperature difference ΔT is determined by the monitored values of thermocouples on the surfaces of heating and cooling units. Those two parameters are the items to be measured in the GHP test. The temperatures of the heating and cooling units are set as 40 °C and 30 °C, respectively. Furthermore, the correction factor of thermal conductivity λ for the apparatus is 0.909. After starting the apparatus, the data record proceeds with a frequency of once a minute. As the heat transfer tends to be steady gradually, the sampling begins automatically and is performed five times with an interval of 5 min and lasts 20 min in total.

$$\lambda = \frac{\Phi d}{2A_{\mathrm{m}} \cdot \Delta T} \tag{1}$$

where Φ is the total heat flux; d is the thickness of the specimen; A_{m} is the metering area; ΔT is the temperature difference between the hot-side and cold-side surfaces of a specimen.

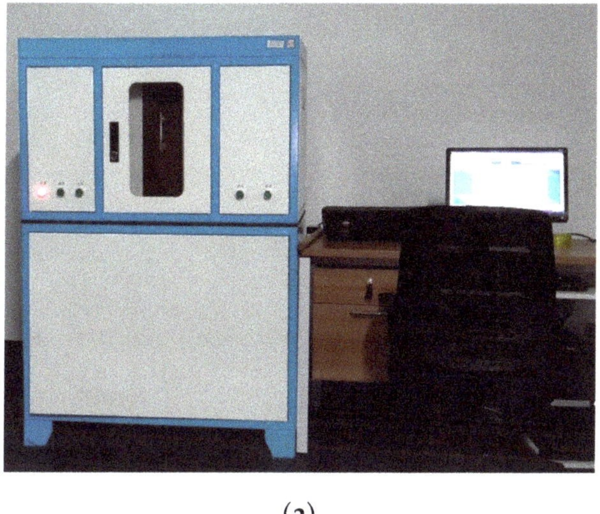

(**a**)

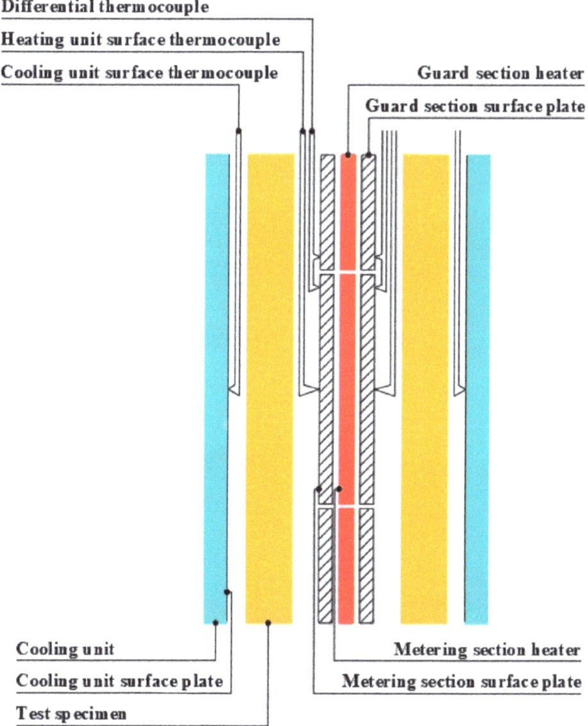

(**b**)

Figure 2. Test apparatus and principle for GHP: (**a**) CD−DR3030 thermal conductivity tester; (**b**) test principle.

Figure 3 shows the test apparatus and principle for TCB-HFM, and the arrangement details of the heat flux meter (HFM) and thermocouples are demonstrated in Figure 4. As shown in the figures, the specimen is embedded in a 1000 mm × 1000 mm × 150 mm rock wool block with a 300 mm × 300 mm square perforation in the center, which is embedded in the specimen frame, and the gaps between the specimen and the surrounding rock wool are sealed off using the nano-silica aerogel felt and polyurethane foam. An HFM with the resolution, measurement range, and standard uncertainty caused by the apparatus resolution of 0.1 W/m^2, 0~2000 W/m^2, and 0.028867 W/m^2 is arranged on the hot side of the specimen to measure the heat flux density q through the specimen center, while three thermocouples for each side with the resolution, measurement range, and standard uncertainty caused by the apparatus resolution of 0.1 °C, −50~120 °C, and 0.028867 °C are arranged around the center to collect the temperature differences ΔT between the two surfaces of the specimen. Vaseline and aluminum foil tapes are used to attach and fix the HFM and thermocouples to the corresponding monitored locations. Different from the test theory of the hot box (Equation (2)), the heating power does not need to be measured to determine the total heat flux Φ in this test since the hot box in Figure 3 is only used to provide a steady and controllable temperature difference condition, while the heat flux density is accurately determined according to the records of the HFM. To generate significant heat flux through the specimen, the air temperatures of the hot side and cold side chambers, which can be adjusted by the number of cooling fans in work and the power of the heating unit, are set as $T_i = 30$ °C and $T_e = 5$ °C, respectively. As the limited record duration of the datalogger is 180 min, the data record starts when the steady state comes, and the record frequency is once a minute, the same as that of the GHP test. The data from the last 20 min are used for statistical analysis and comparison with the results of the GHP test. The λ can be finally determined according to the theory of heat flux meter method expressed by Equation (3), in which the ΔT is the average for three monitored locations of thermocouples shown in Figure 4a. The test theory of TCB-HFM is clear, and the operation is simple. Meanwhile, the TCB-HFM can be used for specimens that are thicker compared with the GHP, is limited by the apparatus, and also can apply to performing the thermal property investigation of large-dimension wall specimens, which is conducive to the full use of experimental equipment resources.

$$\lambda = d \cdot K \tag{2a}$$

$$K = \frac{\Phi}{A \cdot \Delta T} \tag{2b}$$

$$\lambda = \frac{d}{R} \tag{3a}$$

$$R = \frac{\Delta T}{q} \tag{3b}$$

where R is the thermal resistance; q is the heat flux density; K is the thermal transmittance; A is the area of the specimen perpendicular to the direction of heat flux.

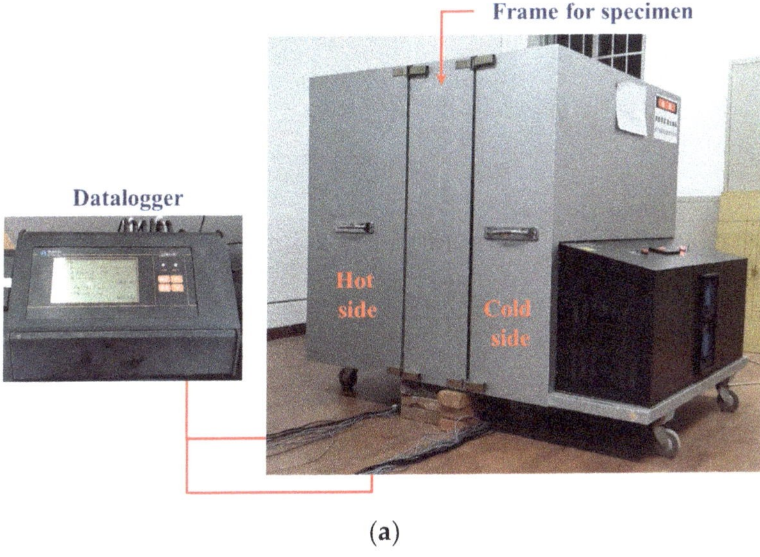

(**a**)

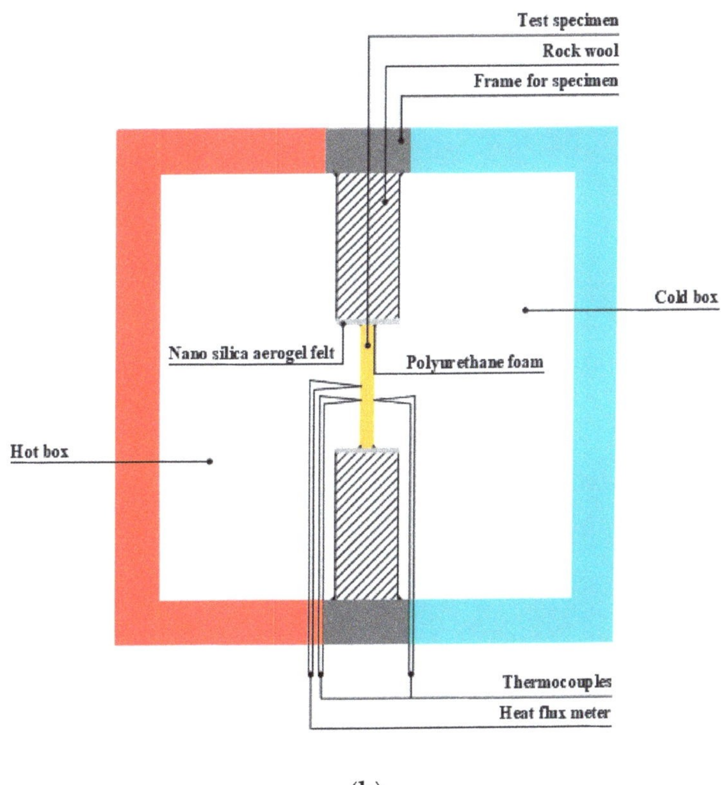

(**b**)

Figure 3. Test apparatus and principle for TCB-HFM: (**a**) JTRG−1 hot box and JTNT−A datalogger connected with the heat flux meter and thermocouples; (**b**) test principle.

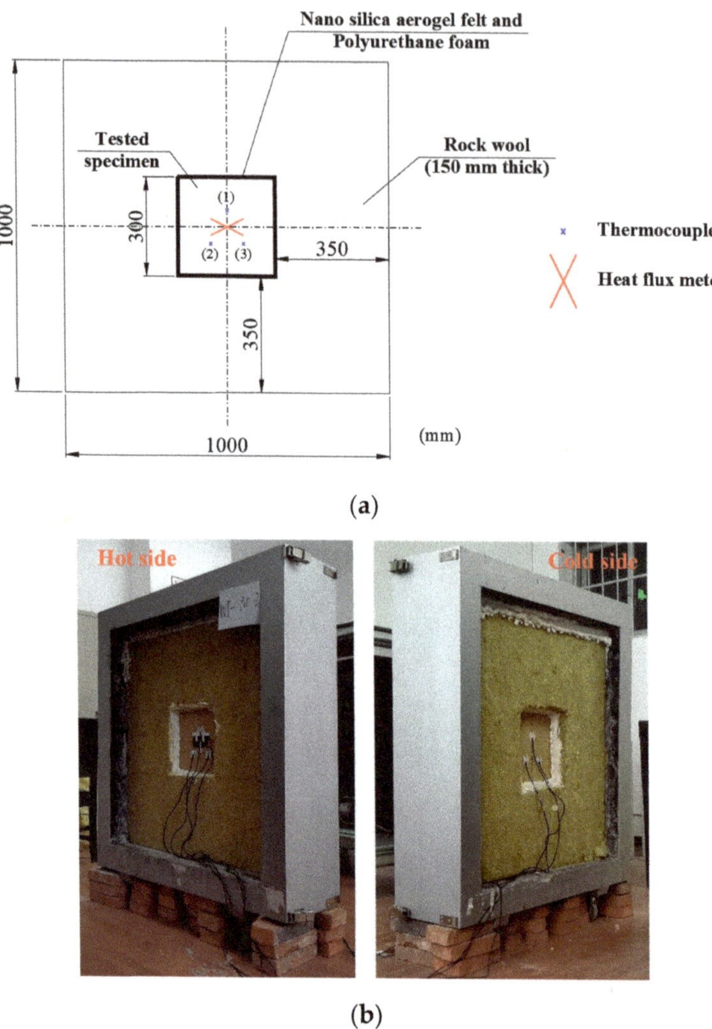

Figure 4. Arrangement details of the heat flux meter and thermocouples: (**a**) schematic diagram; (**b**) actual arrangement.

3. Experimental Results

3.1. Validation of the Temperature Control Box—Heat Flux Meter Method

3.1.1. Experimental Results for the Guarded Hot Plate Method

The sampling results of specimen WT−t25* for the GHP test are summarized in Table 2. Based on the data in the table, Equation (2), and the correction factor, the thermal conductivity λ of WSSB25 is determined to be 0.185 W/(m·K). The measurement uncertainty of λ caused by the measuring apparatus resolution is 0.000263 W/(m·K). To make a better comparison with the results of the TCB-HFM test, the monitored parameters are converted into q and ΔT according to Equation (4), and the two parameters are taken as the uniform items for the following analyses and comparisons, which are shown versus record time t in Figure 5. It can be noticed that the q and ΔT sharply decrease and increase, respectively, at the preliminary stage of the test and then gradually tend to be steady. About 70 min after the test begins, the heat transfer comes to a steady state. The sampling starts from the 121st minute, and we focus on the data for the last 20 min from then on. The statistical results for that duration, such as the coefficient of variation (COV) and maximum deviation from the average (MDA), are listed in Table 3. It is found that the COV and MDA are very small, which indicates that the fluctuation of q and ΔT is slight during the sampling stage.

$$\frac{\Phi d}{2A_{\mathrm{m}} \cdot \Delta T} = \frac{qd}{\Delta T} \qquad (4)$$

Table 2. Sampling results of specimen WT−t25*.

Sample Number	Temperature or Temperature Difference for the Left Specimen (°C)			Temperature or Temperature Difference for the Right Specimen (°C)			Φ (W)
	Metering Section Surface Plate T_{L1}	Cooling Unit Surface Plate T_{L2}	ΔT_{L}	Metering Section Surface Plate T_{R1}	Cooling Unit Surface Plate T_{R2}	ΔT_{R}	
1	39.938	29.979	9.959	40.104	29.979	10.125	3.641
2	39.947	30.009	9.938	40.060	30.002	10.058	3.642
3	39.949	30.000	9.949	40.063	29.996	10.067	3.642
4	39.949	30.019	9.930	40.064	30.011	10.053	3.642
5	39.950	30.040	9.910	40.067	30.035	10.032	3.642
Average	39.947	30.009	9.938	40.072	30.005	10.067	3.642

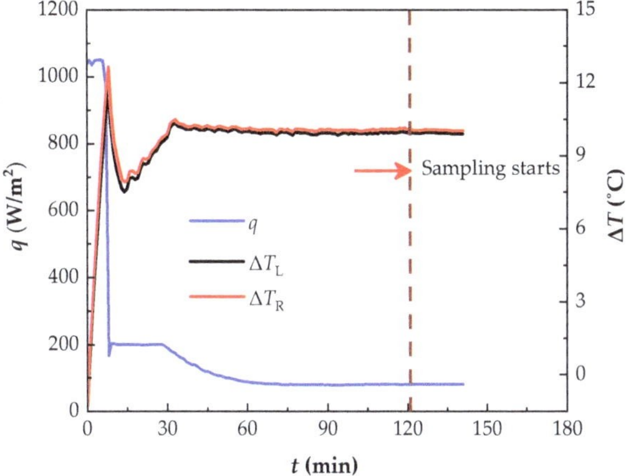

Figure 5. q and ΔT versus t for specimen WT−t25*.

Table 3. Statistical results of q and ΔT for specimen WT−t25*.

Thermal Parameter	Average	Coefficient of Variation (COV)	Maximum Deviation from the Average (MDA)
q	80.932 W/m^2	0.01%	0.03%
ΔT_{L}	9.935 °C	0.21%	0.33%
ΔT_{R}	10.059 °C	0.19%	0.66%

3.1.2. Experimental Results for the Temperature Control Box—Heat Flux Meter Method and Method Validation

The q and ΔT of specimens WT−t25−1 and WT−t25−2 for the TCB-HFM tests are directly monitored by the arranged HFM and thermocouples and presented versus t in Figure 6. To raise the comparability of the test results between GHP and TCB-HFM, the same coordinate ranges in Figure 5 are used in Figure 6. It can be seen that significantly greater q and ΔT of the specimens WT−t25−1 and WT−t25−2 are shown compared with those of the specimen WT−t25*, which attributes to the greater temperature difference between the hot side and the cold side for the initial setting of TCB-HFM tests. In addition, the fluctuation amplitudes of q and ΔT are slightly bigger than those shown in Figure 5, which can be for the reason that, unlike the GHP test, the heating and cooling units are not tightly next to the specimen for the TCB-HFM test. For further comparison, Table 4 shows the statistical results of data recorded during the last 20 min. It is found that the COV and MDA are small, with maximums of 0.70% and 1.41%, respectively, although slightly greater than that of the specimen WT−t25* shown in Table 3. In addition, it is demonstrated that the data fluctuation amplitudes are much smaller than a generally acceptable limitation of 5%, and the fluctuations are not unidirectional (see Figure 7), which verifies the steady state and data credibility of the sampling stage. Based on Equation (1) and the data in Table 4, the average of λ for specimens WT−t25−1 and WT−t25−2 is finally determined as 0.196 W/(m·K), having a 5.9% deviation from the result of the GHP test. The magnitudes of the data fluctuation and deviation from the result of the control test are in a reasonable range, which indicates that the TCB-HFM can be regarded as an alternative test method to conduct investigations on the thermal properties of materials.

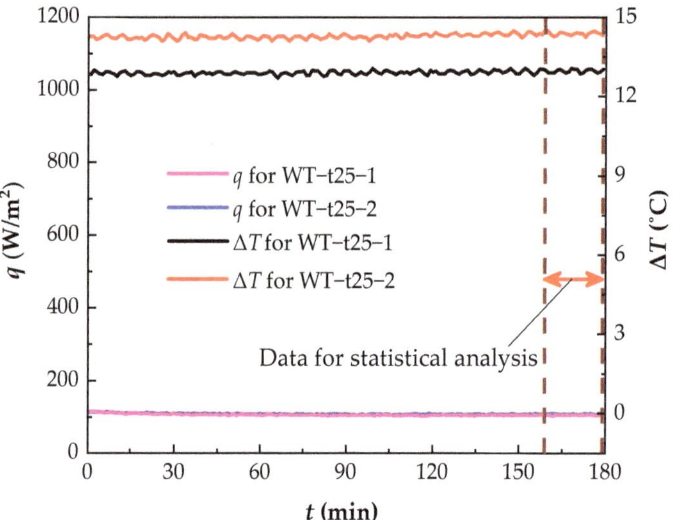

Figure 6. q and ΔT versus t for specimens WT−t25−1 and WT−t25−2.

Table 4. Statistical results of q and ΔT for specimens WT−t25−1 and WT−t25−2.

Specimen	Thermal Parameter	Average	Coefficient of Variation (COV)	Maximum Deviation from the Average (MDA)
WT−t25−1	q	105.729 W/m²	0.68%	1.39%
	ΔT_1	11.943 °C	0.50%	1.20%
	ΔT_2	13.167 °C	0.50%	1.27%
	ΔT_3	13.795 °C	0.67%	0.76%
WT−t25−2	q	108.733 W/m²	0.69%	1.26%
	ΔT_1	13.971 °C	0.40%	0.92%
	ΔT_2	15.214 °C	0.70%	1.41%
	ΔT_3	14.019 °C	0.48%	0.85%

Note: The subscript of "ΔT" corresponds to the number of the thermocouple shown in Figure 4a.

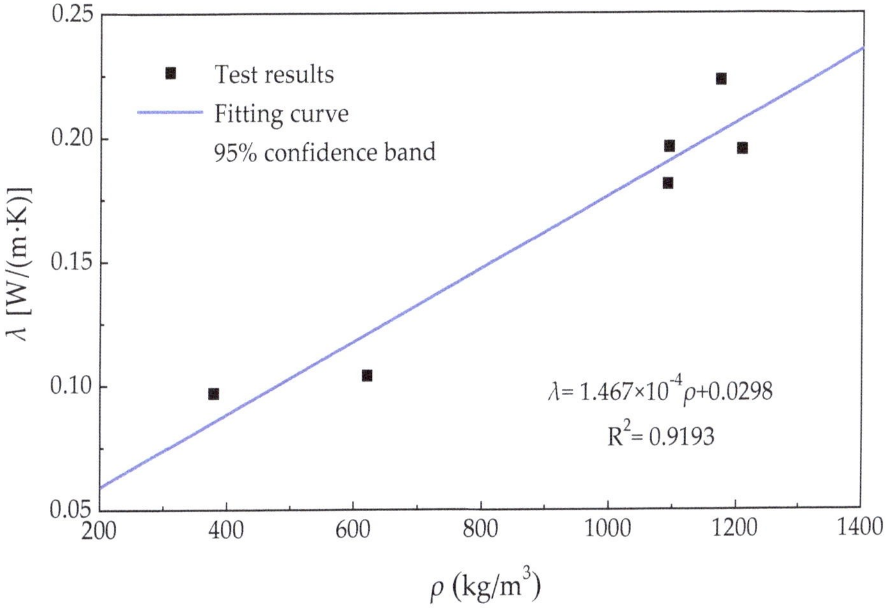

Figure 7. Fitting curve and λ test results versus ρ.

3.2. Thermal Property of Straw Boards

Analogously, the thermal property tests of the straw boards with other specifications were carried out via TCB-HFM. The experimental statistical results are summarized in Table 5. It can be noticed that the data fluctuation amplitudes have no abnormalities and are all within 3%, with the maximum MDA for ΔT and q being 1.95% and 2.77%, respectively. Considering the densities of straw boards of different specifications shown in Table 1, it is found that the λ of WSSB12, WSSB15, WSSB18, and WSSB25, which have higher densities, is obviously greater than that of the low-density WSSB30 and PSB58. Thus, a linear correlation between ρ and λ is assumed, and the proposed fitting formula is expressed as Equation (5). Figure 7 shows the fitting curve and λ test results versus ρ. It can be seen that the data scatters of test results evenly distribute on both sides of the fitting curve and are all in the range of the 95% confidence band. The coefficient of determination R^2 of the fitting formula is equal to 0.9193, which demonstrates the significant correlation between ρ and λ of straw boards in this study. Therefore, the fitting formula can be useful for relevant researchers and designers to understand the thermal conductivities of such materials when thermal tests are not available. It should be noted that the above λ prediction formula is developed based on the test specimens that are conditioned in the lab environment with the temperature and relative humidity of 19 °C and 29%, respectively. In the case that the straw boards are used in an environment that is significantly different from that lab environment, the prediction results could have some deviation from the actual values.

$$\lambda = 1.467 \times 10^{-4}\rho + 0.0298 \tag{5}$$

where ρ is in kg/m^3, and λ is in W/(m·K).

Table 5. Statistical results of thermal parameters derived from the TCB-HFM tests.

Specimen	MDA for ΔT	MDA for q	λ [W/(m·K)]	λ_{avg} [W/(m·K)]
WT−t12−1	1.92%	0.99%	0.177	
WT−t12−2	1.79%	1.40%	0.198	0.195
WT−t12−3	1.94%	1.49%	0.193	
WT−t15−1	1.67%	1.18%	0.180	
WT−t15−2	1.38%	1.11%	0.182	0.181
WT−t18−1	1.18%	0.77%	0.199	
WT−t18−2	1.95%	1.12%	0.231	0.223
WT−t18−3	1.60%	0.90%	0.215	
WT−t25−1	1.27%	1.39%	0.204	
WT−t25−2	1.41%	1.26%	0.189	0.196
WT−t30−1	1.35%	0.97%	0.101	
WT−t30−2	1.02%	0.62%	0.107	0.104
ST−t58−1	1.22%	2.15%	0.095	
ST−t58−2	1.36%	2.77%	0.100	0.097

Note: λ_{avg} refers to the average of λ.

4. Discussion

The advantages of the proposed alternative test method, TCB-HFM, can be concluded by comparing it with the commonly used steady-state test methods stated in the Introduction for investigating the thermal properties of insulation materials. Compared to the heat flux meter method, the hot box used for providing a steady and controllable temperature difference condition in the TCB-HFM tests makes the specimen in an enclosed space. This contributes to avoiding reliance on the indoor and outdoor temperature difference and eliminating the effect of ambient temperature and relative humidity on measuring results of thermal conductivity. Compared to the guarded hot plate method, the TCB-HFM can be used for measuring the thermal conductivity of plate-shaped specimens with relatively small dimensions and applies to the thermal property investigation on large-dimension wall specimens. Thus, such a method makes it more convenient to perform a series of investigations on thermal properties from a material to the corresponding member and is conducive to the full use of experimental equipment resources. In contrast with the hot box method, the monitoring of heat flux in the TCB-HFM no longer relies on recording the heating power but derives from the readings of the heat flux meters. Such a method leaves out the process of considering the error between the heat flux and the heating power and is more concise.

As for the proposed fitting formula for predicting the thermal conductivity of such straw bio-based materials by the density, i.e., Equation (5), it is developed based on the analysis of the material characteristics and ambient factors. The factors that affect the thermal conductivity can be concluded as straw bio-based material characteristics and ambient conditions, including the types of straw and binder, fiber orientation, ambient temperature, relative humidity, and density, as stated in the Introduction. In this study, the types of straw and binder are relatively single, the straw fibers are all scattered without a specific direction, and the effect of ambient temperature and relative humidity is negligible since the tests are conducted in the enclosed apparatus and the same lab environment. Thus, the material density is the most related to the thermal conductivity of such straw bio-based material in this study, and it is reasonable to propose a fitting formula for predicting the thermal conductivity based on the density. Moreover, similar correlations between the density and thermal conductivity of straw bio-based materials have also been concluded in the relative references [7,23].

To compare the thermal conductivities of the straw boards with those of other straw bio-based materials and investigate the applicability of the fitting formula, the data scatters for the λ of straw bales and composites from the previous studies [4,8,9,13,15,25,28] are shown in Figure 8 versus ρ. As stated in the Introduction, the densities of straw bales and composites in previous studies are within a small range. It can be seen from the figure that most of the straw boards concerned in this study have higher densities and λ compared with previous studies, which widens the cover range of the straw bio-based materials of known relation between λ and ρ. In addition, the data from the references are obtained from different lab environments that involve ambient temperatures from 10 °C to 35 °C and relative humidity from 40% to 50%. These two variables mainly affect the moisture content of a specimen tested with non-enclosed apparatus [5] and finally influence the specimen's density. The straw types include rice, wheat, bean, et al. The fiber orientation for the straw bales is random, and the binder types of the straw composites include liquid glass, methylene diphenyl diisocyanate resin, clay, cement, et al. The corresponding data scatters distribute on both sides of the fitting curve and are all in the 95% confidence band range, similar to the data scatters for the test results in this study. It is indicated that the fitting formula has the feasibility to predict the λ of more types of straw bio-based materials. Moreover, the density is dominant among several factors that have effects on the thermal conductivity of straw bio-based materials. The thermal properties of straw bio-based materials with different density specifications need to be studied in a later study to extend the data set. Thus, the formula for λ prediction can be further perfected and used in relevant research and engineering design.

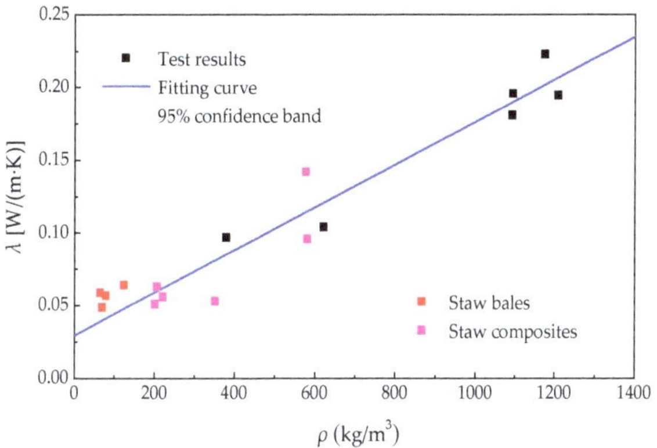

Figure 8. λ of straw bales and composites from the previous studies [4,8,9,13,15,25,28] versus ρ.

5. Conclusions

To validate the TCB-HFM method used for thermal property tests, understand the thermal conductivity of two types of straw boards, and reveal the correlation between straw boards with different specifications and their thermal properties, experimental investigations on two types of boards using both the TCB-HFM and GHP methods were carried out, in which the experimental results of the GHP test were taken as the benchmark. The findings of this study contribute to providing an alternative method to the thermal conductivity test, which can be used for specimens that are thicker compared with the GHP and can also apply to performing the thermal property investigation of large-dimension wall specimens. The proposed λ prediction formula provides a reference for the thermal conductivities of straw bio-based materials when thermal tests are not available to the relevant researchers and designers. This benefits the application of such materials. The main conclusions from this study are summarized as follows.

1. The fluctuation amplitudes of the monitored parameters, including q and ΔT, for the TCB-HFM tests during the steady state are much smaller than a generally acceptable limit of 5%, although slightly bigger than those for the GHP test. Moreover, the 5.9% deviation of the λ between the two test methods is within a reasonable range. The TCB-HFM is verified as an alternative test method to conduct investigations on the thermal properties of materials.

2. The λ of two types of straw boards that include six specifications in total is obtained via the TCB-HFM tests, in which the maximum fluctuation amplitudes for ΔT and q are 1.95% and 2.77% for all specimens during the steady state. Based on the experimental results, the correlation between the ρ and λ for such material is found and expressed by a linear fitting formula with the determination coefficient R^2 of 0.9193.
3. The test results for λ of other straw bio-based materials in previous studies are compared with the fitting curve. It is found that the fitting formula has the feasibility to predict the λ of more types of straw bio-based materials, and the density is dominant among several factors that have effects on the thermal conductivity of straw bio-based materials.

Author Contributions: Conceptualization, K.S. and C.Z.; methodology, K.S.; software, K.S.; validation, K.S.; formal analysis, K.S.; investigation, K.S.; resources, Y.W.; data curation, K.S.; writing—original draft preparation, K.S.; writing—review and editing, C.Z.; visualization, K.S.; supervision, C.Z. and W.Z.; project administration, C.Z. and Y.W.; funding acquisition, C.Z. All authors have read and agreed to the published version of the manuscript.

Funding: This research was funded by the National Key R&D Program of China, grant number 2019YFD1101005, and the China State Construction Engineering Corporation, grant number CSCEC-2020-Z-58-2.

Institutional Review Board Statement: Not applicable.

Informed Consent Statement: Not applicable.

Data Availability Statement: The data presented in this study are available on request from the corresponding author.

Conflicts of Interest: The authors declare no conflict of interest.

References

1. Tlaiji, G.; Ouldboukhitine, S.; Pennec, F.; Biwole, P. Thermal and mechanical behavior of straw-based construction: A review. *Constr. Build. Mater.* **2022**, *316*, 125915. [CrossRef]
2. Shi, L.Y.; Zhang, H.B.; Li, Z.X.; Man, X.; Wu, Y.; Zheng, C.; Liu, J. Analysis of moisture buffering effect of straw-based board in civil defence shelters by field measurements and numerical simulations. *Build. Environ.* **2018**, *143*, 366–377. [CrossRef]
3. Liu, L.; Li, H.; Lazzaretto, A.; Manente, G.; Tong, C.; Liu, Q.; Li, N. The development history and prospects of biomass-based insulation materials for buildings. *Renew. Sustain. Energy Rev.* **2017**, *69*, 912–932. [CrossRef]
4. Cascone, S.; Evola, G.; Gagliano, A.; Sciuto, G.; Baroetto Parisi, C. Laboratory and in-situ measurements for thermal and acoustic performance of straw bales. *Sustainability* **2019**, *11*, 5592. [CrossRef]
5. Ashour, T. The Use of Renewable Agricultural By-Products as Building Materials. Ph.D. Thesis, Benha University, Moshtohor, Egypt, 2003. [CrossRef]
6. Beck, A.; Heinemann, U.; Reidinger, M.; Fricke, J. Thermal Transport in Straw Insulation. *J. Build. Phys.* **2004**, *27*, 227–234. [CrossRef]
7. Vejeliene, J. Processed straw as effective thermal insulation for building envelope constructions. *Eng. Struct. Technol.* **2012**, *4*, 96–103. [CrossRef]
8. Shea, A.; Wall, K.; Walker, P. Evaluation of the thermal performance of an innovative prefabricated natural plant fibre building system. *Build. Serv. Eng. Res. Technol.* **2013**, *34*, 369–380. [CrossRef]
9. Wei, K.; Lv, C.L.; Chen, M.Z.; Zhou, X.; Dai, Z.; Shen, D. Development and performance evaluation of a new thermal insulation material from rice straw using high frequency hot-pressing. *Energy Build.* **2015**, *87*, 116–122. [CrossRef]
10. Conti, L.; Barbari, M.; Monti, M. Steady-State thermal properties of rectangular straw-bales (RSB) for building. *Buildings* **2016**, *6*, 44. [CrossRef]
11. D'Alessandro, F.; Bianchi, F.; Baldinelli, G.; Rotili, A.; Schiavoni, S. Straw bale constructions: Laboratory, in field and numerical assessment of energy and environmental performance. *J. Build. Eng.* **2017**, *11*, 56–68. [CrossRef]
12. Gallegos-Ortega, R.; Magaña-Guzmán, T.; Reyes-López, J.A.; Romero-Hernández, M.S. Thermal behavior of a straw bale building from data obtained in situ. A case in Northwestern México. *Build. Environ.* **2017**, *124*, 116–122. [CrossRef]
13. Sabapathy, K.A.; Gedupudi, S. Straw bale based constructions: Measurement of effective thermal transport properties. *Constr. Build. Mater.* **2019**, *198*, 182–194. [CrossRef]
14. Dušek, J.; Jerman, M.; Podlena, M.; Böhm, M.; Černý, R. Sustainable composite material based on surface-modified rape straw and environment-friendly adhesive. *Constr. Build. Mater.* **2021**, *300*, 124036. [CrossRef]

15. Ratiarisoa, R.V.; Magniont, C.; Ginestet, S.; Oms, C.; Escadeillas, G. Assessment of distilled lavender stalks as bioaggregate for building materials: Hygrothermal properties, mechanical performance and chemical interactions with mineral pozzolanic binder. *Constr. Build. Mater.* **2016**, *124*, 801–815. [CrossRef]

16. Zhou, Y.P.; Trabelsi, A.; Mankibi, M.E. A review on the properties of straw insulation for buildings. *Constr. Build. Mater.* **2022**, *330*, 127215. [CrossRef]

17. ISO 8301:1991; Thermal Insulation—Determination of Steady-State Thermal Resistance and Related Properties—Heat Flow Meter Apparatus. International Organization for Standardization: Geneva, Switzerland, 1991. Available online: https://www.iso.org/standard/15421.html (accessed on 10 July 2023).

18. GB/T 10295-2008; Thermal Insulation—Determination of Steady-State Thermal Resistance and Related Properties—Heat Flow Meter Apparatus. Chinese Standard: Beijing, China, 2008. Available online: https://www.codeofchina.com/standard/GBT10295-2008.html (accessed on 10 July 2023).

19. ISO 8302:1991; Thermal Insulation—Determination of Steady-State Thermal Resistance and Related Properties—Guarded Hot Plate Apparatus. International Organization for Standardization: Geneva, Switzerland, 1991. Available online: https://www.iso.org/standard/15422.html (accessed on 10 July 2023).

20. GB/T 10294-2008; Thermal Insulation—Determination of Steady-State Thermal Resistance and Related Properties—Guarded Hot Plate Apparatus. Chinese Standard: Beijing, China, 2008. Available online: https://www.codeofchina.com/search/default.html?page=1&keyword=GB/T%2010294-2008 (accessed on 10 July 2023).

21. ISO 8990:1994; Thermal Insulation—Determination of Steady-State Thermal Transmission Properties—Calibrated and Guarded Hot Box. International Organization for Standardization: Geneva, Switzerland, 1994. Available online: https://www.iso.org/standard/16519.html (accessed on 10 July 2023).

22. GB/T 13475-2008; Thermal Insulation—Determination of Steady-State Thermal Transmission Properties—Calibrated and Guarded Hot Box. Chinese Standard: Beijing, China, 2008. Available online: https://www.codeofchina.com/search/default.html?page=1&keyword=GB/T%2013475-2008 (accessed on 10 July 2023).

23. Costes, J.-P.; Evrard, A.; Biot, B.; Keutgen, G.; Daras, A.; Dubois, S.; Lebeau, F.; Courard, L. Thermal Conductivity of Straw Bales: Full Size Measurements Considering the Direction of the Heat Flow. *Buildings* **2017**, *7*, 11. [CrossRef]

24. Douzane, O.; Promis, G.; Roucoult, J.; Le, A.D.T.; Langlet, T. Hygrothermal performance of a straw bale building: In situ and laboratory investigations. *J. Build. Eng.* **2016**, *8*, 91–98. [CrossRef]

25. Bakatovich, A.; Davydenko, N.; Gaspar, F. Thermal insulating plates produced on the basis of vegetable agricultural waste. *Energy Build.* **2018**, *180*, 72–82. [CrossRef]

26. Sun, K.; Zheng, C.R.; Wang, X.D. Thermal performance and thermal transmittance prediction of novel light-gauge steel-framed straw walls. *J. Build. Eng.* **2023**, *67*, 2352–7102. [CrossRef]

27. GB/T 21723-2008; Wheat/Rice-Straw Particleboard. Chinese Standard: Beijing, China, 2008.

28. Liuzzi, S.; Rigante, S.; Ruggiero, F.; Stefanizzi, P. Straw based materials for building retrofitting and energy efficiency. *Key Eng. Mater.* **2016**, *678*, 50–63. [CrossRef]

 *sustainability*

Article

A Novel Rectangular-Section Combined Beam of Welded Thin-Walled H-Shape Steel/Camphor Pine Wood: The Bending Performance Study

Chang Wu [1,2,*], Junwei Duan [3], Ziheng Yang [1], Zhijiang Zhao [1] and Yegong Xu [1]

1   Department of Civil Engineering, Lanzhou University of Technology, Lanzhou 730050, China
2   Western Engineering Research Center of Disaster Mitigation in Civil Engineering Ministry of Education, Lanzhou 730050, China
3   China Construction First Construction Group Co., Ltd., Shanghai 201103, China
*   Correspondence: wuchang317@edu.lut.cn

Abstract: At present, the development of green building materials is imminent. Traditional wood structures show low strength and are easy to crack. Steel structures are also prone to instability. A novel rectangular-section composite beam from the welded thin-walled H-shape steel/camphor pine was proposed in this work. The force deformation, section strain distribution law, and damage mechanism of the combined beam were studied to optimize the composite beam design, clarify the stress characteristics, present a more reasonable and more efficient cross-section design, and promote green and environmental protection techniques. Furthermore, the effect of different factors such as steel yield strength, H-type steel web thickness, H-type steel web height, H-type steel flange thickness, H-type steel upper flange covered-board thickness, and combined beam width was investigated. The ABAQUS simulation with the finite element software was also performed and was verified through empirical experiments. According to the results: (1) the damage process of the composite beam was divided into three steps, namely elastic stage, elastic–plastic step, and destruction stage, and the cross-middle section strain met the flat section assumption; (2) additionally, the bond connection was reliable, the two deformations were consistent, and the effect of the combination was significant. The study of the main factors showed that an increase in the yield strength, the H-type steel web height, the steel H-beam upper flange thickness, and the combined beam width caused a significant enhancement in the bending bearing capacity. The combined beam led to high bending stiffness, high bending bearing capacity, and good ductility under bending.

Keywords: composite beam; material performance test; bending performance; theoretical analysis; numerical simulation

Citation: Wu, C.; Duan, J.; Yang, Z.; Zhao, Z.; Xu, Y. A Novel Rectangular-Section Combined Beam of Welded Thin-Walled H-Shape Steel/Camphor Pine Wood: The Bending Performance Study. *Sustainability* **2023**, *15*, 7450. https://doi.org/10.3390/su15097450

Academic Editor: Claudia Casapulla

Received: 21 February 2023
Revised: 26 April 2023
Accepted: 27 April 2023
Published: 30 April 2023

1. Introduction

The gradual shortage of limited resources and energy has become more and more serious due to the rapid development of urbanization and industrialization. Traditional reinforced concrete buildings need to dismantle after the structure expiration, which inevitably produces a large amount of construction waste. The improper disposal of construction waste can cause environmental pollution, low efficiency of construction waste recovery, and difficulties in construction waste recovery. The discovery of a solution to manage construction waste has converted into an urgent problem. Nowadays, human societies search for green environmental protection, energy saving, ecology, and health [1–8]. Because of the need for development in the construction industry, world research for steel–wood combinations has gradually increased. Borri et al. [9] performed 21 double-shear tests to reinforce the bent wood beams and study the ductility, stiffness, and strength response of reinforcement wood beams. Nadir et al. [10] performed the bending tests on carbon fiber and glass fiber paste wood composite beams and concluded that the bending stiffness of the wood beam can be significantly improved, and the bending bearing capacity calculation formula was deduced. Hassanieh et al. [11] investigated the short-term behaviour of innovative steel-timber composite (STC) floors comprising cross-laminated timber panels connected to steel girders by various mechanical fasteners and/or glue. The results of four-point bending tests performed on full-scale STC beams are reported and the structural behaviour of the proposed STC system is studied. Shekarchi et al. [12] performed a three-point bending test on 20 strengthened beams and four un-strengthened (control) beams to investigate the bending performance of the squeezed glass fiber reinforced polymer. Tohid et al. [13] investigated the bending and shear strengths of lightweight composite I-beams made of timber. The result of both experiments and simulation confirmed the carrying capacity of the I-beam beam is greatly improved. Dar [14] performed the bending test on a rectangular thin-walled steel box and wood-filled composite beam and concluded the composite beam showed good bending performance. Pan et al. [15] investigated a new structural cantilever member bolted by steel cores and wood. Moreover, an analytical model of the cantilever members was developed using ANSYS. The result of both experiments and simulation confirmed the desired mechanical properties of the new cantilever members. Fujita et al. [16] performed the bending tests on steel and wood composite and concluded that the fracture of the composite occurred because of the initial crack at the outer edge of the wood on the tensile side of the member. Hassanieh et al. [17] performed four-point bending tests on seven steel-timber composite (STC) beams to investigate the load–deflection curve, stiffness diagram, maximum load curve, and failure mode of STC beams. McConnell et al. [18] carried out four-point bending tests on unreinforced, reinforced, and post-tensioned glulam timber beams. Compared with the unreinforced beam, the bending strength and stiffness of the reinforced beam and the post-tensioned beam improved. Lee Teng-hui [19] proposed a new type of steel–wood composite beam and studied its bending performance and theoretical analysis. Li Yushun et al. [20] investigated the bonding stress and slip distribution of steel–bamboo interface, the interface of two types of steel and bamboo plate is designed, and the calculation method of the interface bearing capacity under short-term load is proposed. Chen Aiguo et al. [21] proposed a new type of steel–wood composite beam was presented and the bending performance test of 9 steel–wood composite beams was conducted. The results show that H-shaped steel and wood boards have good overall working performance and significant combination effects. Wu [22] discussed the ultimate bending stiffness and the ultimate bending bearing capacity of the combined beam of inner-filled square thin-walled rectangular steel pipe. Zhang Yi et al. [23] took the bolt spacing and shear span ratio as the essential factor to conduct the bending test of the I-type combined beam under the action of a concentrated load. The results showed that the bending stiffness of the combined beam increased by 201%. Liu Degui et al. [24] proposed a built-in thin-walled H-shape steel–wood combination beam. The bending test was carried out with the width of the flange board, the bolt spacing, the thickness of the thin wall steel H-beam, the thickness of the flange board, and the height and the thickness of the board at the web. The results represented that the H-shape steel improved the bending bearing capacity, bending stiffness, energy consumption, and combined beam ductility. Zhang et al. [25] performed a hollow glulam beam design with stiffening plates in a hollow rectangular cross-section with high flexural behaviour and stability. To increase the carrying capacity, the bottom of the beam was reinforced with a fiber-reinforced polymer. The results show that the bearing capacity of reinforced glulam beams is slightly improved.

According to the literature review, a new type of rectangular-section combination beam from welded thin-walled steel H-beam/camphor pine was studied. The combination beam of three thin-walled steel plates was welded into an H-type skeleton by argon arc. The sides of the upper flange and web were covered with pine wood using strong epoxy resin AB glue in the rectangular section combination beam. The section of this combined beam is shown in Figure 1.

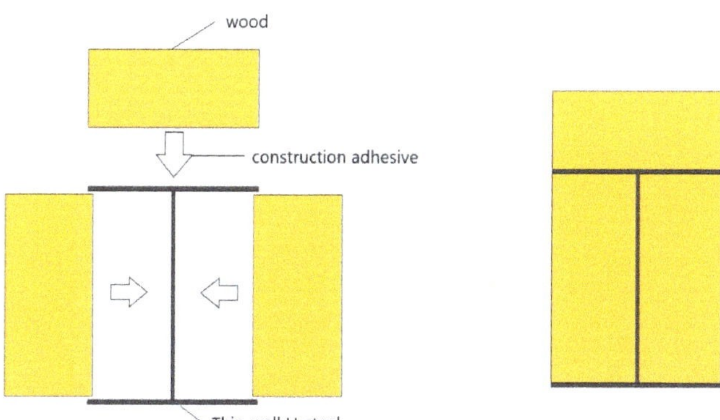

Figure 1. The section of steel and wood composite beam.

The green ecological steel–wood composite beam combines the advantages of wood and steel, and the advantages of the two materials are complementary. The coordinated work not only plays the optimal performance of the two materials, but also resists various stresses with reasonable division of labor, enhances the strength of the structure, and extends the span. In addition, both materials are renewable resources, showing their advantages of sustainable development, and can be widely used in green buildings.

2. Materials and Methods

2.1. Design of Test Specimen

The main parameters for the design of the steel–wood composite beam included the top board thickness, the flange thickness of the added steel, and the steel type on the failure form, bending stiffness, and bending bearing capacity. The values for the thickness of the upper flange-covered board in the steel section were 30 mm, 40 mm, and 50 mm. The thickness of the welded thin-walled H-shape steel flange was considered 1.5 mm, 2 mm, and 2.5 mm. Table 1 represents the parameters including the dimension of specific components.

Table 1. The Parameters of the tests (mm).

Specimen Code	t_f	t_w	h_1	t_1	t_2	t_3	b_1	l	$b \times h$
L-M	-	-	-	-	-	-	-	1700	100×155
L-Z-1	1.5	1.5	125	49.25	49.25	30	100	1700	100×155
L-Z-2	1.5	1.5	125	49.25	49.25	40	100	1700	100×165
L-Z-3	1.5	1.5	125	49.25	49.25	50	100	1700	100×175
L-Z-4	2	1.5	125	49.25	49.25	30	100	1700	100×155
L-Z-5	2.5	1.5	125	49.25	49.25	30	100	1700	100×155

Note: tf, tw, and h1 are the H-type steel flange, web thickness, and height, respectively; t1 is the thickness of the left board of the web, t2 is the thickness of the right board; t3 is the upper flange board; b1 is the upper flange width and the steel H-beam width; l is the total length of the test beam; b is the combined beam width; h is the total height of the combined beam, the thickness of the upper flange board and the height of the H-shape steel.

This composite beam is light in weight. Take the L-Z-2 composite beam as an example, the air-dry density of wood is 0.523 g/cm³, the density of Q235 steel is 7.85 g/cm³, and the density of epoxy resin AB glue is 1.45 g/cm³. The mass of steel required for each meter of composite beam is 3.8 kg, the mass of wood is 7.26 kg, the mass of epoxy resin AB glue is 1.57 kg, and the total mass of one running mass is 12.63 kg.

The structural adhesive used is Deyi brand E44 epoxy resin AB adhesive, in which component A includes: epoxy resin, filler, toughening agent, etc. Group B includes amination/anhydride curing agent polythiol, coupling agent, etc. After curing, the temperature resistance of epoxy resin AB adhesive is 20~80. In the process of using the member, the ambient temperature on the strength after curing can be ignored.

2.2. Test Piece Production

The steps of experiment piece production were as follows:

Stage 1: the welded thin-walled H-shaped steel and wood were polished.

Stage 2: alcohol was used to wipe the steel and wood, remove the dust, and remove impurities, and grease from the wood surface.

Stage 3: the binder (epoxy AB glue) was mixed.

Stage 4: the binder was applied.

Stage 5: the bonding surface remained immobile for complete adhesion. The molded member is shown in Figure 2.

Figure 2. The test piece of steel and wood composite.

2.3. Loading Test

The test adopts the controlled displacement loading method. The speed was 1 mm·min^{-1} and held the load for two minutes to read the number of each instrument. Five linear displacement sensors were arranged on the lower side of the beam in the upper flange middle on the side of the wood with M1, and the web with four sheets M2, M3, M4, and M5. The two strain sheets were adjusted at the edge of the upper and lower flanges of the H-shape steel, numbered G1 and G2. The DH3816 static strain measurement system was used to collect the data. The strain sheets were arranged as shown in Figure 3. The field of loading is shown in Figure 4.

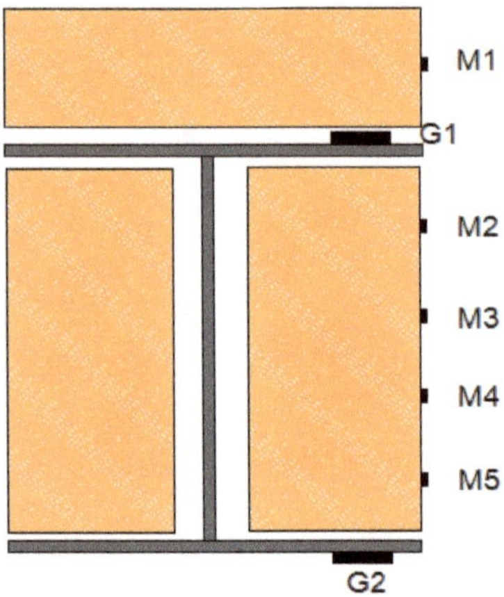

Figure 3. The layout of the strain sheets.

Figure 4. The field of loading.

3. Results and Discussion

3.1. Loading Test and Destruction Process

The beams L-Z-1~L-Z-5 showed similar destructive forms, and the final brittle fracture in the wood on both sides of the H-type steel web reached the ultimate tensile strength. It made the combined beam ineffective. The L-Z-2 composite beam is taken as an example for analysis. At the beginning of loading, the L-Z-2 combination beam was deformed with an increase in the load. When the load measurement showed 35.07 kN, the "click" sound of the wood fiber inside the composite beam and the "crackling sound" of degassing was heard for the first time. The greater the load, the denser and the louder the sound. After the load of 45.19 kN, the primary crack occurred at the right end of the specimen and gradually extended to the span of pure bending (Figure 5a). Figure 5b shows the left end of the composite beam in this stage. When the load value represented 49.31 kN, the crack was gradually enlarged (Figure 5c). As the load reached 56.47 kN, the composite beam entered the elastic–plastic step. The upper flange alongside the right side of welded steel H-beam web is shown in Figure 5d. When the load measurement represented 59.72 kN, there was a 'loud bang' in the zone of the lower flange of the steel H-beam. The lower flange of the steel H-beam reached the yield strength toward the outside, and the steel bent. Therefore, the 45° inclined crack took place in the wood in the same position (Figure 5e), and the composite beam was damaged.

3.2. Mid-Span Cross-Section Strain

Cross-section strain curves across L-Z-2 are shown in Figure 6. The maximum component load value in the figure was taken as the corresponding yield load. The strain value above the neutral axis was negative, indicating that the section above the neutral axis was pressed under the load, making the strain sheet and the value to be negative. The strain value below the neutral axis was positive, indicating that the section below the neutral axis was pulled under the load. It straightened the strain sheet, and the value was positive. The position of the neutral axis in the pure bend section of the component was always unchanged throughout the whole process from the initial load to the yield load, representing that the high-strength epoxy AB glue connection was effective.

3.3. Cross-Span Section Load–Strain Relationship

The cross-sectional load–strain curve for the combined beam L−Z−2 is shown in Figure 7. The values of the left strain abscissa half-axis were negative, representing the sticking position of the strain sheets M1 and G1 in the compression area of the member. The values of the right strain abscissa half-axis were positive, indicating the sticking position of the strain sheets M5 and G2 in the tensile area. From the figure, the strain measurement of each point in the section increased proportionally from the initial load to the yield load. The strain between the pressure area and tensile region was almost symmetric around the ordinate. The strain change of steel and wood in the tensile and compression areas was almost synchronized, showing a very reliable glue connection between the steel and wood.

Furthermore, the two deformations could work together, and the combined effect was significant. Furthermore, the value of strain G1 in the compression area was higher than that of M1 in the compression area. It indicated that the steel–wood combination beam played a complete role in the bending test of the steel.

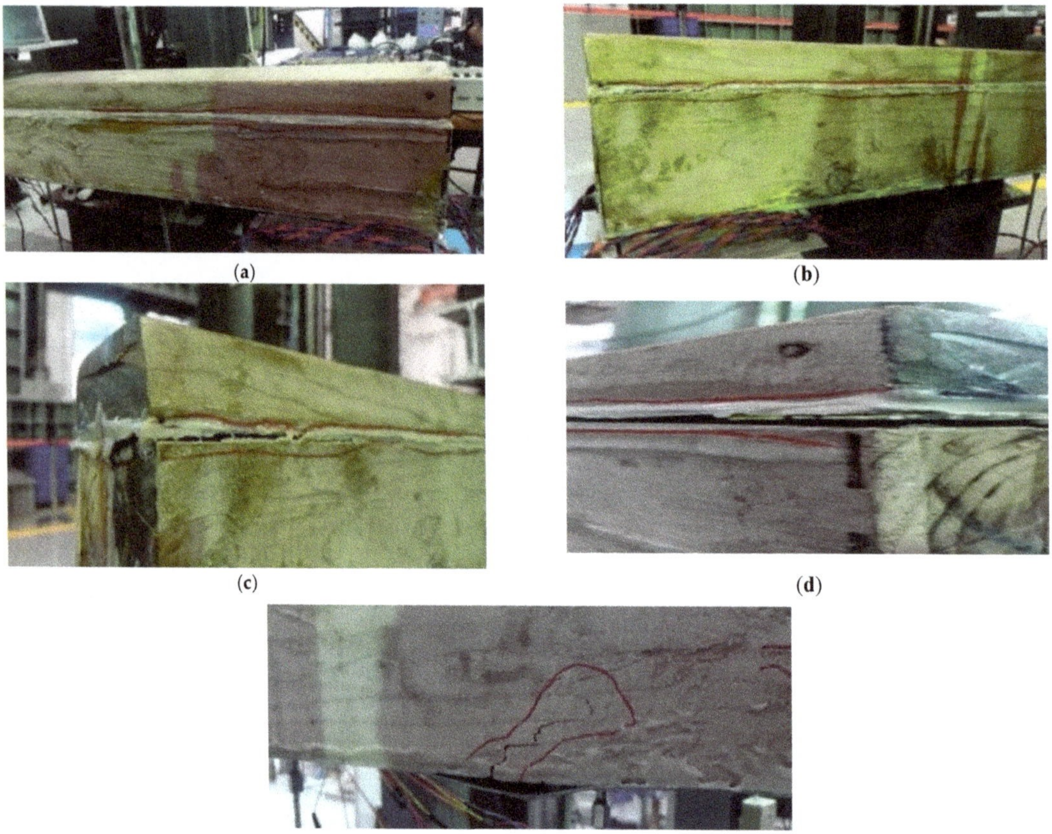

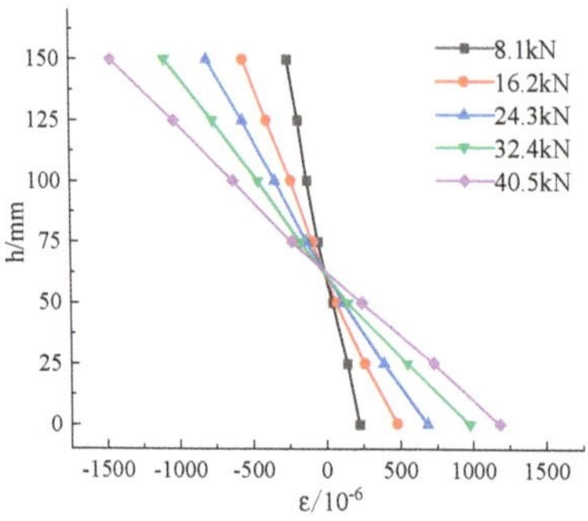

Figure 5. The combination beam L-Z-2 destruction process (**a**) the opened glue at the right upper flange; (**b**) the opened glue at the left upper flange; (**c**) the opened glue and widened at the right upper flange; (**d**) the unwelded H-shape steel web and upper flange; (**e**) the lower flange buckling, wood cracking.

Figure 6. The diagram of cross-section strain for the L-Z-2 span.

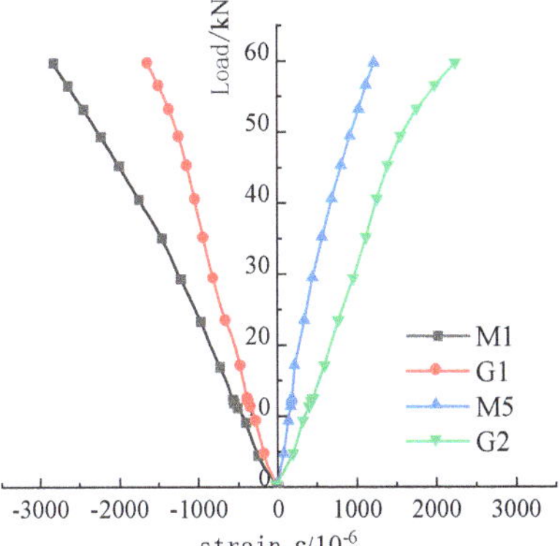

Figure 7. The diagram of cross-section load–strain for the L-Z-2 span.

3.4. Overall Deformation

The curve of overall deformation for the combined beam $L-Z-2$ under different loads is shown in Figure 8. The abscissa in the figure was the net span length of the combined beam, and the ordinate was the corresponding deflection. In the early step of loading, both steel and wood were in the elastic stage, playing a complete role in the tensile performance of H-shape steel and the compression performance of wood. The overall performance was desired, and the overall deformation curve in the whole stress process of components was basically symmetrical. In the early step of loading, the deflection enhancement in each measurement point followed a less linear development, and the overall curve was relatively gentle. With the further increase in the load, the steel entered the yield stage, and the cementing surface was serious, leading to a reduction in the overall stiffness of the combined beam. In the late step of loading, the deflection developed more rapidly than in the early stage, and increased significantly higher, until the component was destroyed. In the whole process of loading, the span deflection of the pure bending section was always the largest.

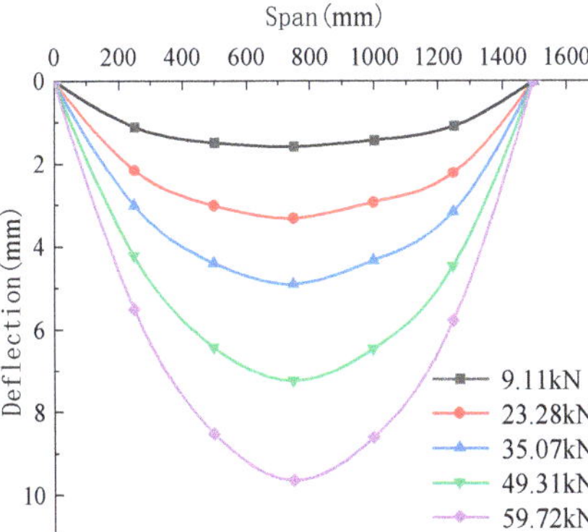

Figure 8. The diagram of overall deformation.

3.5. Load–Cross Deflection Relationship

The load–span median deflection diagram of the L-M, L-Z-1~L-Z-5 specimens are shown in Figure 9. The test period from the loading beginning to the failure was divided into three steps: elastic stage, elastic–plastic step, and destructive stage. Among the samples, the whole stress process of specimens L-M and L-Z-5 was the most representative. In the elastic step, the welding of the wood from the thin-walled H-shape steel and the left and right sides of the web, and the wood at the upper flange, under the strength of the high-strength binder, made the span deflection show an increasing linear trend with the load enhancement and always remained linear before the yield load of the component. The adhesive cracking started when the load exceeded the strength of the binder. At the same time, the welded thin-walled H-shape steel began to yield and entered the elastic–plastic stage. From the figure, there was a clear yield platform, after which the slope of the curve started to decrease and became nonlinear. It meant that the bending stiffness of the component gradually decreased, and the carrying capacity began to decrease. A gradual increase in the load caused a gradual reduction in the stiffness, which entered the destruction step. With an equal amount of load, the deformation expanded deflection rather than in the previous stage. When the load increased gradually, the component was damaged. According to the load–span medium deflection diagram, the final deflection of the member L-M was about 8 mm, and the final damage phenomenon indicated a brittle failure. The final deflection of the remaining combined components exceeded the final deflection of the components L-M, and the final damage phenomenon occurred as ductility damage. The component L-Z-5 was a typical representative component of ductile failure, and its final deflection was 16.87 mm, which also represented that the component delayed the damage when the steel was in the destructive stage. This kind of combined beam resulted in a good deformation ability and ductility under the bending process.

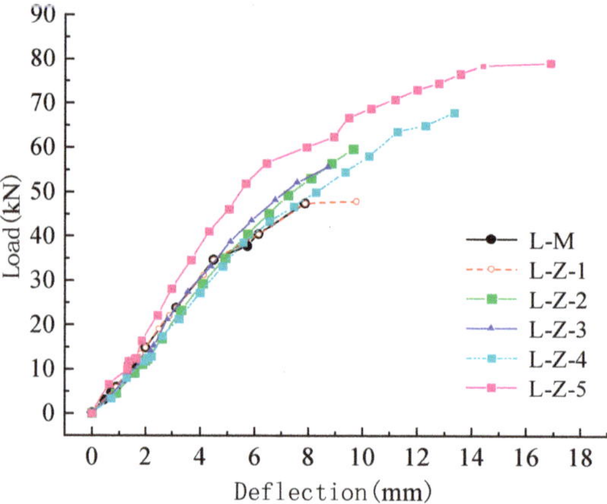

Figure 9. The diagram of beam load–span medium deflection.

4. The Results of the Effective Factors

4.1. Effect of Welding Thin-Walled H-Shape Steel

The influence of welded thin-walled steel H-beam was studied on the bending capacity of the combined beam, as shown in Figure 10 and Table 2. As a comparison between the combined beam L-Z-1 and pure wood beam L-M, the bending bearing capacity of these beams was 47.934 kN and 47.429 kN, respectively. The thin-walled H-shape steel was welded was about 1.94%, and the bending bearing capacity was not much different. The maximum bearing capacity increased only by 0.505 kN, and the enhancement range was about 1.06%. The analysis reasons were that the combination beam L-Z-1 was the first processing and assembly component, with uneven coating, uneven thickness, insufficient bonding strength, maintenance, and not in place, so that the bending bearing capacity was not higher than that of the pure wood beam L-M, and the effect is not obvious. Compared with combined beam L-Z-4 and pure wood beam L-M, the bending capacity of combined beam L-Z-4 was 67.958 kN, that of L-M beam was 47.429 kN, the welded thin-walled H-shape steel was about 3.76%, the bending capacity significantly increased, the maximum bearing capacity increased by 20.529 kN, the increased range was about 42.0%. Compared with combined beam L-Z-5 and pure wood beam L-M, the bending bearing capacity of combined beam L-Z-5 was 79.169 kN, and that of pure wood beam L-M was 47.429 kN. Welded thin-walled H-shape steel was added to about 4.41%. Compared with the combination beam L-Z-5 and pure wood beam L-M, the maximum bearing capacity increased by 31.19 kN and by about 67.87%. It was known from the analysis that after welding, the maximum bending bearing capacity of different sizes was higher than that of the pure wood beam.

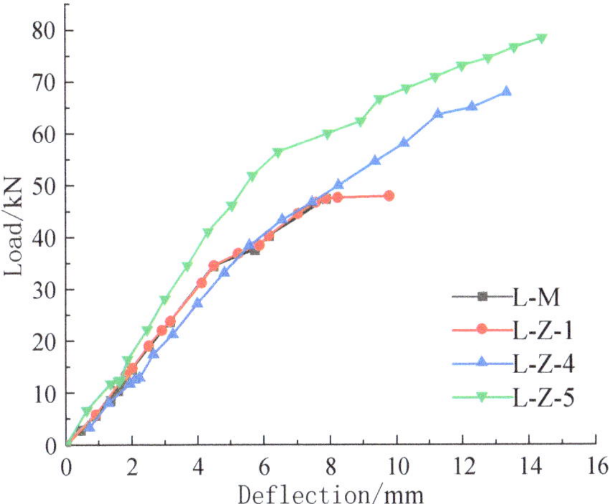

Figure 10. The diagram of load-deflection for the comparison among the beams.

Table 2. Flexural load capacity of specimens.

Specimen Number	P/kN	γ_P/%
L-M	47.429	-
L-Z-1	47.934	1.06
L-Z-4	67.958	42
L-Z-5	79.169	68.87

Note: γ_p is the percentage improvement of built-in thin-walled H-shaped steel and wood composite beam P relative to pure wood beam P.

4.2. Influence of the Flange Thickness of Welded Thin-Walled H-Shape Steel

The effect of the flange thickness was investigated on the bending capacity of the combined beam, as shown in Figure 11 and Table 3. When the thickness of the flange of the beam L-Z-1 increased from 1.5 mm to 2 mm to the thickness of the beam L-Z-4, the bending capacity increased from 47.934 kN of the beam L-Z-1 to 67.958 kN of the beam L-Z-4, and the maximum bending capacity increased by 20.024 kN, with an increase of about 41.77%. When the thickness of beam L-Z-4 increased from 2 mm to 2.5 mm of beam L-Z-5, the maximum bending bearing capacity increased from 67.958 kN of beam L-Z-4 to 79.169 kN of beam L-Z-4, increasing the bending bearing capacity by 11.211 kN by about 16.50%. Consequently, the maximum bending bearing capacity of the corresponding combined beam was also different due to the various thicknesses of the welded thin-walled H-type steel flange. With the gradual increase in the flange thickness of the welded thin-walled H-shape steel, the maximum bending bearing capacity was gradually enhanced, but the increase was gradually smaller. Therefore, the optimum flange thickness was 2 mm.

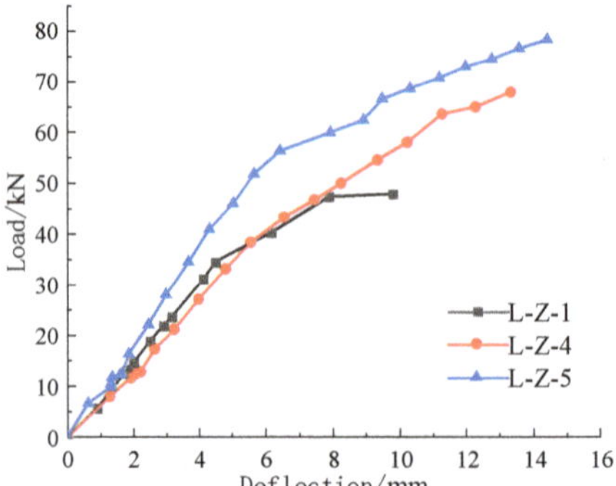

Figure 11. The load-deflection diagram for comparison among the three beams.

Table 3. Flexural load capacity of specimens with different flange thickness.

Specimen Number	H-Beam Flange Thickness/mm	P/kN	γ_P/%
L-Z-1	1.5	47.934	—
L-Z-4	2	67.958	41.77
L-Z-5	2.5	79.169	16.5

Note: γ_p is the percentage improvement of built-in thin-walled H-shaped steel and wood composite beam P relative to composite beam L-Z-1 P.

4.3. Effect of the Upper Flange Thickness

The combined beam load-mid-span deflection curve is shown in Figure 12. Compared with the L-Z-1 combined beam and L-Z-2 combined beam, the thickness of the upper flange covered-board of H-shape steel increased from 30 mm to 40 mm, and the corresponding bending bearing capacity was enhanced from 47.934 kN to 59.720 kN, with an increase by 24.59%. Compared with the beam L-Z-2 and the beam L-Z-3, when the thickness of the upper flange covered-board increased from 40 mm to 50 mm, the bending bearing capacity was not much different. The reasons might be due to the processing error of the component and the uneven coating on the adhesive surface. However, from the three curves in the figure, the overall bending stiffness of combination beam L-Z-3 was constantly greater than that of combination beam L-Z-2. Before the mid-span displacement reached 8.77 mm, the bending bearing capacity of combined beam L-Z-3 was steadily higher than that of combined beam L-Z-2. Overall, the bending bearing capacity is improved. According to the analysis, the thickness of the flange of the steel H-beam caused a significant influence on the bending bearing capacity of the combined beam. After the increase in the board thickness, the overall height of the composite beam was enhanced, and the bending bearing capacity also gradually increased.

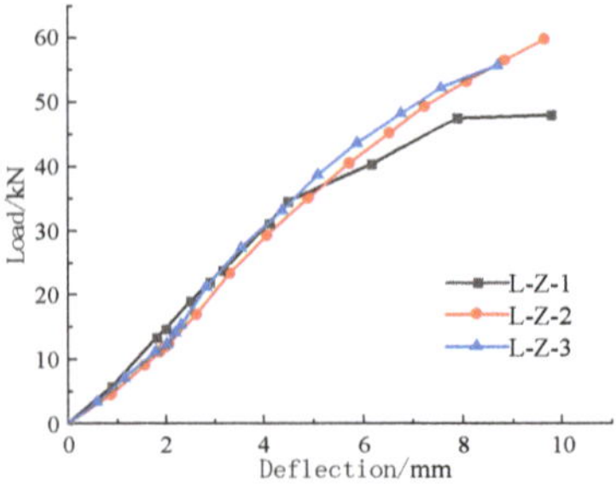

Figure 12. A comparison of load-deflection curves.

5. Finite Element Parameter and Results Comparison

5.1. Finite Element Model Establishment

5.1.1. Basic Assumptions and Material Constitutive Relationship

(1)   Basic assumption

1.   The wood when pulled and pressed; There was no relative sliding between;
2.   Steel and wood; without considering the influence of slip;
3.   The influence of residual stress and residual deformation was not considered during steel welding;
4.   The natural defects, creep, and stress relaxation of the wood itself were not considered.

(2)   Steel constitutive structure

The failure of the combined beam was due to the failure of the wood on both sides of the H-shape steel web after reaching the tensile strength, and the steel failed to play the ultimate tensile strength role. The steel conformed to the ideal elastic–plastic model. Its constitutive relationship is shown in Figure 13.

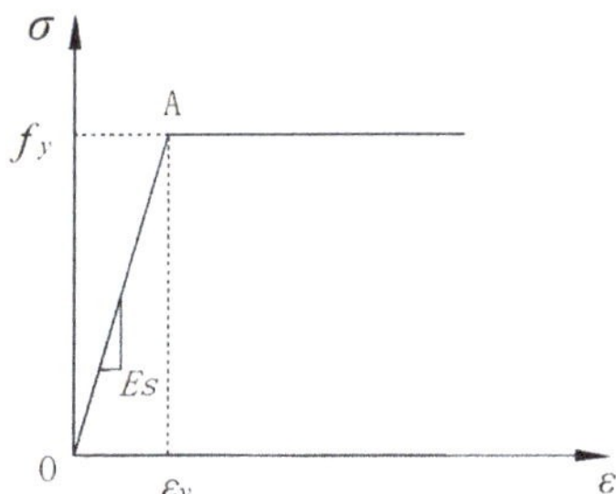

Figure 13. The diagram of steel constitutive relationship.

(3)   Wood constitutionality

Wood constitutive followed the Chen [26] constitutive relation as an ideal elastoplastic model as shown in Figure 14.

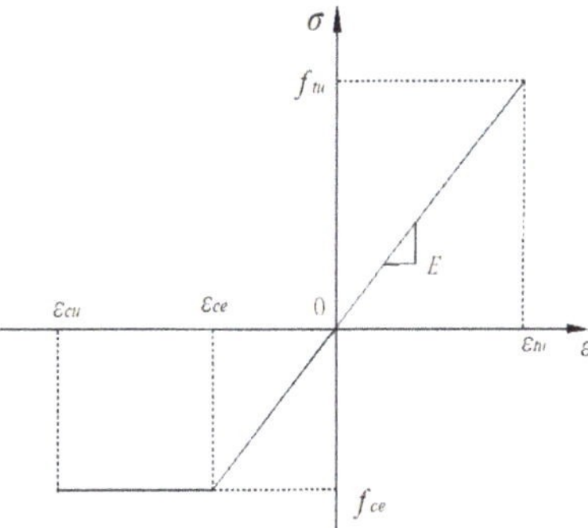

Figure 14. The diagram of wood constitutive relationship.

5.1.2. Material Properties

Wood belonged to the orthogonal anisotropic material. The longitudinal direction (L) was the wood length direction, as indicated by the main direction 3. Radial (R) was indicated by primary direction 2, and tangential (T) was represented by primary direction 1. The engineering constant was selected in the elastic type. The timber property index parameters of the wood were inputted as shown in Table 4. The material direction of the wood was assigned.

Table 4. The parameters of wood materials performance elasticity.

E_L (MPa)	E_R (MPa)	E_T (MPa)	μ_{RT}	μ_{LR}	μ_{LT}	G_{LT}	G_{LR}	G_{RT}
11,022	1102.2	551.1	0.37	0.022	0.37	661.32	826.65	198.40

The properties included Q235B steel, elastic modulus 2.03×105 MPa, Poisson ratio 0.3, yield strength 285.64 MPa. The steel pad block was set as a rigid body.

5.1.3. Model Establishment

The components were divided into three categories for modeling, including wood, steel, and pads. The overall model of the combined beam is shown in Figure 15a.

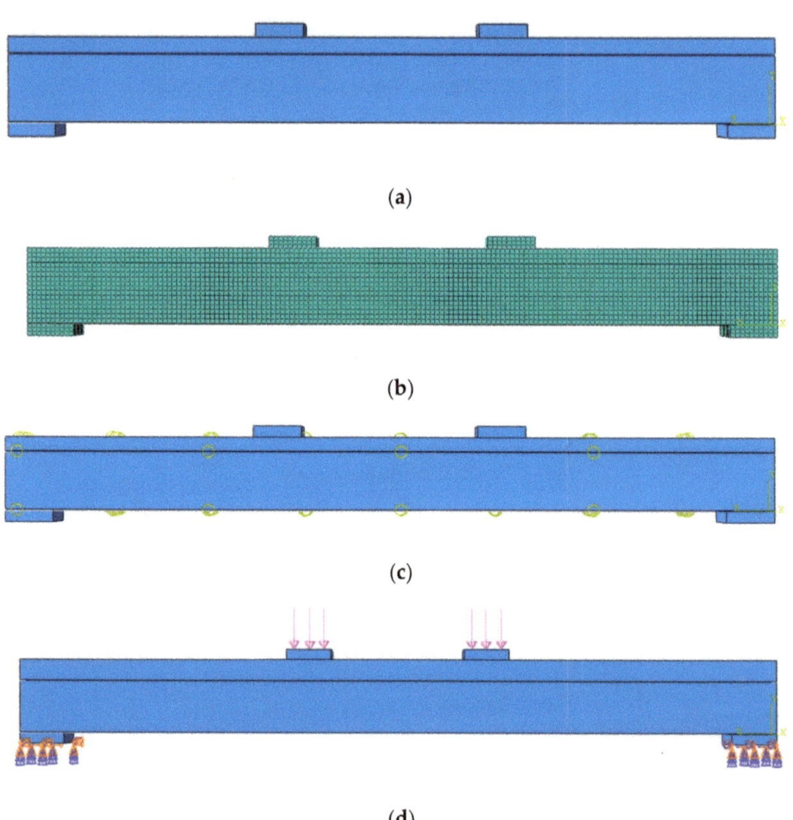

(a)

(b)

(c)

(d)

Figure 15. Model establishment. (**a**) Overall model of the combined beam; (**b**) Meshing of the combined beam; (**c**) The diagram of interface interaction; (**d**) The boundary conditions and loads.

5.1.4. Grid Division

A structured partition geometric model was used. The model grid size was 10 mm. The components were gridded by swept-through arranging seeds on the geometric model (Figure 15b).

5.1.5. Definition of Interface Interactions

This paper uses the overall combination beam as the research object, and the connection between the components is not the focus of this paper. The connections at different sites used the "Tie" binding constraints. The connection of each part was made to achieve the purpose of common force and coordinated deformation. After the interface, the interaction is shown in Figure 15c.

5.1.6. Boundary Conditions and Loads

The model could be reduced to boundary condition constraints in the form of simple branch beams. Constraints at the support were $Ux = Uy = Uz = 0$ and $URx = URy = 0$ at one end. The other end formed sliding support with $Ux = Uy = 0$ and $URx = URy = 0$. Loading on the upper surface of the two upper steel pads was applied in an equivalent pressure form (Figure 15d).

5.2. Results of Finite Element Calculation

5.2.1. The Stress Process of Components

Figure 16 shows a comparison of the overall deformation stress cloud map of the combined beam and the test damage map. From Figure 16, the two deformations and damage were consistent. Figure 17 shows the load–span deflection curve of the ABAQUS simulation member L-Z-1. Three points A, B, and, C are important nodes during loading. A indicates how well the lower flange at the steel pad just reached its yield. At this time, the wood at the upper flange of the steel H-beam and the wood on both sides of the steel H-beam web did not approach the yield strength. Point B was that after the upper and lower flanges of H-shape steel reached the yield strength, the wood at the upper flange of H-shape steel approached the tensile yield strength, and the wood tensile area on both sides of the H-shape steel web did not reach the tensile yield strength. At point C, both the upper and lower flange and the web of the H-shape steel approached the yield strength, and the wood at the upper flange reached the smooth grain compressive yield strength. At the same time, the wood tensile area on both sides of the H-shape steel web approached the tensile yield strength.

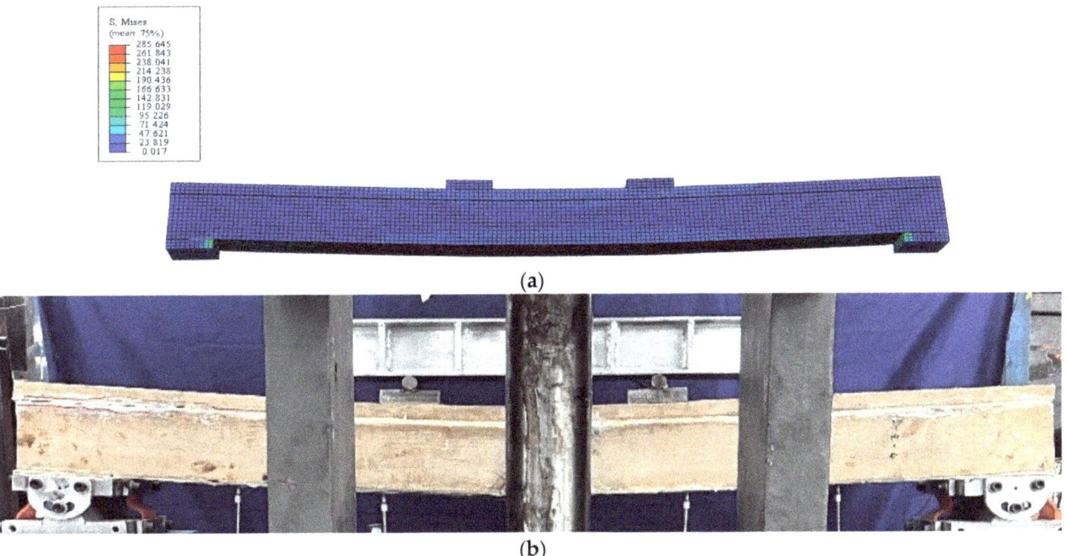

Figure 16. A comparison between (**a**) the cloud diagram of combined beam overall deformation stress; (**b**) the failure test diagram.

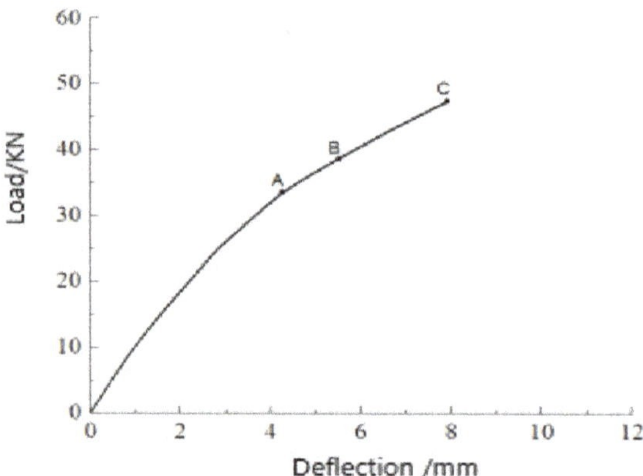

Figure 17. The diagram of load–span medium deflection of beam L-Z-1.

Before point A, neither the lower flange nor the web of the H-shaped steel reached the yield strength. The stress cloud chart is shown in Figure 18, and the lower side of the H-shape steel web at the lower cushion block support did not reach the yield, and the stress was 260.70 MPa. The reason was that with an increase in load, the H-shape steel web in the lower pad support. Therefore, the stress was too large but did not approach the yield stress. At the moment, only the wood at the position of the upper flange pad on the H-type steel showed the largest compression stress, and the maximum stress was 4.20 MPa. It was not 1/10 of the compression yield stress. The wood on both sides of the H-type steel web did not yield and the compressive stress is much lower than that of the wood at the upper flange pad.

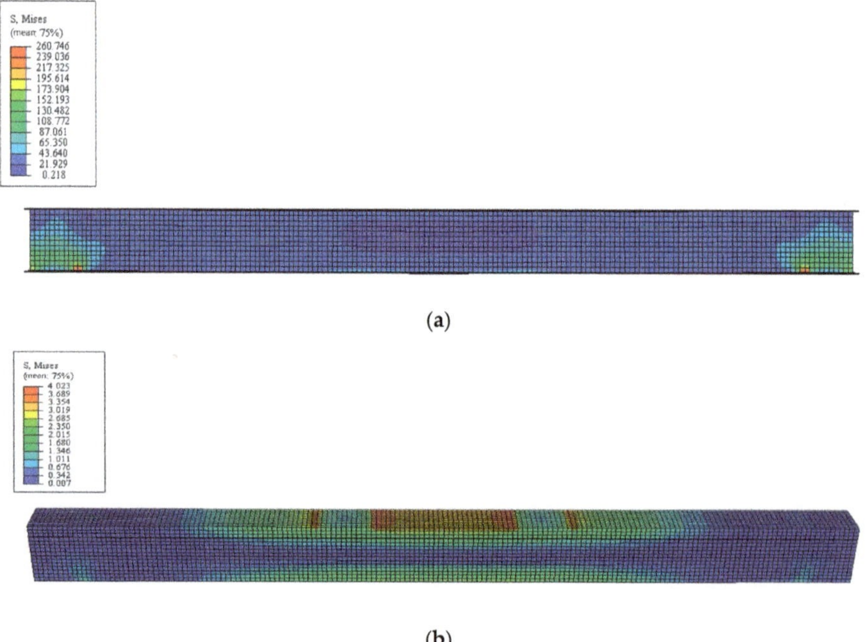

Figure 18. The cloud diagram of the stress distribution during early loading for (a) H-shape steel; (b) wood.

Figure 19 represents a cloud diagram of the stress distribution for the H-type steel and wood at point A. At point A, the H-shape steel webs of the lower flange span of the H-shape steel and the support pad also reached the yield strength of the steel, and the stress was 285.60 MPa. At this time, the wood at the upper flange and the wood on the sides of the steel H-beam web did not reach the yield strength, the maximum compression stress at the wood pad at the upper flange was 18.52 MPa, and the maximum compression stress on the sides of the steel H-beam web was approximately 7.73 MPa. The maximum tensile stress of the wood area on both sides of the H-type steel web was about 13.90 MPa.

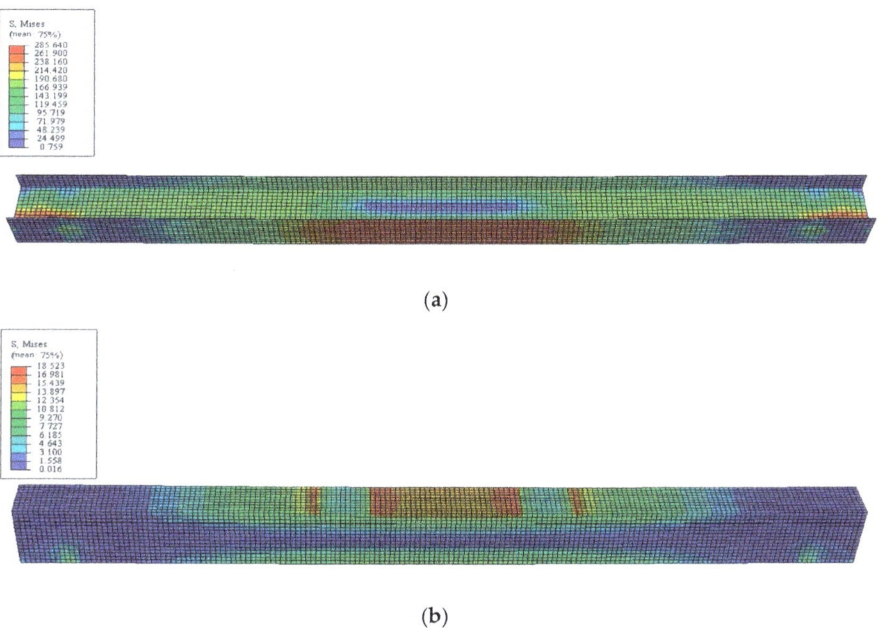

(a)

(b)

Figure 19. The cloud map of stress distribution at point A for (**a**) steel H-beam; (**b**) wood.

Figure 20 shows a cloud map of the stress distribution for the H-type steel and wood at point B. After the further increase in the load, the stress generated by the member also increased immediately. When point B was reached, the overall H-type steel approached the yield strength, and the stress was 285.60 MPa. At this time, the wood compressive strength at the upper flange reached the yield compressive strength of 25.52 MPa, and the maximum wood compressive stress on both sides of the H-shape steel web was about 17.02 MPa. The maximum tensile stress of the wood area on both sides of the H-type steel web was about 19.14 MPa.

Figure 21 represents a cloud map of the stress distribution for steel and wood H-beam. When point C was reached, the H-shape steel web and the upper and lower flange approached the yield strength, and the stress was 285.6 MPa. The wood at the upper flange of the H-shape steel reached the compressive strength of 25.52 MPa, and the wood could continue to bear it despite the yield strength. The wood tensile area on both sides of the H-shape steel web also achieved tensile yield strength, and the stress was 46.43 MPa. According to the test phenomenon, the wood on both sides of the member H-type steel web was a brittle fracture, and the member was damaged.

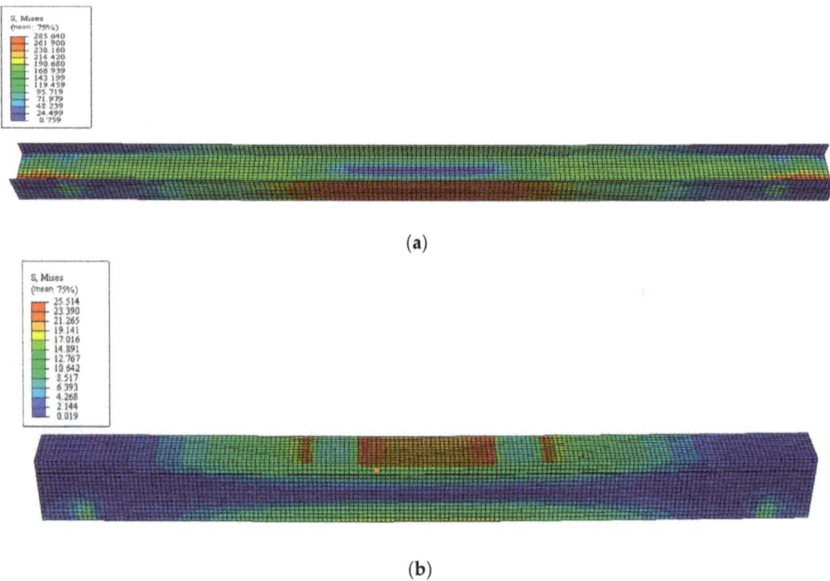

(a)

(b)

Figure 20. The cloud diagram of the stress distribution at point B for (**a**) H-type steel; (**b**) wood.

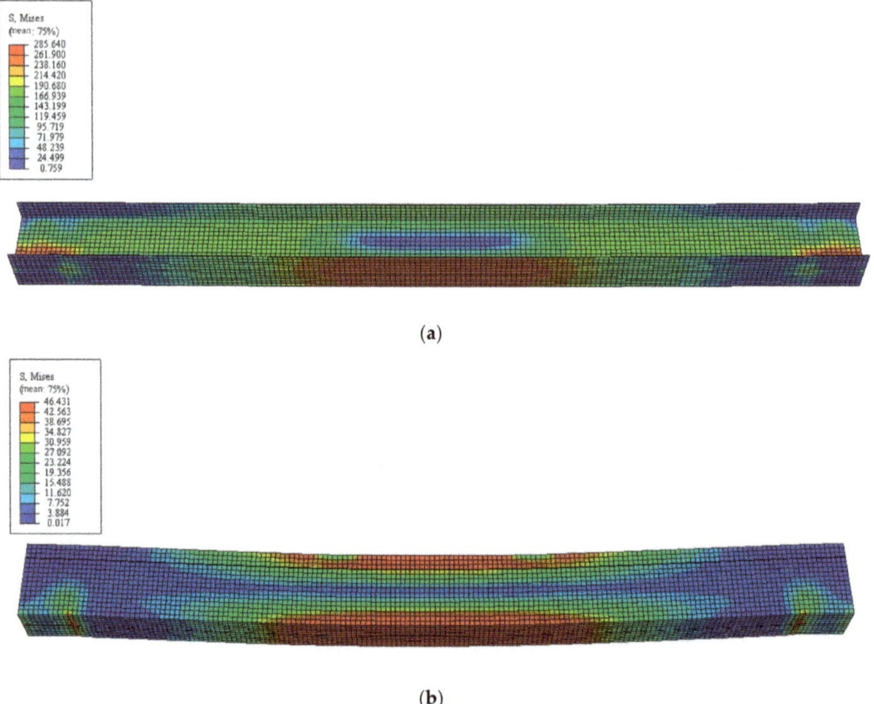

(a)

(b)

Figure 21. The cloud map of the stress distribution at point C for (**a**) steel H-beam; (**b**) wood.

5.2.2. The Change in Component Deflection in the Span

Figure 22 shows the diagram of the interspan deflection of composite beam L-Z-1 as an ABAQUS simulation component under the load. Figure 22a represents the graph corresponding to the important node A in Figure 17, namely the cross-medium deflection graph corresponding to the end of the elastic phase. When the component was in the elastic–plastic phase, the span medium deflection increased nonlinearly. Figure 22b shows the cross-medium deflection diagram corresponding to the main node of B in Figure 17. A transmedian deflection diagram was observed at the end of the elastoplastic phase. After the maximum bearing capacity was achieved in the damage stage, the failure occurred. Figure 22 represents the cross-medium deflection diagram corresponding to the important node of C in Figure 17, i.e., the cross-medium deflection diagram corresponding to the breaking moment.

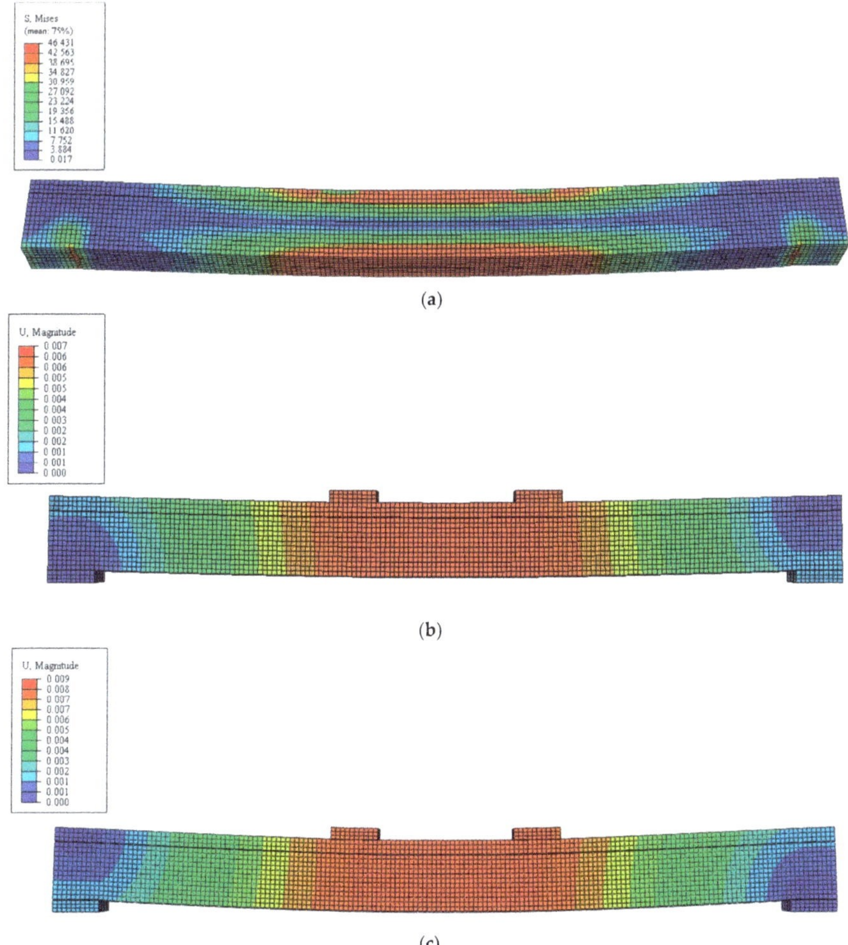

Figure 22. The Medium deflection diagram of (**a**) the A-point span; (**b**) the B-point span; (**c**) the C-point span.

5.2.3. The Combined Beam Load–Span Medium Deflection during the Simulation and Experiments

The relationship between the simulation results and the experiment results is represented in Figure 23. The overall situation change of each combination beam was the same, and the overall error was small. It showed that the simulation value confirmed the test value and that the finite element simulation was reliable. In the elastic phase, the stiffness of the simulated combination beam was greater than that of the test combination beam. In the elastic range, the curve of load and deflection was linear. After the load value reached about 1/2 of the maximum load, the curve gradually offset the straight trend development. The overall stiffness decreased from the elastic stage. The growth rate of the maximum deflection in the span was gradually reduced until the ultimate tensile strength of the wood was reached. It resulted in the overall destruction of the combined beam and the loss of bearing capacity.

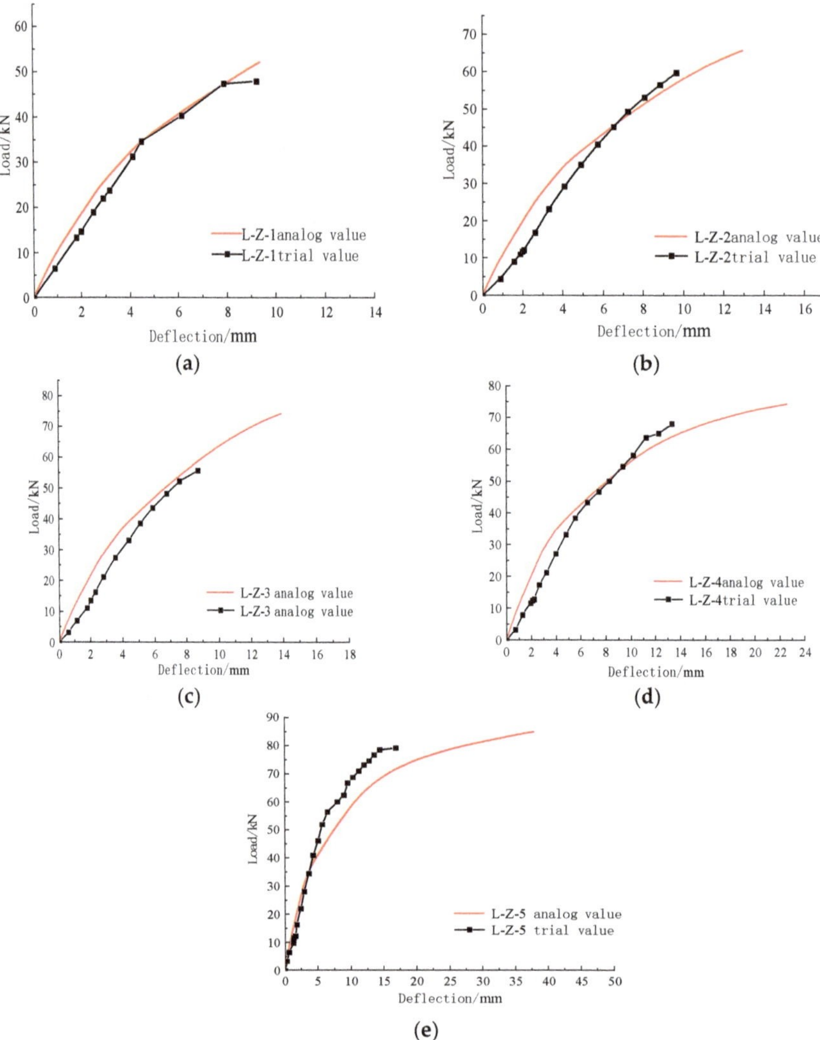

Figure 23. The comparison of the curves of combined beam load–span medium deflection for (**a**) L-Z-1; (**b**) L-Z-2; (**c**) L-Z-3; (**d**) L-Z-4; (**e**) L-Z-5.

5.3. Main Factors in the Combined Beam Bending Performance

5.3.1. Influence of Steel Yield Strength

The yield strength of steel types Q235B, Q345B, and Q355B was selected for simulation. The remaining parameters of the combined beam remained unchanged, and the load–cross range deflection curve is shown in Figure 24. In the elastic range, the three curves almost coincided. When the yield strength increased from Q235B to Q345B, the carrying capacity was enhanced by 42.22%. When the yield strength increased from Q345B to Q355B, the carrying capacity was enhanced by 19.22%. Therefore, the bending bearing capacity increased with the yield strength, but the load enhancement rate showed a gradually decreasing trend. Although the yield strength of steel improved the bearing capacity to a certain extent, the improvement rate in the bearing capacity gradually became smaller.

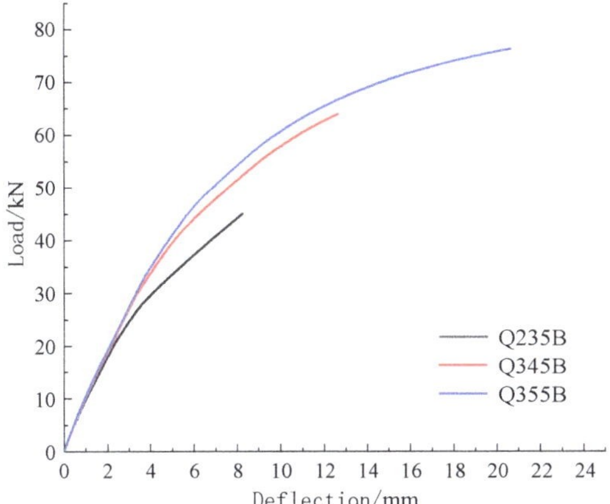

Figure 24. A comparison of load–span deflection curves for different yield strengths.

5.3.2. Influence of the Flange Thickness of Type Steel H-Beam

The combined beam of three flange thicknesses of 1.5 mm, 2 mm, and 2.5 mm was selected for simulation analysis. The effect of steel thickness on the load deflection is shown in Figure 25. When the flange thickness increased from 1.5 mm to 2 mm, the maximum bearing capacity was enhanced by 22.12 kN, and the maximum bearing capacity increased by 42.38%. When the flange thickness was enhanced from 2 mm to 2.5 mm, the maximum bearing capacity increased by 10.81 kN, and the maximum bearing capacity was enhanced by 14.55%. Therefore, the maximum bearing capacity increased with the thickness of the H-shape steel flange. The thickness of the H-shape steel flange improved the maximum bearing capacity significantly.

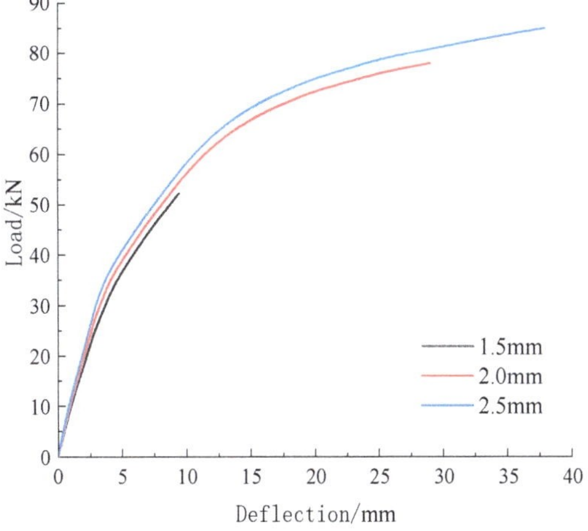

Figure 25. The comparison of load–span medium deflection curves for different flange thicknesses.

5.3.3. Effect of the H-Type Steel Web Thickness

The values of 1.5 mm, 2 mm, and 2.5 mm were considered to study the H-type steel thickness influence on the bearing capacity of the combined beam. A comparison diagram of the load versus deflection is shown in Figure 26. According to the figure, the increase in the maximum bearing capacity was smaller. The maximum bearing capacity of the combined beam with a web thickness of 1.5 mm and 2 mm was 52.20 kN and 64.86 kN, respectively. The maximum bearing capacity increased by 24.25%. The load-deflection curve of the combined beam with a web thickness of 2 mm and 2.5 mm was very close, the deformation trend was consistent, and the maximum bearing capacity increased by 5%. From the perspective of the steel quantity of H-shape steel, the thickness of the web was adjusted from 1.5 mm to 2 mm, the steel volume increased by 33.33%, the maximum bearing capacity was enhanced by 24.25%, the steel volume increased by 1%, and the maximum bearing capacity was enhanced by 0.73%. When the thickness of the web was changed from 2 mm to 2.5 mm, the steel volume increased by 25%, and the maximum bearing capacity was enhanced by 5%. The amount of steel used increased by 1%, and the maximum bearing capacity was enhanced by 0.2%. Consequently, the thickness of the H-type steel web was 2 mm as the optimized web thickness.

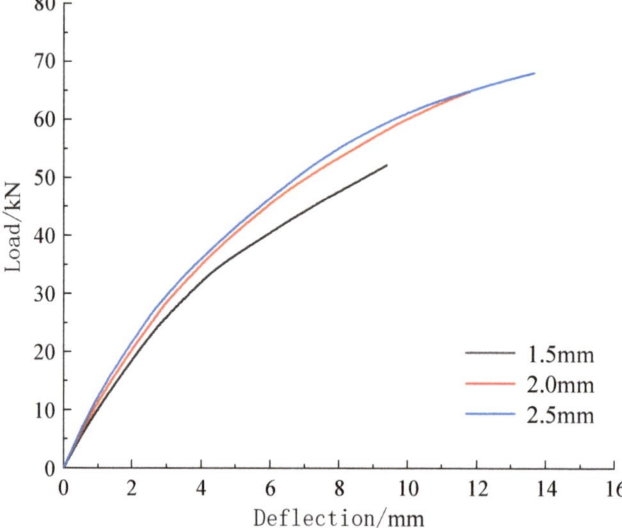

Figure 26. A comparison of load–span medium deflection curves under different web thicknesses.

5.3.4. Influence of the H-Type Steel Web Height

The effect of the three web heights of 125 mm, 150 mm, and 175 mm was investigated on the bearing capacity of the combined beam. The curve of load–span medium deflection is shown in Figure 27. According to the figure, with a gradual increase in the height of the H-shape steel web, the maximum capacity of the corresponding combined beam was enhanced significantly. The height of the web increased from 125 mm to 150 mm, and the maximum bearing capacity was enhanced by 64.56%. The height of the H-shape steel web increased from 150 mm to 175 mm, and the maximum bearing capacity was enhanced by 27.93%. The maximum bearing capacity of the combined beam was analyzed from the amount of type steel. When the height of the H-type steel web increased from 125 mm to 150 mm, the amount of steel H-beam of the composite was enhanced by 20.49%, the maximum bearing capacity increased by 64.56%, the amount of H-shape steel was enhanced by 1%, and the maximum bearing capacity increased by 3.15%. When the height of the H-type steel web was enhanced from 150 mm to 175 mm, the amount of H-type steel for the combined beam increased by 17.00%, the maximum bearing capacity was enhanced by 27.93%, the amount of H-type steel increased by 1%, and the maximum bearing capacity was enhanced by 1.64%. Finally, an increase in the height of the H-shape steel web significantly improved the bearing capacity of the combined beam, and the optimum web height should be 150 mm.

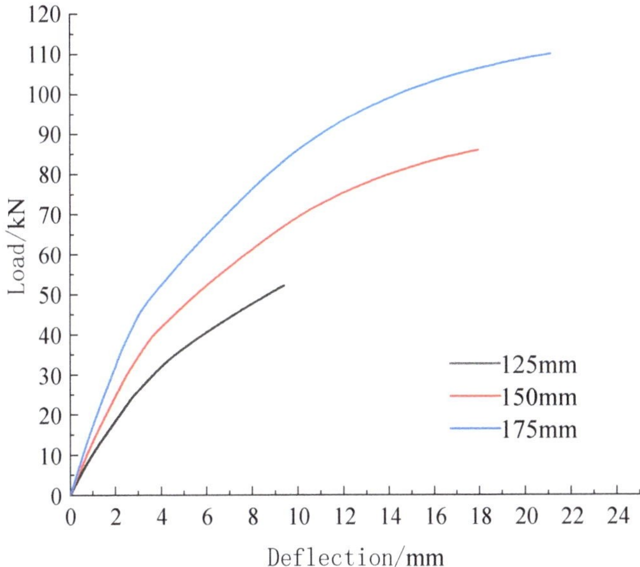

Figure 27. The comparison of load–span medium deflection curves for different web heights.

5.3.5. The Thickness of Steel H-Beam

The thickness of 30 mm, 40 mm, and 50 mm represented the maximum bearing capacity of the combined beam. The load–cross deflection curve pair is shown in Figure 28. The three curves changed consistently. As the thickness of the upper flange of the steel H-beam gradually increased, the maximum bearing capacity also was enhanced. When the board thickness of the upper flange cover of H-shape steel increased from 30 mm to 40 mm, the corresponding maximum bearing capacity was enhanced by 15.52 kN and 25.90%. The thickness of the upper flange-covered board of H-shape steel was enhanced from 40 mm to 50 mm, corresponding to the maximum bearing capacity increase of 8.53 kN and 12.98%. As the thickness of the flange of the H-type steel increased step by step, the maximum bearing capacity was enhanced significantly, and the thickness of the flange of the H-type steel indirectly increased the overall height of the combined beam, and the maximum bearing capacity was enhanced immediately. In summary, the thickness of the upper flange of the H-shape steel composite beam was 40 mm.

5.3.6. Effect of the Composite Beam Section Width

The influence of setting the width of 100 mm, 120 mm, 120 mm, and 140 mm was studied. The curve of load–span deflection is shown in Figure 29. The three curves resulted in the same trend. As the width of the combined beam gradually increased, the maximum carrying capacity occurred. The width of the combined beam was enhanced from 100 mm to 120 mm, corresponding to the maximum bearing capacity increased by 17.86 kN. It was enhanced by 34.21%. The width of the combined beam increased from 120 mm to 140 mm, corresponding to the maximum bearing capacity of 17.64 kN, and was enhanced by 25.18%. Therefore, the maximum bearing capacity increased step-by-step with the width of the combined beam.

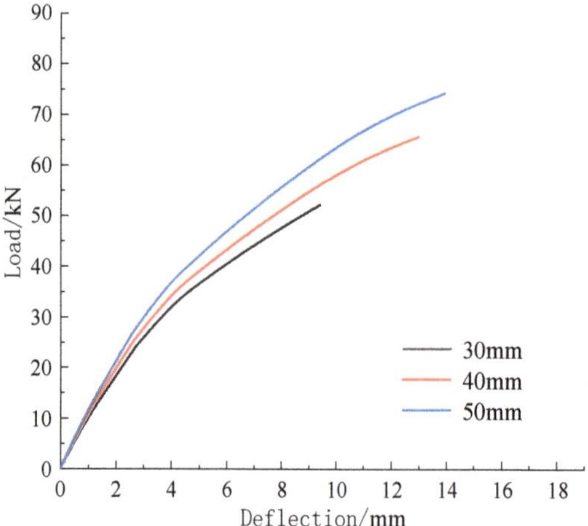

Figure 28. A Comparison of load–span deflection curves for the upper flange covered with different board thicknesses.

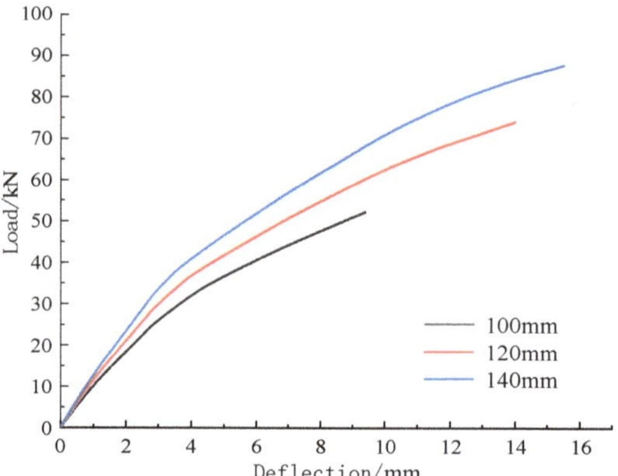

Figure 29. The comparison of the load–span medium deflection curves at different widths.

In this thesis, the combination of experimental and numerical simulations of composite beams has shown good performance, and the addition of steel sections to wood can effectively improve the flexural load capacity of pure wood beams. In countries where wood resources are abundant, such as Japan and Canada, the practical application of composite beams in residential housing is gradually realized, and the research in this paper has some value.

6. Conclusions

Based on the experimental and numerical simulation method, this paper studied the novel rectangular-section combined beam of welded thin-walled H-shape steel/camphor pine wood under the bending performance study The conclusions are as follows:

(1) The combined beam can be roughly divided into three stages from loading to destruction: elastic stage, elastic–plastic stage, and destruction stage. The final damage is caused by the yielding of the steel section, the yielding of the wood in the compression zone, and the maximum tensile strength of the wood in the tension zone, and the wood is damaged by the tensile break.

(2) The overall deformation curve of the combined beam shows that the deflection in the span of the pure bending section is always the largest during the loading process. The

gluing connection between steel and wood is very reliable, the deformation of both is coordinated, and the combined effect is significant.

(3)   Improving the yield strength of steel can effectively improve the flexural load-carrying capacity of the combined beam. the thickness of the H-beam flange has a significant improvement in the maximum load-carrying capacity. The effect of web thickness is smaller. the height of the H-beam web has obvious improvement on the combined beam bearing capacity. The maximum bearing capacity is increased by increasing the thickness of the cladding plate on the flange of the H-beam. The width of the combined beam has less effect on the bearing capacity.

In summary, this test combination beam performs better in the test and simulation, there will be some problems in the research, but with the depth of the subsequent research, gradually overcome the shortcomings, and provide a reference for the research in related fields.

Author Contributions: Conceptualization, C.W.; Methodology, C.W.; Software, J.D. Writing—original draft, J.D.; Writing—review & editing, Z.Y., Z.Z. and Y.X. All authors have read and agreed to the published version of the manuscript.

Funding: This work is supported by the National Natural Science Foundation of China (52268029) and the Open Fund of Shock and Vibration of Engineering Materials and Structures Key Laboratory of Sichuan Province (Grant No. 20kfgk03).

Institutional Review Board Statement: Not applicable.

Informed Consent Statement: Not applicable.

Data Availability Statement: Not applicable.

Conflicts of Interest: The authors declare no conflict of interest.

References

1.   Chen, S. Talk about green and environmental protection building materials. *Sci. Technol. Entrep. Mon.* **2010**, *23*, 118–119.

2.   Zhou, R.; Qing, L. Green environmental building materials and their application. *Resour. Inf. Eng.* **2017**, *32*, 82–83.

3.   Zhang, J.; Gu, L. The development status, challenges and policy suggestions of green building in China. *Energy China* **2012**, *34*, 19–24.

4.   Chau, C.K.; Hui, W.K.; Ng, W.Y.; Powell, G. Assessment of CO_2 emissions reduction in high-rise concrete office buildings using different material use options. *Resour. Conserv. Recycl.* **2012**, *61*, 22–34. [CrossRef]

5.   Wang, S.; Zhou, C.; Li, G.; Feng, K. CO_2, economic growth, and energy consumption in China's provinces: Investigating the spatiotemporal and econometric characteristics of China's CO_2 emissions. *Ecol. Indic.* **2016**, *69*, 184–195. [CrossRef]

6.   Ramage, M.H.; Burridge, H.; Busse-Wicher, M.; Fereday, G.; Reynolds, T.; Shah, D.U.; Wu, G.; Yu, L.; Fleming, P.; Densley-Tingley, D.; et al. The wood from the trees: The use of timber in construction. *Renew. Sustain. Energy Rev.* **2017**, *68*, 333–359. [CrossRef]

7.   Akbarnezhad, A.; Jianzhuang, X. Estimation and Minimization of Embodied Carbon of Buildings: A Review. *Buildings* **2017**, *7*, 5. [CrossRef]

8.   Chiniforush, A.A.; Akbarnezhad, A.; Valipour, H.; Xiao, J. Energy implications of using steel-timber composite (STC) elements in buildings. *Energy Build.* **2018**, *176*, 203–215. [CrossRef]

9.   Borri, A.; Corradi, M. Strengthening of timber beams with high strength steel cords. *Compos. Part B Eng.* **2011**, *42*, 1480–1491. [CrossRef]

10.   Nadir, Y.; Nagarajan, P.; Ameen, M.; Arif, M.M. Flexural stiffness and strength enhancement of horizontally glued laminated wood beams with GFRP and CFRP composite sheets. *Constr. Build. Mater.* **2016**, *112*, 547–555. [CrossRef]

11.   Hassanieh, A.; Valipour, H.; Bradford, M. Experimental and numerical investigation of short-term behaviour of CLT-steel composite beams. *Eng. Struct.* **2017**, *144*, 43–57. [CrossRef]

12.   Shekarchi, M.; Oskouei, A.V.; Raftery, G.M. Flexural behavior of timber beams strengthened with pultruded glass fiber reinforced polymer profiles. *Compos. Struct.* **2020**, *241*, 112062. [CrossRef]

13.   Ghanbari-Ghazijahani, T.; Russo, T.; Valipour, H.R. Lightweight timber I-beams reinforced by composite materials. *Compos. Struct.* **2020**, *233*, 111579. [CrossRef]

14.   Dar, A.R. An experimental study on the flexural behavior of cold-formed steel composite beams. *Mater. Today Proc.* **2020**, *27 (Pt 1) Pt 1*, 340–343. [CrossRef]

15.   Pan, F.T.; Kou, L.; Wang, C.Y. Mechanical Properties Analysis of Steel-Timber Combined Member. *Adv. Mater. Res.* **2012**, *347–353*, 1477. [CrossRef]

16.   Fujita, M.; Iwata, M. Bending Test of the Composite Steel-Timber Beam. *Appl. Mech. Mater.* **2013**, *351–352*, 415–421. [CrossRef]

17. Hassanieh, A.; Valipour, H.R.; Bradford, M.A. Experimental and numerical study of steel-timber composite (STC) beams. *J. Constr. Steel Res.* **2016**, *122*, 367–378. [CrossRef]
18. McConnell, E.; McPolin, D.; Taylor, S. Post-tensioning of glulam timber with steel tendons. *Constr. Build. Mater.* **2014**, *73*, 426–433. [CrossRef]
19. Denghui, L. Study on Bending Properties of Steel-Wood Composite Beam. Master's Thesis, Beijing Jiaotong University, Beijing, China, 2016.
20. Yushun, L.; Jialiang, Z.; Xiuhua, Z.; Yu, W.; Jun, G. Study on bond stress and bond slip of bamboo-steel interface under static load. *J. Build. Struct.* **2015**, *36*, 114–123.
21. Aiguo, C.; Teng-Hui, L.; Chao, F.; Qingguo, Z.; Jihui, X. Experimental study on flexural behavior of H-shaped steel-wood composite beams. *J. Build. Struct.* **2016**, *37* (Suppl. 1), 261–267.
22. Xianzhe, W. Study on Flexural Capacity of Thin-Walled Flitch Filled Rectangular Steel Tube Composite Beams. Master's Thesis, Southwest University of Science and Technology, Mianyang, China, 2018.
23. Ye, Z.; Feiyang, X.; Hongda, Y.; Xinmiao, M.; Ying, G. Experimental Study and Finite Element Analysis of Steel and Wood Combined Beam under Centralized load. *J. Beijing For. Univ.* **2021**, *43*, 127–136.
24. Degui, L.; Yuhao, W.; Yong, W.; Wei, Z.; Tao, W. Study on the bending properties of thin-wall H-steel-wood composite beam. *J. Build. Struct.* **2022**, *43*(05), 149–163.
25. Zhang, X.; Zhang, Y.; Xie, X. Experimental and analytical investigation of the flexural behaviour of stiffened hollow glulam beams reinforced with fibre reinforced polymer. *Structures* **2023**, *50*, 810–822. [CrossRef]
26. Chen, Y.X. Flexural Analysis and Design of Timber Strengthened with High-Strength Composites. Ph.D. Thesis, Rutgers The State University of New Jersey, New Brunswick, NJ, USA, 2003.

 *sustainability*

Article

Intelligent Assessment Method of Structural Reliability Driven by Carrying Capacity Sustainable Target: Taking Bearing Capacity as Criterion

Guoliang Shi [1,2], Zhansheng Liu [1,2,*], Dengzhou Xian [3] and Rongtian Zhang [1]

[1]   Faculty of Architecture, Civil and Transportation Engineering, Beijing University of Technology, Beijing 100124, China; guoliangshi@emails.bjut.edu.cn (G.S.); zhangrt123@emails.bjut.edu.cn (R.Z.)
[2]   The Key Laboratory of Urban Security and Disaster Engineering of the Ministry of Education, Beijing University of Technology, Beijing 100124, China
[3]   Hebei Construction Building Group Co., Ltd., Shijiazhuang 050051, China; xdz@hbjgjt.cn
*   Correspondence: liuzhansheng@bjut.edu.cn

Abstract: Large-scale building structures are subject to numerous uncertain loads during their service life, leading to a decrease in structural reliability. Real-time analysis and accurate prediction of structural reliability is a key step to improve the bearing capacity of buildings. This study proposes an intelligent assessment method for structural reliability driven by a sustainability target, which incorporated digital twin technology to establish an intelligent evaluation framework for structural reliability. Under the guidance of the evaluation framework, the establishment method of a structural high-fidelity twin model is formed. The mechanical properties and reliability analysis mechanism are established based on the high-fidelity twin model. The theoretical method was validated by experimental analysis of a rigid cable truss construction. The results showed the simulation accuracy of the high-fidelity twin model formed by the modeling method is up to 95%. With the guidance of the proposed evaluation method, the mechanical response of the structure under different load cases was accurately analyzed, and the coupling relationship between component failure and reliability indicators was obtained. The twinning model can be used to analyze the reliability of the structure in real time and help to set maintenance measures of structural safety. By analyzing the bearing capacity and reliability index of the structure, the safety of the structure under load is guaranteed. The sustainability of structural performance is achieved during the normal service period of the structure. The proposed reliability assessment method provides a new approach to improving the sustainability of building bearing capacity.

Keywords: sustainable structure; digital twin; mechanical response; reliability analysis; experimental validation

Citation: Shi, G.; Liu, Z.; Xian, D.; Zhang, R. Intelligent Assessment Method of Structural Reliability Driven by Carrying Capacity Sustainable Target: Taking Bearing Capacity as Criterion. *Sustainability* **2023**, *15*, 10655. https://doi.org/10.3390/su151310655

Academic Editors: Zhihua Chen, Qingshan Yang, Yue Wu and Yansheng Du

Received: 14 March 2023
Revised: 15 May 2023
Accepted: 27 June 2023
Published: 6 July 2023

1. Introduction

Large public buildings use spatial and prestressed structures [1] that are exposed to wind loads and temperature effects during their service life, significantly affecting their serviceability and bearing capacity [2]. The frequent collapse of structures poses a serious threat to life and property. The collapse of the World Trade Center in 2001 [3] and the unexpected collapse of a terminal building at Paris Charles de Gaulle Airport in 2004 were examples of these events. Therefore, reliability analysis of structures has consistently been a crucial area of research in civil engineering [4].

Regarding reliability analysis of structures, most existing studies use finite element simulation and experimental validation [5]. Lu et al. [6] studied the failure mechanism of a 6 m-span plane trusses string structure through experiments and numerical simulations, analyzing the effects of different cable rupture types on structural reliability. Vaezzadeh et al. [7] discovered the progressive collapse mode of cable-net structures through experimental analysis, with the main factor being constraint failure. Ozer et al. [8] conducted modal analysis of a bridge structure by establishing a finite element model (FEM), comparing frequency and vibration mode changes to judge reliability. Bojorquez et al. [9] evaluated the impact of corrosion on the reliability of reinforced concrete structures using Monte Carlo simulation and stochastic modeling methods, highlighting its important role in structural failure and the necessity for consideration in reliability design.

The rapid development of smart buildings and intelligent construction has put forward new requirements for the bearing capacity of structures [10]. Against this backdrop, improving the sustainability of building bearing capacity is an urgent issue [11]. Traditional reliability analysis methods cannot achieve continuous evaluation and rely on data generated by experiments. Real-time analysis and accurate prediction of structural reliability are significant for improving the sustainability of buildings. Digital twin (DT) technology plays a role in the interactive mapping of deficiency and excess and spatiotemporal evolution analysis in structural health monitoring [12]. DT technology realizes the interactive mapping between structural entities and virtual models. The establishment of the twin model of the structure provides a powerful tool for the intelligent evaluation of reliability. Driven by the twin model, the sustainability of structural bearing capacity can be effectively improved. [13].

Wang et al. [14] proposed a damage identification method for cable dome structures by combining DT and deep learning. This method can identify damage type, location, and degree intelligently with high accuracy and strong robustness. Ritto et al. [15] explored an ensemble method based on finite element modeling and machine learning under the twin concept. This method effectively integrates engineering reality with virtual models, offering reliable decisions for structural safety control. Zhu et al. [16] proposed a DT-based safety assessment method for dams. This assessment method makes automated and efficient monitoring of dam conditions possible. Liu et al. [17] proposed a DT-based intelligent evaluation method in order to enhance the constructional safety degree of prestressed steel structures. This method ensures a safety level for structures and elevates the intelligent level of construction safety evaluation.

Driven by sustainability goals and the DT concept, this study proposes an intelligent assessment method for structural reliability. The safety of the structure under load is ensured by analyzing the bearing capacity and reliability index of the structure. The sustainability of structural performance is achieved during the normal service period of the structure. DT uses a high-fidelity dynamic virtual model to simulate the state and behavior of physical entities. As a link between the real physical world and the virtual digital space, it is a key enabling technology for structural safety assessment. By establishing an HF twin model to map the mechanical state of the structure in real time, the method accurately evaluates the failure mode and reliability development of the structure. The method provides decision support for structural bearing capacity maintenance. Integrating DTs into structural reliability evaluation can allow analysis of the mechanical response under extreme operating conditions in the twin model, reducing the cost of experimental research. More importantly, this research method realizes real-time analysis and accurate prediction of structural reliability, effectively improving the sustainability of structural bearing capacity. The organizational structure of this paper is as follows.

The second section puts forward the theoretical method of reliable evaluation. In this chapter, the framework and process of evaluation are formed. Driven by the evaluation process, the establishment method of the twin model is first formed. Based on the establishment of the twin model, the mechanical properties and reliability of the structure are analyzed. Analyzing the mechanical properties and reliability of the structure provides a basis for improving the sustainability of the carrying capacity.

Based on the theoretical method in Section 2, Section 3 is verified by experiments. During testing, a high-fidelity virtual model of the structure was first established. The simulation ability is improved by modifying the model. The unfavorable load conditions and failure modes are obtained in the multi-span and multi-stage loading process. Finally, the reliability index of the structure is analyzed in different extreme load conditions.

Section 4 is the conclusion of this study.

2. Methodology

To lift the sustainability of building bearing capacity, this study proposes an intelligent reliability assessment method for structures. By using DT technology, an HF twin model of structures was created, which enables analysis of structural mechanical properties and reliability.

2.1. Intelligent Evaluation Framework Driven by DTs

During normal service life, a structure is affected by external factors such as temperature and wind loads, which threaten its safety. During the lifetime of the structure, internal causes such as component relaxation and component corrosion will also affect its reliability. Real-time analysis and accurate prediction of structural reliability have become essential scientific issues. The physical parameters, geometric parameters and boundary parameters which represent the stiffness and mass of the structure are all time-varying functions [18], making the analysis of structural reliability a multidimensional mechanical problem that couples time and space [19]. DTs, driven by multidimensional virtual models and integrated data, can monitor, simulate, predict, optimize, and meet practical functional services and application requirements through closed-loop interactions between the virtual and real worlds [20]. DTs encompass information technologies such as the IoT and AI, fully considering the spatiotemporal evolution characteristics of structural mechanical states, and realizing HF mapping of the structure. Multidimensional, multiple process, multiple parallel operation time simulations, and integrated management of deficiency and excess are achieved under the support of the five-dimensional space of DTs. The structural reliability intelligent evaluation framework driven by DTs is shown in Figure 1. Establishing an HF twin model based on the state of the actual structure is the first step to achieve reliability assessment. Comparing the measured values of structural mechanical properties with the simulated values, the basic parameters of the structure in the model are modified to improve the simulation ability of the model. Based on the establishment of the virtual model, the mechanical properties of the structure are analyzed. In the virtual model, the mechanical response corresponding to the applied load of different spans is analyzed to obtain unfavorable load conditions. The failure mode of the structure is analyzed under the unfavorable load condition. For extreme load conditions of the structure, such as component failure, the reliability index of the structure is analyzed in the virtual model. The residual bearing capacity and reliability of the structure are obtained by analyzing the mechanical properties of the structure, which provides a basis for ensuring the safety of the structure.

Under the guidance of DT, this study achieves intelligent evaluation of structural reliability by considering the spatiotemporal evolution of the structural system and its interaction with the virtual reality interaction. First, an HF twin model is established for interactive mapping in the same physical space, forming DT data for structural mechanical performance and reliability analysis. By integrating DT technology with the effects and mechanical performance parameters of the structure, a multiple-scale model of the longitudinal dimension of the structural system is realized. Based on the real-time evolution of the load during the service life of the structure, an operational system of structural dynamic collaboration driven by DT is established.

Virtual models were established according to the physical structure to accurately map the structural mechanical state. In this study, the structural mechanical performance was analyzed through the multi-span and multi-stage distribution of loads. Extreme operating conditions were set in the virtual model to analyze the structural failure probability and obtain the structural reliability index ultimately. Based on the analysis of mechanical performance and reliability, maintenance strategies were developed to effectively improve the sustainability of structural bearing capacity.

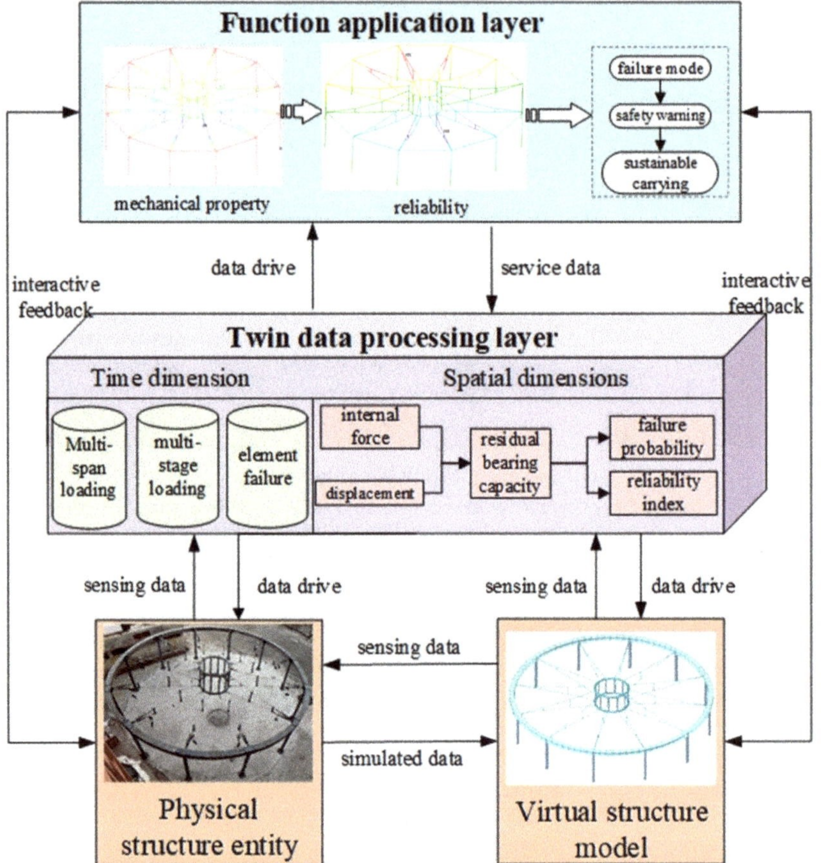

Figure 1. The structural reliability intelligent evaluation framework driven by DTs.

2.2. Establishment of the DT Model

Establishing a structural HF twin model is the first step in achieving intelligent reliability assessment. This section analyzes the traditional finite element modeling process and forms a model correction method. Following the establishment of the DT model, the structural reliability can be analyzed in real time and the development trend of the reliability index under extreme operating conditions can be accurately predicted. With the DT model established, it effectively reduces the dependence on experimental data.

2.2.1. Construction of FEM

Through the study of structural performance and reliability, the traditional simulation method relies on FEMs [21]. The FEM analysis software ANSYS is used to analyze the structural performance, with the following modeling process.

(1) Assign material properties and real constants to the corresponding components.
(2) Arrange key joints in the software, establish corresponding component units, and form a structural model.
(3) Apply corresponding constraints to the connection joints of the structure.
(4) Apply the loads that the structure bears at the designated joints.
(5) Obtain the structural mechanical response under the action of the loads.

2.2.2. Modification of Virtual Model

On the basis of establishing the FEM, this study proposes a physical property optimization method based on a genetic algorithm [22,23]. In the optimization process, physical attributes such as component size are used as optimization parameters. The lower and upper limits are determined according to the machining and construction accuracy of physical properties. The error ($\Delta\mu$) between the displacement simulation value of the structure in its formed shape and the measured value is taken as the optimization objective. The optimization process of attribute parameters in the virtual model is expressed as Equation (1).

$$\begin{cases} a_1 \leq \delta_1 \leq b_1 \\ a_2 \leq \delta_2 \leq b_2 \\ \cdots \\ a_n \leq \delta_n \leq b_n \end{cases} \xRightarrow{\text{iteration}} \min(\Delta\mu) \xRightarrow{\text{select}} \begin{cases} \delta_1^{opt} \\ \delta_2^{opt} \\ \cdots \\ \delta_n^{opt} \end{cases} \tag{1}$$

where δ_i equals the *i*th optimization variable, such as the prestress of a component, and a_i and b_i denote the lower and upper limits of the *i*th optimization variable. The optimization algorithm is used to adjust the basic parameters of the model to minimize the error between the simulated and measured vertical displacement values, while δ_i^{opt} is the optimal design parameter corresponding to the minimum error. By modifying the structural physical properties in the model, the accuracy of model simulation is significantly improved. In this model, the mechanical state of the structure is analyzed in real time and the safety risk is predicted, which provides support for the intelligent evaluation of structural mechanical performance and reliability. The basic parameter optimization process of the virtual model is shown in Figure 2. The parameters are optimized in Python. The minimum error between the simulated value and the measured value is taken as the objective function, and the value range of each basic parameter (size, prestress) is taken as the constraint condition. Driven by the genetic algorithm, the parameters are optimized to improve the simulation ability of the virtual model.

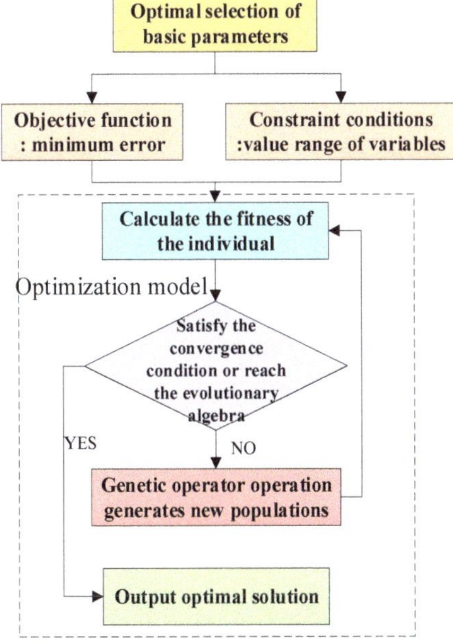

Figure 2. Model modification process based on the genetic algorithm.

2.3. Analysis of Structural Mechanical Performance and Reliability

For the analysis of structural performance under different load cases, this study developed a twin model of the experimental model formed by the theoretical method in Section 2.1. In the formed shape, the HF of the twin model is verified by comparing the measured and simulated values of the structural mechanical parameters. Driven by the genetic algorithm, the basic parameters in the model are optimized to improve the simulation ability of the model. A variety of static load conditions are arranged in the test model and the twin model to obtain the mechanical response of the structure. By setting multi-span loads, the most unfavorable load on the structural performance is obtained. Multi-stage loading is performed for the most unfavorable load to obtain the failure mode of the structure. The reliability analysis of the structure is carried out in the twin model for extreme loads, such as component failure. By analyzing the mechanical properties and reliability of the structure, the bearing capacity of the structure is evaluated to improve the sustainability of the structure. This process is shown in Figure 3. In the analysis of structural mechanical performance and reliability, this study used a tension meter and a total station to capture the changes in cable force and displacement. The twin model of the structure is established by ANSYS.

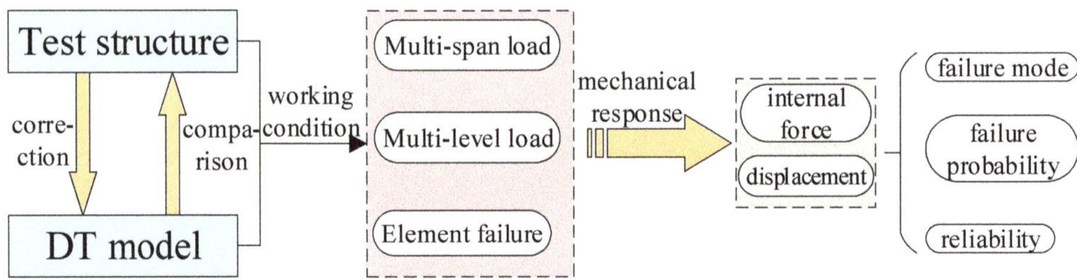

Figure 3. Analysis process of structural mechanical performance and reliability.

This study conducted an analysis of the internal forces and displacements of the structure under multi-stage loading to evaluate the failure mode of the structure. The loads corresponding to the ultimate limit state (ULS) and serviceability limit state (SLS) of the structure were obtained. The reliability analysis of the structure after component failure was carried out in the twin model. Response surface methodology and the Monte Carlo method were applied in the reliability analysis of the structure. Firstly, the response equation was obtained using the response surface method, and then 1 million sampling simulations were performed with Latin hypercube sampling (LHS) to obtain the structural failure probability.

In the DT model, the effect of structural failure on the ULS and SLS was studied. For the ULS, the loading condition (LS) is expressed in Equation (2) [24], and the function of the failure is expressed in Equation (3).

$$L(ULS) = 1.3 \times dead\ load + 1.5 \times live\ load \tag{2}$$

$$f(ULS) = [\sigma] - \sigma_{max} \tag{3}$$

where $[\sigma]$ represents the limit value of the component stress, which is equal to the material strength divided by the safety factor, σ_{max} is the maximum stress of the components in the structure, and $f(ULS) < 0$ refers to structural failure.

For the SLS, the LS is expressed in Equation (4) [24], and the failure function is expressed in Equation (5).

$$L(SLS) = 1.0 \times dead\ load + 1.0 \times live\ load \tag{4}$$

$$g(SLS) = [\mu] - \mu_{max} \tag{5}$$

where $[\mu]$ represents the limit value of the structural displacement, which is equal to the structural span divided by the safety factor, μ_{max} represents the maximum displacement of the structure, and $g(SLS) < 0$ represents structural failure. By analyzing the structural failure function, the structural failure probability was obtained, and the development trend of the structural reliability index was captured. Based on the structural mechanical performance and reliability analysis, this study provides a basis for the safety maintenance of the structure, effectively ensuring the sustainability of the structural bearing capacity.

3. Experimental Verification

To demonstrate the feasibility of the research method, an experimental verification was carried out on a rigid cable truss construction. In the experimental model, the span of the cable truss was 6 m. The model consists of 12 bays of cable truss, inner ring beam, outer ring beam, and outer column. The radial cables include upper radial cables (URCs) and lower radial cables (LRCs) with a diameter of 9 mm, and the cable forces of the URCs and LRCs in formed shape are 15 kN and 16 kN, respectively. The single-bay cable truss is connected to the radial cable by two struts, the inner ring beam includes upper and lower ring beams, which are connected by inner struts. The height of the inner strut is 0.75 m, and the inner radius is 0.5 m. The height of the outer column is 1.5 m. The experimental model of this study is shown in Figure 4. The mechanical properties of cable members are the focus of this study. Tension meters are arranged in each cable to collect the change of cable force. The displacement of each node under self-weight condition is collected by total station. By comparing the measured and simulated values of the displacement, the size and prestress of the component are corrected. The modified virtual model can accurately map the mechanical properties of the structure. The feasibility of the model is verified by comparing the measured and simulated values of the cable force. On the basis of establishing a high-fidelity twin model, the state of the structure under various load conditions can be accurately analyzed in the model. By analyzing the mechanical properties of the structure under extreme conditions, the reliability of the structure is judged, which provides a basis for improving the bearing capacity of the structure. The structure is formed by adjusting the sleeve in the test model. The vertical load of the joint is applied by arranging weight blocks on the joint.

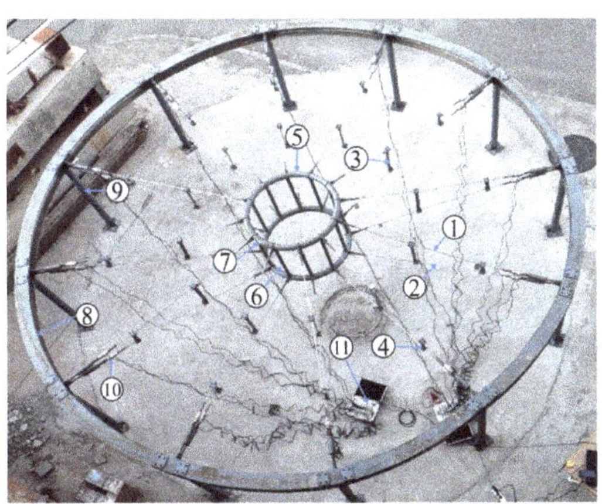

Figure 4. The experimental model: (① URC, ② LRC, ③ middle strut, ④ outer strut, ⑤ upper ring beam, ⑥ lower ring beam, ⑦ inner strut, ⑧ outer ring beam, ⑨ outer column, ⑩ tension meter, ⑪ collecting box).

3.1. Structural Twin Modeling

Establishing an HF twin model is the first step of intelligent assessment. Based on the modeling method in Section 2.2, this study first established the FEM of the structure. The cable components in the structure are built using Link10 elements, while the rigid components are built using Beam188 elements. The boundary conditions for the cable components are based on three-direction articulation. The base of the structure's columns is a rigid constraint. The cable elements are prestressed by applying initial strain. The vertical load is set in the model to simulate the actual structure loading process.

Based on the FEM, the structural vertical displacement in its formed shape was obtained. The error between the simulation and measured values of the vertical displacement was taken as the optimization objective. The physical properties optimized were the dimensions and prestress of the cables. The optimization parameters were iterated under the drive of a genetic algorithm, as shown in Figure 5. The optimal physical properties were obtained after 12 iterations. The error between the obtained vertical displacement in the twin model and the measured value was within 5%. In the process of virtual model optimization, the values of key parameters of genetic algorithm are shown in Table 1.

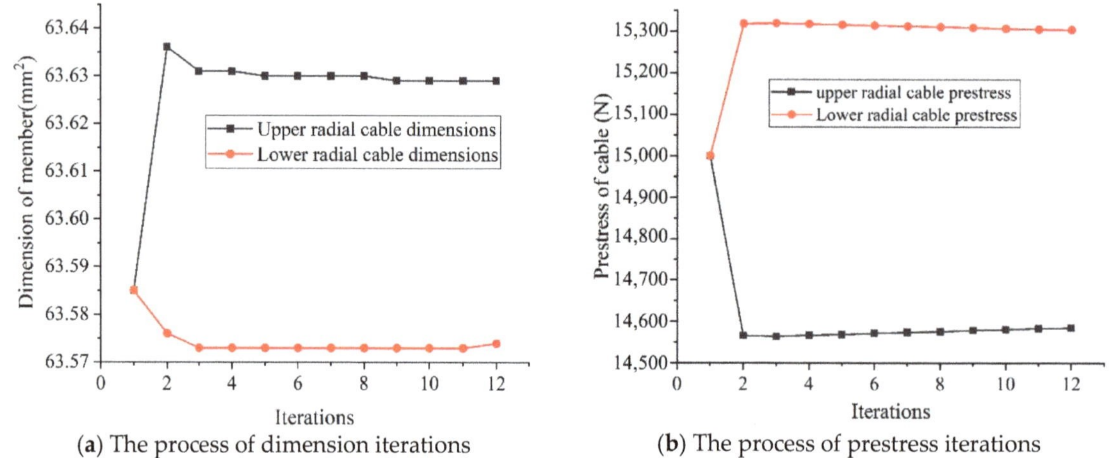

(**a**) The process of dimension iterations

(**b**) The process of prestress iterations

Figure 5. The iteration process of physical properties.

Table 1. Optimization parameters of the model.

Parameter	Adjustable Range	Value	Meaning
η	$[0, 1]$	0.01	learning rate
γ	$[0, \infty]$	500	genetic algebra
ς	$[0, \infty]$	50	population size

Based on the optimized physical properties, an HF structural twin model was established, as shown in Figure 6. To verify the effectiveness of the twin model, the simulation and measured values of the radial cable forces were obtained for the structure in its formed shape. The comparison showed that the fidelity of the cable force analysis by the DT model was also within 5%. The comparison of the cable force simulation and measured values is shown in Table 2. Since the structure is symmetric in geometry and loading, Table 2 shows the comparison of the radial cables of a single bay.

Figure 6. The HF twin model of the structure.

Table 2. The comparison of cable force between simulation value and measured value.

Component No.	Measured Value (N)	Simulation Value (N)	Error Value
URC 1	12,821	13,424	4.7%
LRC 1	14,588	14,034	−3.8%

3.2. Mechanical Response of Obtained Structure Based on DTs

A DT model can accurately and effectively simulate the mechanical properties of the results. This section analyzes the mechanical response of the structure under different span loads using the DT model. The structure's failure modes were obtained through multi-stage loading according to the most unfavorable load cases. The structural reliability index was obtained for extreme operating conditions.

3.2.1. Mechanical Performance of Different Load Conditions

Four load cases were analyzed in this study in static load analysis, as shown in Table 3. According to the literature [25], the constant load was taken as 0.15 kN/m^2, and the surface load was converted to an equivalent joint load [26], where the joint load on the inner strut was 33 N, on the middle strut was 92.3 N, and on the outer strut was 146 N. The specific LSs are shown in Figure 7.

Table 3. The load cases.

	Conditions of Load Combination
LS 1	1/4-span constant loads
LS 2	1/2-span constant loads
LS 3	3/4-span constant loads
LS 4	Full span constant loads

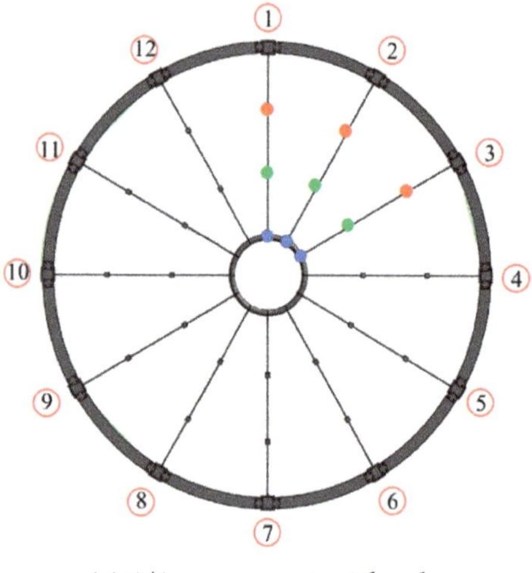

(**a**) 1/4-span constant loads

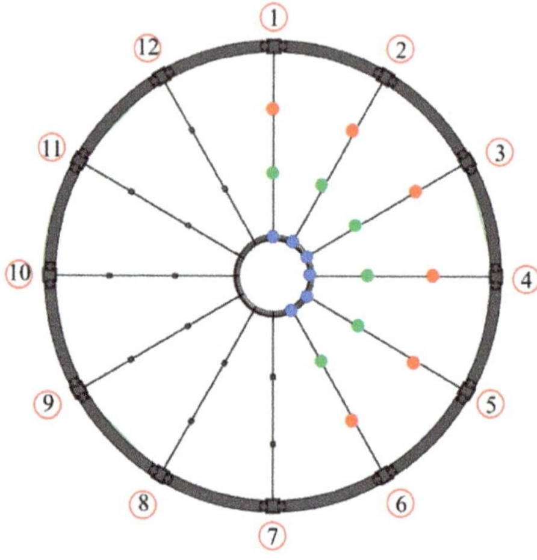

(**b**) 1/2-span constant loads

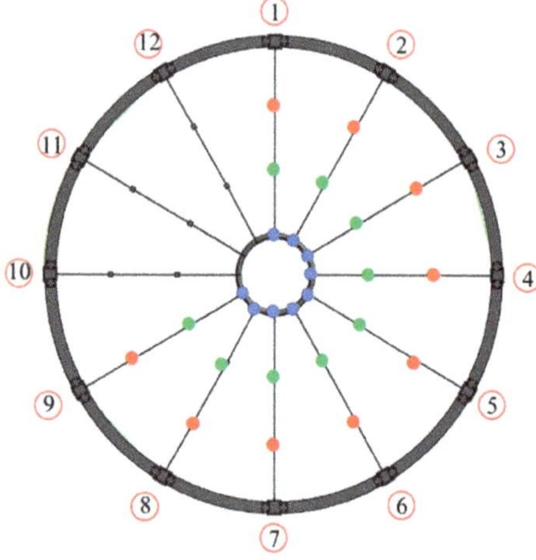

(**c**) 3/4-span constant loads

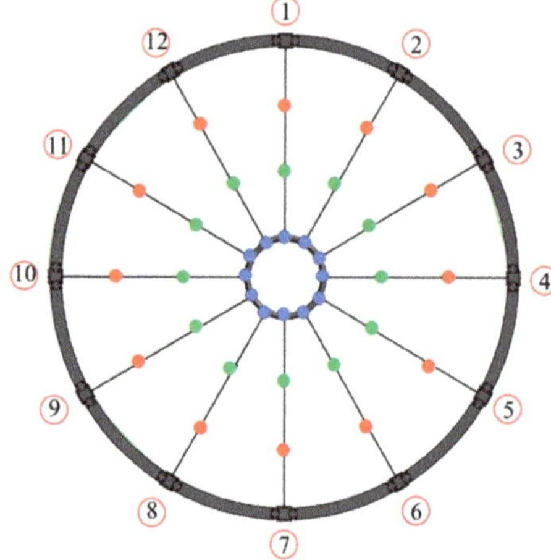

(**d**) Full span constant loads

Figure 7. The specific LSs (red is 146 N, green is 93.2 N, and blue is 33 N, The number in the figure indicates the axis position of the cable truss).

(1)   Quarter-span constant loads

Quarter-span loads were applied to the first, second, and third bays of radial cables in the experimental model to load the structure. The DT model's simulation analysis of internal forces and displacements is shown in Figure 8. The vertical coordinate axis in Figures 8, 9, 10 and 11b represents the vertical displacement, which gradually decreases from inside to outside. Under the action of load, the displacement of the external strut is maximal, so the corresponding curve is on the inside.

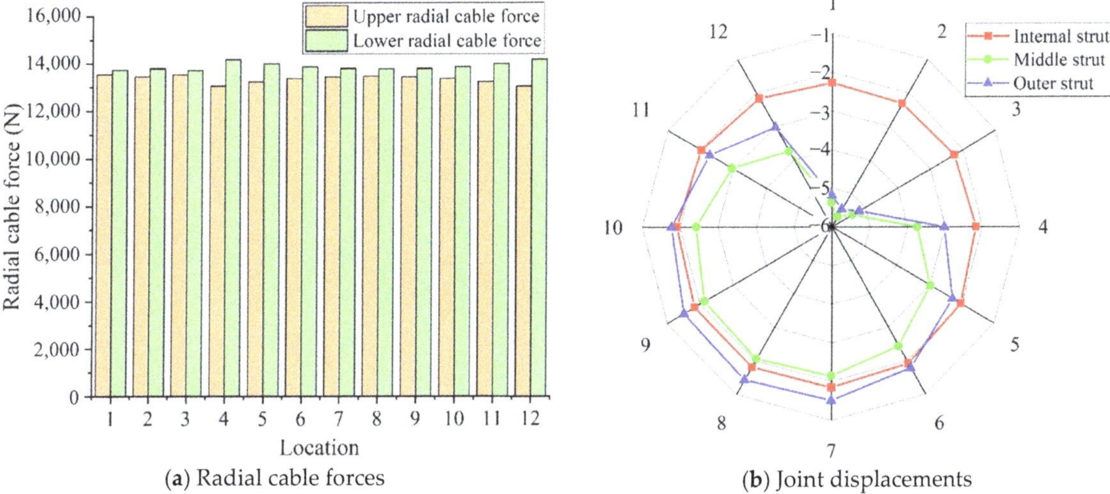

(**a**) Radial cable forces

(**b**) Joint displacements

Figure 8. Simulation analysis of the mechanical properties of the structure.

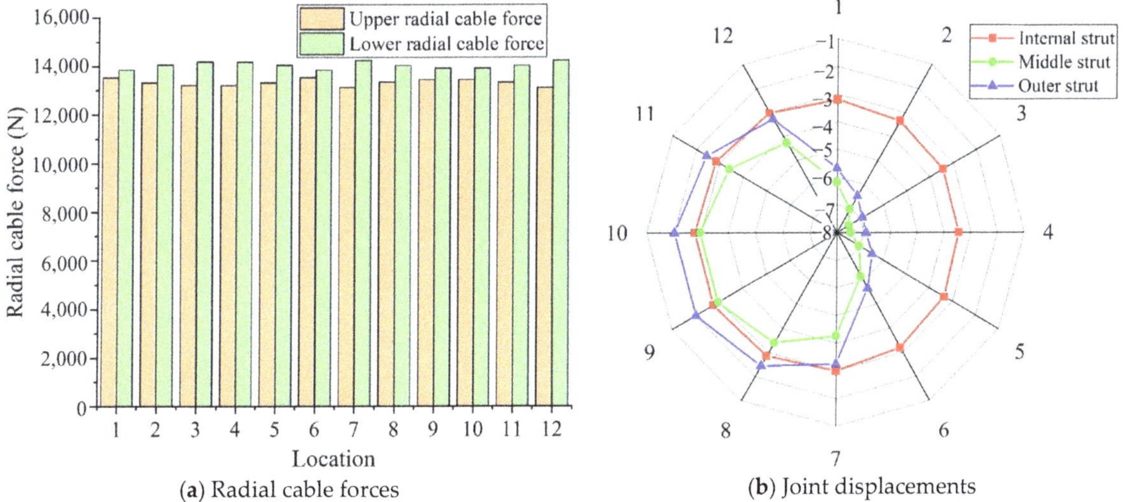

(**a**) Radial cable forces

(**b**) Joint displacements

Figure 9. Simulation analysis of the mechanical properties of the structure.

Compared with the structure in formed shape, when a 1/4-span load was applied, the force in the URC increased, while that in the LRC decreased. The force in the URC of the first and third bays increased the most, increasing from 13,424 N to 13,555 N, a variation rate of 1%. The force in the LRC of the first and third bays decreased the most, decreasing from 14,034 N to 13,745 N, a variation rate of 2.1%. The vertical displacement at the loading point changed the most, and the joint displacement of the middle strut was the largest. When a 1/4-span load was applied, the joint displacement of the middle strut in the second bay was the largest, at −5.68 mm.

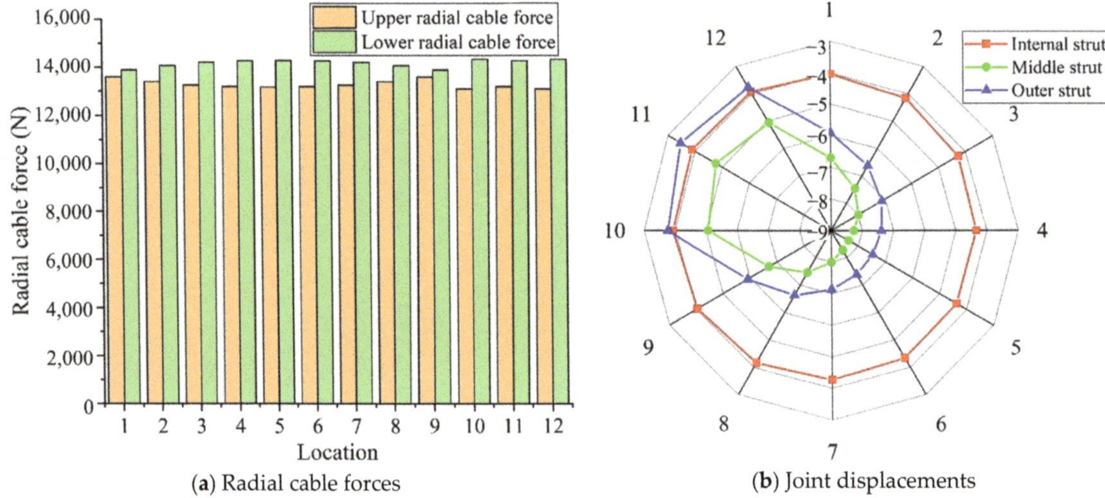

(**a**) Radial cable forces

(**b**) Joint displacements

Figure 10. Simulation analysis of the mechanical properties of the structure.

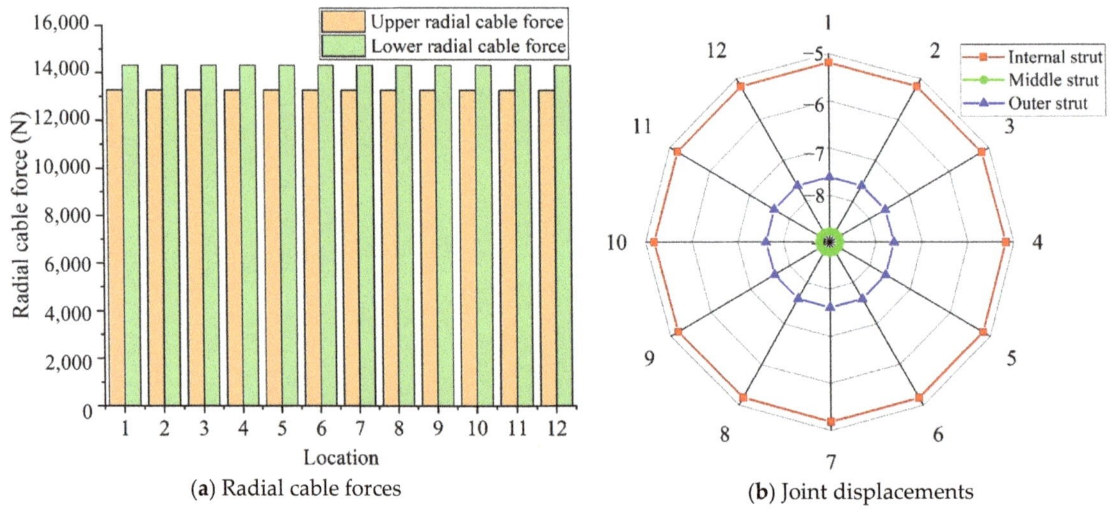

(**a**) Radial cable forces

(**b**) Joint displacements

Figure 11. Simulation analysis of the mechanical properties of the structure.

(2) Half-span constant loads

One to six bays of radial cables in the test model were loaded to apply a 1/2-span load to the structure. The internal forces and displacements obtained from the twin-model simulation analysis are shown in Figure 9.

When a 1/2-span load was applied, the maximum force of the URC for the first and sixth bays was 13,540 N, while the maximum force of the LRC for the third and fourth bays was 14,168 N. Compared with the formed shape, the upward cable force of the third and fourth bays changed the most, with a change rate of 1.6%. The LRC force of the seventh and twelfth bays changed the most, with a change rate of 1.5%. The vertical displacement at the loading position changed the most, and the joint displacement of the middle strut was the largest, reaching −7.48 mm when a 1/2-span load was applied to the third and fourth bays.

(3)   Three-quarter-span constant loads

Bays 1–9 of radial cables in the test model were loaded to apply a 3/4-span load to the structure. The internal forces and displacements obtained from the twin-model simulation analysis are shown in Figure 10.

When a 3/4-span load was applied, the maximum force of the URC for the first and ninth bays was 13,599 N, while the maximum force of the LRC for the fifth bay was 14,291 N. Compared with the formed shape, the upward cable force of the tenth and twelfth bays changed the most, with a change rate of 2.3%. The LRC force of the tenth and twelfth bays changed the most, with a change rate of 2.2%. The vertical displacement at the loading position changed the most, and the joint displacement of the middle strut was the largest, reaching −8.36 mm when a 3/4-span load was applied to the fifth bay.

(4)   Full-span constant loads

When subjected to full-span constant load, the internal forces and displacements of the test model were analyzed, and the results are shown in Figure 11 using the twin-model simulation.

When the full-span load was applied, the force on the URC was 13,271 N, and that on the LRC was 14,326 N. The force on the URC decreased by 1.1% as a whole, while the force on the LRC increased by 2.1% as a whole. The maximum joint displacement of the middle strut was −8.78 mm, and the minimum joint displacement of the inner strut was −5.18 mm.

3.2.2. Structural Failure Mode

By analyzing the structural mechanical performance under different span LSs, it was found that the maximum internal force and displacement of the structure occurred under full-span constant load. The failure modes of the cable truss structure are transfinite deformation, cable breaking, strut destabilization, and cable relaxation [27,28]. In the twin model, multi-level loading with full-span constant load was carried out to analyze the trend of changes in the structure's vertical displacement and internal forces. The mechanical response of the structure as the load increases is shown in Figure 12. Under the full-span load, the displacement of the upper and lower joints of the same strut is the same (Figure 12a). The same cable has the same stress in all sections (Figure 12b).

According to the cable structure technical specification [29–33], the limit value of vertical displacement under normal service conditions is 1/250 of the span, that is, 24 mm. In this experiment, transfinite deformation occurred when loading level 15 constant load. The tensile strength of the cable and the strength of the steel component used in the experiment were 1670 MPa and 345 MPa, respectively. As shown in Figure 12, when loading up to level 95 constant load, the LRC force first reaches the breaking force. Among the struts, the stress change in the outer strut is the most obvious. Through the analysis of the twin model, the failure mode of the structure's ultimate bearing capacity state is identified as cable breaking.

3.2.3. Reliability of Components after Failure

In order to improve the sustainability of structural bearing capacity, a reliability analysis of the structure was conducted. In this paper, the Monte Carlo method based on response surface was used to analyze the damage situation of the cable truss structure. In this method, the structural response surface is first fitted using the sampling method, and then the response surface equation is used to replace the structure's FEM. After that, the structure is analyzed for reliability using the Monte Carlo method.

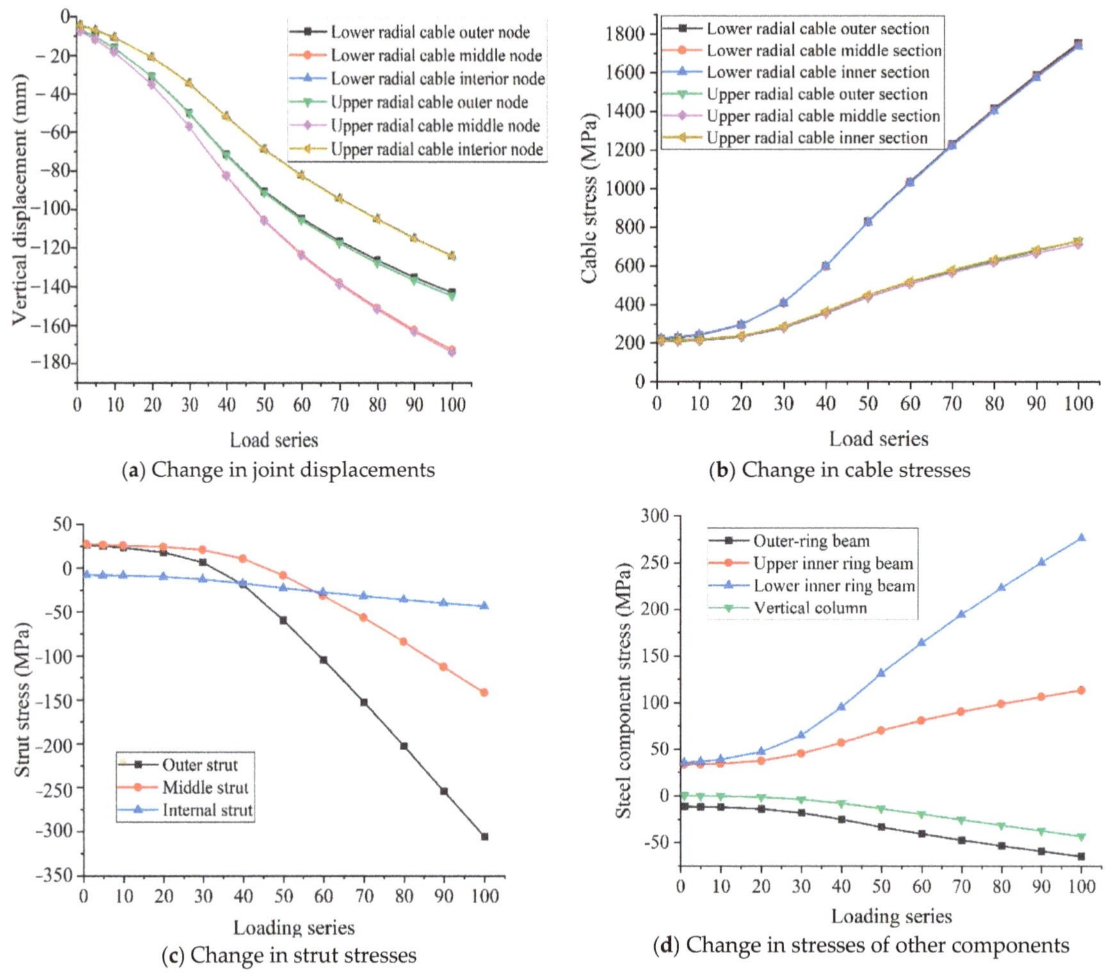

Figure 12. Structural mechanical response of multi-stage loading process.

Structural failure probability under different conditions was calculated using Equations (3) and (5). The displacement limit value was set to 24 mm. According to the analysis in Section 3.2.2, cable breaking is the failure mode of the ULS. In Equation (5), the limit value of the component stress is the cable's tensile strength divided by the safety factor. According to the cable structure technical specification [29–33], the safety factor is 2.0. The extreme value of stress is 835 MPa. Under the guidance of the theoretical method in Section 2.3, this study conducted an evaluation of the structural reliability under four typical operating conditions [34–36], as shown in Table 4.

As shown in Table 4, the failure of the LRC has the greatest impact on the structural reliability, and the reliability index of the SLS is close to 0. Therefore, during the structural service life, the stress state of the LRC should be closely monitored to prevent component relaxation or excessive stress. In addition to typical operating conditions, the structural reliability can be analyzed in real time for other operating conditions using the DT model. Structural bearing capacity maintenance can be carried out based on the changes in the reliability index.

Table 4. Evaluation of structural reliability.

Components Failure	SLS		ULS	
	Failure Probability	Reliability Index	Failure Probability	Reliability Index
Single URC failure	3.71887×10^{-1}	0.3269	7.59028×10^{-3}	2.4280
Single LRC failure	4.26797×10^{-1}	0.1845	8.49569×10^{-3}	2.3869
Single middle strut failure	3.13185×10^{-1}	0.4868	6.14143×10^{-3}	2.5039
Single outer strut failure	2.81372×10^{-1}	0.5788	5.79581×10^{-3}	2.5243

3.3. Discussion

Using the twin-modeling method, an HF twin model of the test structure was established. The simulation accuracy of the twin model was found to be over 95% through comparison, which fully demonstrates the feasibility of the DT modeling method.

The structural mechanical performance was analyzed during multi-span loading. It was found that the URC on the edge span had the maximum cable force, and the LRC on the middle span had the greatest change in cable force. The joint displacement of the middle strut was greater than that of the outer and inner struts, and the joint displacement was the largest for the loading on the middle span. The sensitive areas of the structure were determined by analyzing its response to multi-span loading.

The structural failure modes and structural bearing capacity were obtained through multi-stage loading. The structure exceeded the deformation limit when subjected to a constant load at level 15. When loaded to level 95, the LRC force reached its breaking point. Under the ultimate load, the cable force of the lower radial cable is 2.2 times that of the upper radial cable. Therefore, the stress performance of the lower radial cable should be emphasized in the loading process. Through the analysis of multi-stage loading, it is found that grade 15 constant load corresponds to the normal service limit state of the structure, and grade 95 constant load corresponds to the ultimate state of the structure's bearing capacity.

The importance of various components was obtained by analyzing the structure's reliability. The order of importance was LRC > URC > outer strut > middle strut. In the DT model, structural reliability under various conditions can be analyzed accurately, providing the basis for improving the sustainability of structural bearing capacity. In the normal service limit state, the reliability index of the failure structure of a single lower radial cable is close to 0, which is the most unfavorable extreme condition of the structure. When a single member fails, the reliability index corresponding to the lower radial cable is a third of that of the outer strut.

4. Conclusions

This study proposes an intelligent reliability assessment method for improving the sustainable carrying capacity of the structure by integrating DT technology. Through the establishment of an HF twin model, real-time analysis and accurate prediction of structural reliability were achieved. On the basis of analyzing the mechanical performance and reliability of the structure, strategy support was provided to ensure the sustainability of the bearing capacity of the structure. The main conclusions are as follows.

(1) The establishment method of the HF twin model was formed by combining the genetic algorithm and finite element modeling theory. By comparing the simulated and measured values of structural mechanical parameters, the physical properties of the model are modified. The DT model can accurately map the structural mechanical state with an error within 5%. Additionally, the DT model can analyze the structural mechanical response under various operating conditions in real time, reducing the dependence on experimental data.

(2) Using the DT model, the analysis method of the structural mechanical performance and reliability was formed. The DT model can accurately analyze the structural carrying capacity and the development trend of reliability indicators. The structural safety status can be accurately judged based on the reliability index, which provides the basis for improving the sustainability of structural bearing capacity.

(3) During the experiment, the advantages of the proposed intelligent assessment method were verified. Under the guidance of the theoretical method, the mechanical behavior law of the structure under multi-span load arrangement is obtained, and the key stress nodes and adverse conditions are captured. By analyzing the development trend of displacement and internal force under multi-stage loads, the ultimate bearing capacity of the structure is obtained. By reaching component failure, the reliability index and key components are obtained by simulation in the twin model. Real-time analysis of the structural mechanical performance and reliability effectively ensured the safety of the structure, providing ideas for improving the sustainable bearing capacity of the structure.

This study focused on the reliability and bearing capacity of cable structures. It provides ideas for other types of structural safety assessment. By obtaining adverse working conditions, failure modes, and reliability indices in the twin model, it provides a basis for improving the sustainability of structural bearing capacity. For concrete structure and masonry structure, it is necessary to further study the influence of material ratio and cooperative work between materials on structural reliability. The large volume of existing buildings necessitates higher requirements for structural health monitoring. It is important to improve the sustainability of existing building carrying capacity by updating the twin model with the dynamic and static mechanical data of the fusion structure driven by twin data and realizing accurate identification of structural damage. Driven by the sustainability goal of structural carrying capacity, it is advised that future studies focus on forming efficient maintenance measures based on the reliability index of structures. Taking into account the whole life cycle of the building, low energy consumption in the operation and maintenance of the structure is also a future research direction.

Author Contributions: Conceptualization, G.S.; methodology, G.S.; software, G.S.; validation, G.S., Z.L., D.X. and R.Z.; writing—original draft preparation, G.S.; writing—review and editing, Z.L.; project administration, Z.L. and D.X.; funding acquisition, D.X. All authors have read and agreed to the published version of the manuscript.

Funding: The research was funded by the Research and Development Project of the Ministry of Housing and Urban-Rural Development of China (project 2020-K-106).

Institutional Review Board Statement: Not applicable.

Informed Consent Statement: Not applicable.

Data Availability Statement: The data presented in this study are available on request from the corresponding author. The data are not publicly available for confidentiality reasons.

Acknowledgments: The authors would like to thank Beijing University of Technology and Hebei Construction Building Group Co., Ltd., for their support throughout the research project. This work was supported by Yufen Zhang (Hebei University of Technology), Tianping Zhang (Hebei Construction Building Group Co., Ltd.), Liya Zhao (Hebei Construction Building Group Co., Ltd.), Jinsong Xian (COX Architecture Pty Ltd.), Liyong Jia (Hebei Construction Building Group Co., Ltd.).

Nomenclature

Formula symbol	physical meaning
DT	digital twin
HF	high fidelity
F_{DT}	evaluation framework driven by DTs
L_{PSE}	physical structure entity
L_{VSM}	virtual structure model
L_{TDP}	twin data processing layer
L_{FA}	functional application layer
C_N	connections between each component
FEM	finite element model
$\Delta\mu$	displacement error
δ_i	ith optimization variable
a_i	lower limits of the ith optimization variable
b_i	upper limits of the ith optimization variable
δ_i^{opt}	optimal design parameter corresponding to the minimum error
LHS	Latin hypercube sampling
ULS	ultimate limit state
SLS	serviceability limit state
LS	loading condition
$[\sigma]$	limit value of the component stress
σ_{max}	maximum stress of the components in the structure
$[\mu]$	limit value of the structural displacement
μ_{max}	maximum displacement of the structure
URC	upper radial cable
LRC	lower radial cable

References

1. Markosian, N.; Tawadrous, R.; Mastali, M.; Thomas, R.J.; Maguire, M. Performance Evaluation of a Prestressed Belitic Calcium Sulfoaluminate Cement (BCSA) Concrete Bridge Girder. *Sustainability* **2021**, *13*, 7875. [CrossRef]
2. Chen, Z.; Ge, H.; Chan, S. Modelling, Test and Practice of Steel Structures. *Metals* **2022**, *12*, 1212. [CrossRef]
3. Omika, Y.; Fukuzawa, E.; Koshika, N.; Morikawa, H.; Fukuda, R. Structural responses of world trade center under aircraft attacks. *J. Struct. Eng.* **2005**, *131*, 6–15. [CrossRef]
4. Fischer, E.; Biondo, A.E.; Greco, A.; Martinico, F.; Pluchino, A.; Rapisarda, A. Objective and Perceived Risk in Seismic Vulnerability Assessment at an Urban Scale. *Sustainability* **2022**, *14*, 9380. [CrossRef]
5. Liu, H.; Li, B.; Chen, Z.; Ouyang, Y. Damage analysis of aluminum alloy gusset joints under cyclic loading based on continuum damage mechanics. *Eng. Struct.* **2021**, *244*, 112729. [CrossRef]
6. Lu, J.; Zhang, H.; Wu, X. Experimental study on collapse behaviour of truss string structures under cable rupture. *J. Constr. Steel Res.* **2021**, *185*, 106864. [CrossRef]
7. Vaezzadeh, A.; Dolatshahi, K.M. Progressive collapse resistance of cable net structures. *J. Constr. Steel Res.* **2022**, *195*, 107347. [CrossRef]
8. Ozer, E.; Feng, M.Q. Structural Reliability Estimation with Participatory Sensing and Mobile Cyber-Physical Structural Health Monitoring Systems. *Appl. Sci.* **2019**, *9*, 2840. [CrossRef]
9. Bojórquez, J.; Ponce, S.; Ruiz, S.E.; Bojórquez, E.; Reyes-Salazar, A.; Barraza, M.; Chávez, R.; Valenzuela, F.; Leyva, H.; Baca, V. Structural reliability of reinforced concrete buildings under earthquakes and corrosion effects. *Eng. Struct.* **2021**, *237*, 112161. [CrossRef]
10. Callcut, M.; Cerceau Agliozzo, J.-P.; Varga, L.; McMillan, L. Digital Twins in Civil Infrastructure Systems. *Sustainability* **2021**, *13*, 11549. [CrossRef]
11. Selvam, J.; Vajravelu, A.; Nagapan, S.; Arumugham, B.K. Analyzing the Flexural Performance of Cold-Formed Steel Sigma Section Using ABAQUS Software. *Sustainability* **2023**, *15*, 4085. [CrossRef]
12. Bonney, M.S.; de Angelis, M.; Dal Borgo, M.; Wagg, D.J. Contextualisation of information in digital twin processes. *Mech. Syst. Signal Process.* **2023**, *184*, 109657. [CrossRef]
13. Mahmoodian, M.; Shahrivar, F.; Setunge, S.; Mazaheri, S. Development of Digital Twin for Intelligent Maintenance of Civil Infrastructure. *Sustainability* **2022**, *14*, 8664. [CrossRef]
14. Wang, L.; Liu, H.; Chen, Z.; Zhang, F.; Guo, L. Combined digital twin and hierarchical deep learning approach for intelligent damage identification in cable dome structure. *Eng. Struct.* **2023**, *274*, 115172. [CrossRef]

15. Ritto, T.G.; Rochinha, F.A. Digital twin, physics-based model, and machine learning applied to damage detection in structures. *Mech. Syst. Signal Process.* **2021**, *155*, 107614. [CrossRef]

16. Zhu, X.; Bao, T.; Yeoh, J.K.W.; Jia, N.; Li, H. Enhancing dam safety evaluation using dam digital twins. *Struct. Infrastruct. Eng.* **2021**, *19*, 904–920. [CrossRef]

17. Liu, Z.; Shi, G.; Jiao, Z.; Zhao, L. Intelligent Safety Assessment of Prestressed Steel Structures Based on Digital Twins. *Symmetry* **2021**, *13*, 1927. [CrossRef]

18. Liu, Y.-F.; Liu, X.-G.; Fan, J.-S.; Spencer, B.; Wei, X.-C.; Kong, S.-Y.; Guo, X.-H. Refined safety assessment of steel grid structures with crooked tubular members. *Autom. Constr.* **2019**, *99*, 249–264. [CrossRef]

19. Carneiro, G.D.N.; António, C.C. Dimensional reduction applied to the reliability-based robust design optimization of composite structures. *Compos. Struct.* **2020**, *255*, 112937. [CrossRef]

20. El Bazi, N.; Mabrouki, M.; Laayati, O.; Ouhabi, N.; El Hadraoui, H.; Hammouch, F.-E.; Chebak, A. Generic Multi-Layered Digital-Twin-Framework-Enabled Asset Lifecycle Management for the Sustainable Mining Industry. *Sustainability* **2023**, *15*, 3470. [CrossRef]

21. Kaewunruen, S.; Ngamkhanong, C.; Xu, S. Large amplitude vibrations of imperfect spider web structures. *Sci. Rep.* **2020**, *10*, 19161. [CrossRef] [PubMed]

22. Lee, J.; Hyun, H. Multiple Modular Building Construction Project Scheduling Using Genetic Algorithms. *J. Constr. Eng. Manag.* **2019**, *145*, 04018116. [CrossRef]

23. Del Savio, A.A.; de Andrade, S.A.L.; da Silva Vellasco, P.C.G.; Martha, L.F. Genetic algorithm optimization of semi-rigid steel structures. In Proceedings of the Eighth International Conference on the Application of Artificial Intelligence to Civil, Structural and Environmental Engineering, Rome, Italy, 30 August–2 September 2005; pp. 24.1–24.16.

24. *GB 50009–2012*; Load Code for the Design of Building Structures. National Standard of the People's Republic of China: Beijing, China, 2012.

25. *GB 50017-2017*; Code for Design of Steel Structure. National Standard of the People's Republic of China: Beijing, China, 2018.

26. Triller, J.; Immel, R.; Harzheim, L. Topology optimization using difference-based equivalent static loads. *Struct. Multidiscip. Optim.* **2022**, *65*, 226. [CrossRef]

27. Wadee, M.A.; Hadjipantelis, N.; Bazzano, J.B.; Gardner, L.; Lozano-Galant, J.A. Stability of steel struts with externally anchored prestressed cables. *J. Constr. Steel Res.* **2020**, *164*, 105790. [CrossRef]

28. Yan, R.; Chen, Z.; Wang, X.; Liu, H.; Xiao, X. A new equivalent friction element for analysis of cable supported structures. *Steel Compos. Struct.* **2015**, *18*, 947–970. [CrossRef]

29. *JGJ 257-2012*; Technical Specification for Cable Structure. Ministry of Housing and Urban-Rural Development: Beijing, China, 2012.

30. *GB/T232-2010*; Metallic Materials—Bend Test. Standards Press of China: Beijing, China, 2010.

31. *GB/T228.1-2010*; Metallic Materials—Tensile Testing—Part 1: Method of Test at Room Temperature. Standards Press of China: Beijing, China, 2010.

32. *ASCE/SEI 19-10*; Structural Applications of Steel Cables for Buildings. American Society of Civil Engineers (ASCE): Reston, VA, USA, 2010.

33. *EN 1993–1–11*; Eurocode 3: Design of Steel Structures—Part 1–11: Design of Structures with Tension Components. CEN: Paris, France, 1993.

34. Liu, Z.; Li, H.; Liu, Y.; Wang, J.; Tafsirojjaman, T.; Shi, G. A novel numerical approach and experimental study to evaluate the effect of component failure on spoke-wheel cable structure. *J. Build. Eng.* **2022**, *61*, 105268. [CrossRef]

35. Musarat, M.A.; Alaloul, W.S.; Cher, L.S.; Qureshi, A.H.; Alawag, A.M.; Baarimah, A.O. Applications of Building Information Modelling in the Operation and Maintenance Phase of Construction Projects: A Framework for the Malaysian Construction Industry. *Sustainability* **2023**, *15*, 5044. [CrossRef]

36. Tonelli, D.; Beltempo, A.; Cappello, C.; Bursi, O.S.; Zonta, D. Reliability analysis of complex structures based on Bayesian inference. *Struct. Health Monit.* **2023**. [CrossRef]

sustainability

Article

Machine Learning-Based Prediction of Elastic Buckling Coefficients on Diagonally Stiffened Plate Subjected to Shear, Bending, and Compression

Yuqing Yang [1,*], Zaigen Mu [1,*] and Xiao Ge [2]

[1]   School of Civil and Resource Engineering, University of Science and Technology Beijing, Beijing 100083, China
[2]   School of Civil and Transportation Engineering, Hebei University of Technology, Tianjin 300401, China
*   Correspondence: yqyang@ustb.edu.cn (Y.Y.); zgmu@ces.ustb.edu.cn (Z.M.)

Citation: Yang, Y.; Mu, Z.; Ge, X. Machine Learning-Based Prediction of Elastic Buckling Coefficients on Diagonally Stiffened Plate Subjected to Shear, Bending, and Compression. *Sustainability* **2023**, *15*, 7815. https://doi.org/10.3390/su15107815

Academic Editors: Jurgita Antuchevičienė, Qingshan Yang, Zhihua Chen, Yansheng Du and Yue Wu

Received: 15 March 2023
Revised: 27 April 2023
Accepted: 2 May 2023
Published: 10 May 2023

Abstract: The buckling mechanism of diagonally stiffened plates under the combined action of shear, bending, and compression is a complex phenomenon that is difficult to describe with simple and clear explicit expressions. Predicting the elastic buckling coefficient accurately is crucial for calculating the buckling load of these plates. Several factors influence the buckling load of diagonally stiffened plates, including the plate's aspect ratio, the stiffener's flexural and torsional rigidity, and the in-plane load. Traditional analysis methods rely on fitting a large number of finite element numerical simulations to obtain an empirical formula for the buckling coefficient of stiffened plates under a single load. However, this cannot be applied to diagonally stiffened plates under combined loads. To address these limitations, several machine learning (ML) models were developed using the ML method and the SHAP to predict the buckling coefficient of diagonally stiffened plates. Eight ML models were trained, including decision tree (DT), k-nearest neighbor (K-NN), artificial neural network (ANN), random forest (RF), AdaBoost, LightGBM, XGBoost, and CatBoost. The performance of these models was evaluated and found to be highly accurate in predicting the buckling coefficient of diagonally stiffened plates under combined loading. Among the eight models, XGBoost was found to be the best. Further analysis using the SHAP method revealed that the aspect ratio of the plate is the most important feature influencing the elastic buckling coefficient. This was followed by the combined action ratio, as well as the flexure and torsional rigidity of the stiffener. Based on these findings, it is recommended that the stiffener-to-plate flexural stiffness ratio be greater than 20 and that the stiffener's torsional-to-flexural stiffness ratio be greater than 0.4. This will improve the elastic buckling coefficient of diagonally stiffened plates and enable them to achieve higher load capacity.

Keywords: diagonally stiffened plates; combined action; elastic buckling coefficient; machine learning; Shapley Additive exPlanations

1. Introduction

Stiffened plates are widely utilized in various fields such as civil engineering, automotive, shipbuilding, and aerospace [1–3]. The current research focuses on the buckling stability of stiffened plates under in-plane loading [4–8]. In structural engineering, stiffeners are used in beams, plates, shear walls, and other major structural members to improve structural performance [9]. The common forms of stiffened plates include transverse, longitudinal, and diagonal. In 1960, Timoshenko and Gere [10] systematically studied the buckling behavior of rectangular plates and derived the equilibrium equations of rectangular plates under single loads such as compression, bending, and shear. In addition, the relationship between flexure rigidity of stiffeners and critical buckling stress was analyzed for transversely and longitudinally stiffened plates, and the limit of the stiffness ratio was given, thus establishing the foundation for the research of buckling stability of stiffened plates. Mikami [11] and Yonezawa [12] investigated the buckling behavior of diagonally stiffened plates simply supported on four edges. Yuan et al. [13] considered the flexural rigidity of the open diagonal stiffeners, and studied the shear behavior of diagonally stiffened stainless-steel girders using the finite element (FE) software. Martins and Cardoso [14] also investigated how the torsional stiffness of closed-section stiffeners affects the shear buckling behavior of diagonally stiffened plates.

Most current research focuses on the shear buckling behavior of diagonally stiffened plates, and there is a lack of sufficient research on the combined action of multiple loads. This discourages the design of stiffened plates. There is an urgent need to propose a simple and convenient model for the calculation of the buckling performance of diagonally stiffened plates under combined compression-bending-shear action.

On the other hand, new types of composite structures have emerged in recent years. Because many variables affect their mechanical properties, it is difficult to solve their analytical solutions directly. It is also hard to fully reflect the influence of different parameters that affect a structure's performance when using semi-theoretical and semi-experiential methods. Machine learning (ML) can reveal the hidden relationship between input features and prediction results. It can be used to predict the load-bearing capacity, failure mode, and damage assessment of members with excellent accuracy [15–17]. Using ML to predict the bearing capacity of complex structures is one of its important research objectives, such as the punching strength of concrete slabs [18], axial compressive strength of concrete-filled steel tube (CFST) columns [19], buckling loads of perforated steel beams [20], material bond strength [21], concrete splitting and tensile strength [22], and rebound modulus properties [23]. The nonlinear hysteretic response behavior of steel plate shear walls can be simulated using a deep learning method [24]. Furthermore, the SHAP (Shapley Additive exPlanations) approach [25] can effectively explain how each feature contributed to the results and increase the credibility of the ML model [26].

There is a lack of research on the performance of diagonally stiffened plates under complex actions, especially under the coupled action of compression, bending, and shear. There is also a lack of relevant theoretical and design basis. Therefore, the central problem of this study is to determine the elastic buckling coefficients of diagonally stiffened plates under the combined action of bending, compression, and shear. In this paper, the data set affecting the buckling performance of diagonally stiffened plates was established, the performance metrics of different machine learning algorithms in predicting the elastic buckling coefficient were compared, and the optimal model was interpreted using the SHAP method. The results show that integrated learning exhibits a good generalization ability and can quickly predict the elastic buckling coefficients of diagonally stiffened plates under the combined action of compression–bending–shear.

2. Methods

2.1. Buckling Coefficient of Diagonally Stiffened Plate

Timoshenko and Gere [10] studied the elastic stability of unstiffened plates using the energy method and found that the aspect ratio γ of the plate significantly affects the critical buckling stress τ_{cr} of the structure, as given in Equation (1), where k_{cr} is a function with respect γ. Then, according to Bleich [27], the buckling coefficients k_{cr} for a four-edge simply supported plate under shear and bending loads alone are calculated by the approximations Equation (2) and Equation (3), respectively. For example, for a rectangular plate with four-sided simple support, the shear buckling coefficient is 9.34 and the bending buckling coefficient is 23.9.

$$\tau_{cr} = \frac{k_{cr}\eta^2 D}{H^2 t} = \frac{k_{cr}\eta^2 E}{12(1-v^2)\lambda^2}, \text{ where } D = \frac{Et^3}{12(1-v^2)}, \lambda = \frac{H}{t} \tag{1}$$

$$\text{For shear}: k_{cr} = \begin{cases} 5.34 + 4/\gamma^2 & \gamma \geq 1 \\ 4 + 5.34/\gamma^2 & \gamma < 1 \end{cases} \tag{2}$$

$$\text{For bending}: k_{cr} = \begin{cases} 15.87 + 1.87/\gamma^2 + 8.6\gamma^2 & (\gamma < 2/3) \\ 23.9 & (\gamma \geq 2/3) \end{cases} \tag{3}$$

$$\gamma = \frac{L}{H} \tag{4}$$

where L, H, and t are the plate's length, height, and thickness, respectively; D is the plate's flexural stiffness; E is the elastic modulus of the plate; k_{cr} is the elastic buckling coefficient; ν is Poisson's ratio; and λ is the slenderness ratio of the plate.

Tong [28] analyzed the buckling stability of a four-edge simply supported plate under combined shear and non-uniform compression, resulting in the interaction formula for the plate combined action of axial compression, bending, and shear as Equation (5). It is possible to separate the non-uniform compression into a combined bending and axial compression action. A coefficient is defined to represent the form of the distribution of non-uniform axial compression. The distribution factor of non-uniform compression ζ is shown in Equation (7). More information is detailed later in Section 3.3.

$$\frac{\sigma}{\sigma_{cr}} + \left(\frac{\sigma_b}{\sigma_{bcr}}\right)^2 + \left(\frac{\tau}{\tau_{cr}}\right)^2 = 1 \tag{5}$$

$$\begin{cases} \sigma = (1 - 0.5\zeta)\sigma_{max} \\ \sigma_b = 0.5\zeta\sigma_{max} \end{cases} \tag{6}$$

$$\zeta = \frac{\sigma_{max} - \sigma_{min}}{\sigma_{max}} \tag{7}$$

where σ, σ_b and τ are the stress of axial compression, bending, and shear, respectively; σ_{cr}, σ_{bcr} and τ_{cr} are the critical stresses in axial compression, bending, and shear alone, respectively; and σ_{max} and σ_{min} are the maximum and minimum stresses in non-uniform compression, respectively.

Diagonally stiffened plates can be classified as either tension-type or compression-type, depending on the direction of the diagonal stiffeners and the shear force. Under pure shear, stiffeners are subjected to pressure and are called compression-type, while those subjected to tension are called tension-type. Compared to unstiffened plates, diagonally stiffened plates have a larger buckling coefficient, with compression type having the highest buckling coefficient. Mikami [11] assumed that the diagonally stiffened edge is ideally restrained from out-of-plane displacement, meaning the stiffeners have infinite flexural stiffness. Based on the resultant curves of the numerical analysis, equations were fitted to calculate the buckling coefficients of diagonally stiffened plates in pure shear and bending alone. The shear buckling coefficient of diagonally stiffened plates can be approximately expressed by Equation (8), and the bending buckling coefficient of diagonally stiffened plates can be approximated by Equation (9).

$$\text{For shear}: k_{cr} = \begin{cases} 11.9 + 10.1/\gamma + 10.9/\gamma^2 & \text{compression} \\ 17.2 - 22.5/\gamma + 16.7/\gamma^2 & \text{tension} \end{cases} \tag{8}$$

$$\text{For bending}: k_{cr} = 22.5 + 4.23\gamma + 2.75\gamma^2 \tag{9}$$

Considering the flexure rigidity of stiffeners, the critical elastic buckling stress obtained from the FE analysis was compared to the structure's theoretical buckling stress [13]. As a result, the equation for estimating the elastic buckling coefficient of the diagonally stiffened plate was modified as Equation (10), where I_s is the moment of inertia of the stiffeners.

$$k_{cr} = \begin{cases} \left(\frac{24.2}{\gamma} - 2\right) \cdot \sqrt[3]{\frac{I_s}{h \cdot t^3}} + \frac{6.57}{\gamma^2} + 4.92 \; for \; \frac{I_s}{h \cdot t^3} < 1 \wedge \gamma < 1 \\[6pt] \left(\frac{24.2}{\gamma} - 2\right) \cdot \sqrt[3]{\frac{I_s}{h \cdot t^3}} + \frac{4.92}{\gamma^2} + 6.57 \; for \; \frac{I_s}{h \cdot t^3} < 1 \wedge \gamma \geq 1 \\[6pt] \left(\frac{9}{\gamma} - 2\right) \cdot \sqrt[3]{\frac{I_s}{h \cdot t^3}} + \frac{6.57}{\gamma^2} + \frac{15.2}{\gamma} + 4.92 \; for \; \frac{I_s}{h \cdot t^3} \geq 1 \wedge \gamma < 1 \\[6pt] \left(\frac{9}{\gamma} - 2\right) \cdot \sqrt[3]{\frac{I_s}{h \cdot t^3}} + \frac{4.92}{\gamma^2} + \frac{15.2}{\gamma} + 6.57 \; for \; \frac{I_s}{h \cdot t^3} \geq 1 \wedge \gamma \geq 1 \end{cases} \tag{10}$$

Martins and Cardoso [14] investigated the effect of the torsional rigidity of stiffeners on the elastic buckling stress. Based on a nonlinear regression analysis of a large number of numerical simulation results, Equations (11) and (12) were provided to predict the shear buckling coefficients of open and closed diagonally stiffened plates, respectively. In these equations, η_y is the relative flexural rigidity around the strong axis, η_z is the relative flexural rigidity around the weak axis, and φ_x is the relative torsional rigidity for the stiffeners.

$$k_{cr} = \begin{cases} 4.00 + \frac{5.34}{\gamma^2} + \frac{11.50}{\sqrt[3]{\eta_y}} + 5.23 \cdot \frac{\eta_y^{0.404} \cdot \eta_z^{0.038}}{\gamma^{1.371}} \\[4pt] \qquad for \; \eta_y \leq 12(1 - v^2) \wedge 0.5 < \gamma \leq 1.0 \\[8pt] 5.34 + \frac{4.00}{\gamma^2} + \frac{4.07}{\sqrt[3]{\eta_y}} + 6.15 \cdot \frac{\eta_y^{0.428} \cdot \eta_z^{0.020}}{\gamma^{1.434}} \\[4pt] \qquad for \; \eta_y \leq 12(1 - v^2) \wedge 1.0 < \gamma \leq 2.0 \\[8pt] 18.35 + \frac{5.67}{\gamma} + \frac{3.35}{\gamma^2} - \frac{17.96}{\sqrt[3]{\eta_y}} + 10.16 \cdot \frac{\eta_y^{0.026} \cdot \eta_z^{0.076}}{\gamma^{1.788}} \\[4pt] \qquad for \; \eta_y > 12(1 - v^2) \wedge 0.5 < \gamma \leq 1.0 \\[8pt] 12.03 + \frac{9.96}{\gamma} + \frac{11.45}{\gamma^2} - \frac{14.29}{\sqrt[3]{\eta_y}} + 6.92 \cdot \frac{\eta_z^{0.375}}{\gamma^{1.632} \cdot \eta_y^{0.036}} \\[4pt] \qquad for \; \eta_y > 12(1 - v^2) \wedge 1.0 < \gamma \leq 2.0 \end{cases} \tag{11}$$

$$k_{cr} = \begin{cases} 4.00 + \frac{5.34}{\gamma^2} + \frac{5.03}{\sqrt[3]{\eta_y}} + \frac{1.46}{\sqrt[3]{\phi_x}} + 5.95 \cdot \frac{\eta_y^{0.344} \cdot \eta_z^{0.086}}{\gamma^{1.557}} \\[4pt] \qquad for \; \eta_y \leq 12(1 - v^2) \wedge 0.5 < \gamma \leq 1.0 \\[8pt] 5.34 + \frac{4.00}{\gamma^2} \frac{3.85}{\sqrt[3]{\eta_y}} - \frac{1.25}{\sqrt[3]{\phi_x}} + 6.81 \cdot \frac{\eta_y^{0.395} \cdot \eta_z^{0.011}}{\gamma^{1.083}} \\[4pt] \qquad for \; \eta_y \leq 12(1 - v^2) \wedge 1.0 < \gamma \leq 2.0 \\[8pt] 24.49 + \frac{13.62}{\gamma^2} - \frac{17.01}{\sqrt[3]{\eta_y}} + \frac{56.50}{\sqrt[3]{\eta_z}} - \frac{64.40}{\sqrt[3]{\phi_x}} + 2.19 \cdot \frac{\eta_z^{0.641}}{\gamma^{1.942} \cdot \eta_y^{0.211}} \\[4pt] \qquad for \; \eta_y > 12(1 - v^2) \wedge 0.5 < \gamma \leq 1.0 \\[8pt] 17.34 + \frac{16.48}{\gamma^2} - \frac{10.70}{\sqrt[3]{\eta_y}} + \frac{25.62}{\sqrt[3]{\eta_z}} - \frac{29.04}{\sqrt[3]{\phi_x}} + 2.95 \cdot \frac{\eta_z^{0.563}}{\gamma^{1.578} \cdot \eta_y^{0.170}} \\[4pt] \qquad for \; \eta_y > 12(1 - v^2) \wedge 1.0 < \gamma \leq 2.0 \end{cases} \tag{12}$$

In the study of the buckling coefficient of diagonally stiffened plates, researchers have progressed from assuming that the stiffeners are rigid to considering the flexural rigidity and torsional rigidity. This consideration of additional factors brings the analysis results closer to reality. Through extensive finite element parameter analysis, scholars have derived formulas for calculating shear buckling coefficients. However, the loading conditions of the structure, in reality, are complex and can involve the combined action of bending, shear, and compression. Therefore, studying the buckling coefficient of stiffened plates under combined loads is an urgent problem.

2.2. Buckling Mode of Diagonally Stiffened Plate Subjected to Pure Shear

A diagonally stiffened plate with closed cross-section stiffeners, as illustrated in Figure 1, where L is the width of the plate, H is the height of the plate, t is the thickness of the plate, b is the height of the flange of closed cross-section stiffener, b_s is the width of the web of closed cross-section stiffener, and t_s is the thickness of closed cross-section stiffener.

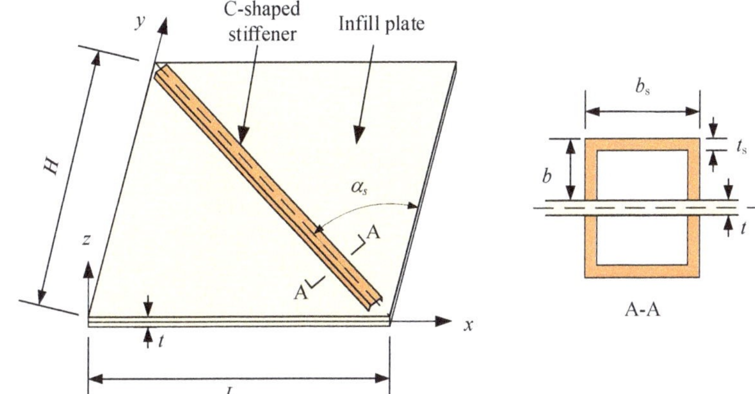

Figure 1. Diagonally stiffened plate with closed cross-section stiffeners.

The stiffener-to-plate flexural stiffness ratio η is used to quantify the relationship between the out-of-plate relative flexural rigidity of the stiffener and the plate, defined as Equation (13). The stiffener's torsional-to-flexural rigidity ratio K is defined as Equation (14).

$$\eta = \frac{E_s I_s}{D L_e}, \text{ where } L_e = H \sin \alpha_s + L \cos \alpha_s \tag{13}$$

$$K = \frac{G_s J_s}{E_s I_s} \tag{14}$$

Previous work [29] analyzed square plates with diagonal stiffeners and found the flexural rigidity of the stiffeners affects the buckling mode of the diagonally stiffened plate (Figure 2). For the compressive type, the buckling coefficient increases continuously with an increasing stiffener-to-plate flexural stiffness ratio η. When η is small, the stiffened plate buckles as a whole, and its elastic shear buckling coefficient is lower than the value $k_{cr0} = 32.9$ calculated by Yonezawa's formula Equation (8). As stiffener stiffness increases, the shear buckling coefficient gradually increases, and the buckling mode shifts to local buckling. The stiffness of the stiffeners that changes the structure from overall buckling to local buckling is called the threshold stiffness η_{Th}. Designers need to ensure that the stiffener has sufficient stiffness to avoid the overall buckling of the structure. When the stiffener stiffness exceeded η_{Th}, the buckling load of the structure continues to increase, and this contribution is provided by the stiffener. In addition, for the tensile type, the stiffener is subjected to tensile stresses and there is no stability problem, so the threshold stiffness η_{Th} is not meaningful for the tensile type diagonally stiffened plate. When the ratio η exceeds a certain value, the structure changes from overall to local buckling and the buckling load remains stable and stops increasing.

Previous research has mainly focused on the buckling behavior of diagonally stiffened plates under shear loading. However, the buckling behavior of diagonally stiffened plates under combined loading depends on several variables, such as geometrical dimensions and loading conditions, which are difficult to express by a simple formula. For this reason, a machine learning algorithm based on data-driven is used to solve this problem and reveal the potential mapping relationships between the influencing factors and the target.

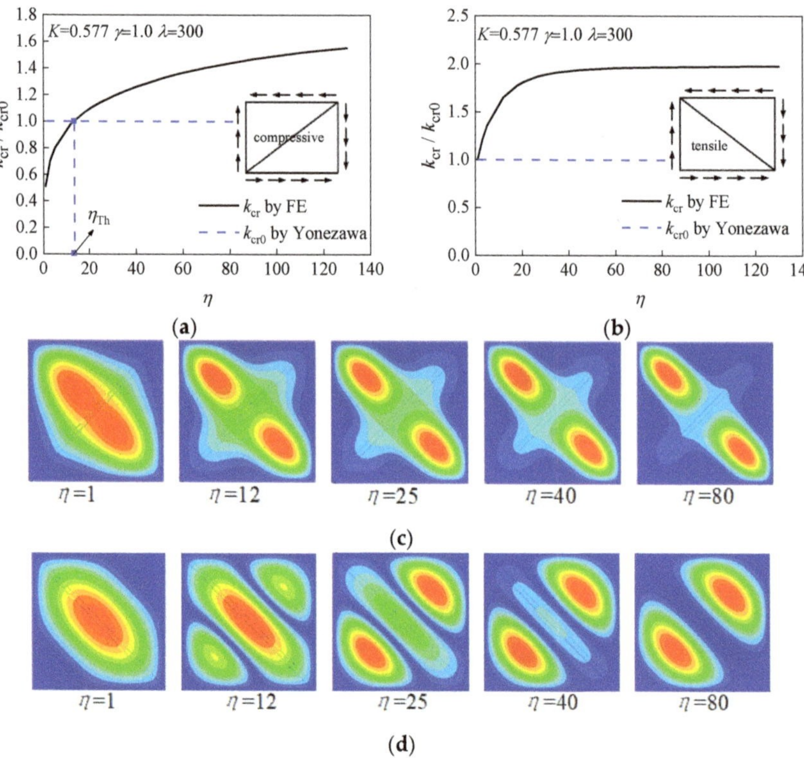

Figure 2. Variation of shear buckling coefficient and buckling mode with η for a diagonally stiffened plate [29]. (**a**) curve of the compressive type; (**b**) curve of the tensile type; (**c**) buckling mode of the compressive type; (**d**) buckling mode of the tensile type.

2.3. Machine Learning Algorithms

The machine learning architecture for predicting the elastic buckling coefficient is illustrated in Figure 3. Several machine learning algorithms were used, including decision tree (DT), random forest (RF), k-nearest neighbor (K-NN), boosting ensemble learning (AdaBoost, XGBoost, LightGBM, and CatBoost), and artificial neural network (ANN). The dataset was divided into a test set and a training set. The training set was used to train the ML models, while the test set was used to evaluate the performance. Furthermore, the SHAP method is utilized to quantify the contribution of the features and to reveal the pattern of their influence on the predictions.

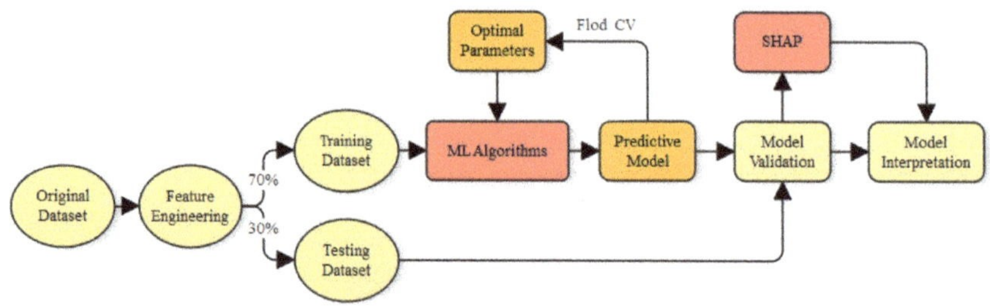

Figure 3. Machine learning architecture.

(1) Decision tree

A decision tree (DT) is a typical supervised machine learning algorithm, including the root node, decision nodes, and terminal nodes. The feature space is recursively partitioned based on a splitting attribute up until all samples correspond to the same class or there are no features that need to be split. The depth of the tree has a significant impact on the DT model's computation time and complexity. Therefore, the maximum depth is the key hyperparameter that needs to be adjusted in the DT model. Furthermore, the maximum number of leaf nodes, the minimum number of segmented samples, and other key hyperparameters are also included.

(2) K-nearest neighbor

K-nearest neighbor (K-NN) is a nonparametric machine learning algorithm [21]. It predicts output values by interpolating the output values of k nearest neighbors in the training set, and the output value can be measured by the Minkowski metric, where $p = 1$ and 2 correspond to Euclidean and Manhattan distances, respectively. Therefore, one key hyperparameter that needs to be fine-tuned in the K-NN model is the number of nearest neighbors.

(3) Artificial neural network

An artificial neural network (ANN) is a mathematical model that simulates the neural system of the human brain [30], also called Multilayer Perceptron (MLP). It consists of one input layer, one or more hidden layers, and one output layer. Multiple neurons are present in each layer. The behavior of each neuron unit is defined by the weights assigned to it. The weights of the neurons are modified by iteration over training data to minimize the error between the predicted and the actual output. The value of the transfer function then checks whether the node should transmit data to the output layer by an activation function. The hyperparameters of the ANN include the number of hidden layers, the number of neurons in each hidden layer, the type of activation function (such as sigmoid, tanh, and relu), and the learning rate. The computational accuracy of the model can be enhanced by adding more hidden layers and neurons, but the cost of computing also rises.

(4) Random forest

Random forest (RF) is a bagging method of ensemble learning. Ensemble learning can generate multiple prediction models and then combine them into a strong learner according to certain rules, which has been proven to be significantly better than single learning methods such as K-NN and DT. RF is a method, consisting of multiple DTs, which randomly selects features to construct independent trees and averages the results of all trees in the forest [31]. RF can be trained more quickly than DT and also lowers the risk of overfitting in DT.

(5) Boosting algorithm

Boosting is another ensemble learning algorithm. There are four main boosting algorithms, namely Adaptive Boosting (AdaBoost) [32], Light Gradient Boosting Machine (LightGBM) [33], Extreme Gradient Boosting (XGBoost) [34], and Categorical Boosting (CatBoost) [35]. In the initial training step of AdaBoost, all instances are weighted equally. Then, the weights of samples misclassified by the previous basic classifier are increased, while the weights of correctly classified samples are decreased and used again to train the next basic classifier. These single models are weighted together to create the final integrated model. Both LightGBM and XGBoost are based on the loss of the gradient vector function to correct DT. LightGBM performs tree splitting based on leaves, which can reduce the error by detecting key points and stopping the computation, improving accuracy and speed. The loss function of XGBoost adds a regularization term to the gradient boosting tree to control the complexity of the model and avoid overfitting. The base model used by CatBoost is an oblivious tree, which is characterized by using the same features for each layer to split and is capable of handling gradient bias and prediction bias, hence avoiding the overfitting problem and obtaining a higher accuracy.

2.4. Evaluation

Machine learning is primarily used to solve classification and regression problems. Evaluation metrics for regression focus on the difference between predicted and true values. Common evaluation metrics for regression include root mean square error (RMSE), mean absolute error (MAE), mean absolute percent error (MAPE), coefficient of determination (R^2), and so on. These metrics are listed in Table 1.

Table 1. Evaluation metrics for machine learning models.

Metrics	Formula	Equation No.	
Root mean squared error (RMSE)	$RMSE = \sqrt{\frac{1}{n}\sum_{i=1}^{n}(y_i - \hat{y}_i)^2}$	(15)	Lower is best
Mean absolute error (MAE)	$MAE = \frac{1}{n}\sum_{i=1}^{n}\lvert y_i - \hat{y}_i\rvert$	(16)	Lower is best
Mean absolute percentage error (MAPE)	$MAPE = \frac{100\%}{n}\sum_{i=1}^{n}\left\lvert\frac{y_i-\hat{y}_i}{y_i}\right\rvert$	(17)	Lower is best
Coefficient of determination (R^2)	$R^2 = 1 - \frac{\sum_{i=1}^{n}(y_i-\hat{y}_i)^2}{\sum_{i=1}^{n}(y_i-\bar{y})^2}$	(18)	Higher is best
adjusted R^2	$Adj.R^2 = 1 - (1 - R^2)\frac{n-1}{n-p-1}$	(19)	Higher is best
Performance index (PI)	$PI = \frac{1}{\bar{y}}\frac{RMSE}{\sqrt{R^2+1}}$	(20)	Lower is best
Variance account for (VAF)	$VAF = \left(1 - \frac{\mathrm{var}(y_i-\hat{y}_i)}{\mathrm{var}(y_i)}\right) \times 100$	(21)	Higher is best
Scatter index (SI)	$SI = \frac{RMSE}{\bar{y}}$	(22)	Lower is best

where, n is the number of samples, y_i is the target value, and $\hat{y}_i$ is the predicted value.

2.5. Explanation of ML Model by SHAP

Explainable ML models can help us understand the mechanisms involved in ML models and improve the credibility of the developed ML models [23,36]. SHAP is a game theoretic approach to explain the output of any machine learning model [25]. SHAP performs model explanation by calculating that each sample or feature contributes to the corresponding predicted value. The SHAP value can be written as

$$g(z') = \phi_0 + \sum_{i=1}^{M} \phi_i z_i'$$

(23)

where the value of z' is 0 or 1, 1 means the feature is the same as the value in the explanation and 0 means missing. ϕ_0 is a constant, which means the average prediction. ϕ_i is the SHAP contribution value of each feature. M is the number of input features.

3. Influencing Features

3.1. Geometric Properties

The diagonally stiffened plate's geometric parameters included the plate's width L, height H, the thickness t, and the closed cross-section stiffeners' web width b, the flange width b_s, and the thickness t_s, as shown in Figure 1. If these six geometric parameters are taken as input features, then the dimension of ML input parameters will be large. High-dimensional datasets have a negative impact on machine learning. On the other hand, during finite element analysis, several parameters are usually fixed as constants and other parameters as variables. It is also convenient to get the relationship between input and output. According to the shear buckling performance of diagonally stiffened plates, the main influence parameters could be the aspect ratio γ, the stiffener-to-plate flexural stiffness ratio η, and the stiffener's torsional-to-flexural rigidity ratio K. These three parameters are calculated by Equations (4), (13) and (14), respectively. These three ratio features are functions of six geometric parameters, which retain the characteristic information of geometric parameters and reduce the input feature's dimension.

3.2. Mechanical Characteristics

The material characteristics that affect the buckling load of the stiffened plate are the modulus of elasticity for the plate E and the stiffeners E_s, and the Poisson's ratio v. This paper takes steel as an example, with E set to 206,000 MPa and Poisson's ratio set to 0.3. The modulus of elasticity of the stiffeners E_s will change, as reflected by the stiffener-to-plate flexural stiffness ratio η according to Equation (13).

3.3. Load Properties

The combined action of bending and axial compression can be regarded as non-uniform compression action [10,28]. As illustrated in Figure 4, the stress distribution under the combined action of bending and axial compression is expressed by the non-uniform compression distribution factor ζ, as expressed by Equation (7). The factor $\zeta = 0.0$ indicates uniform compression and $\zeta = 2.0$ indicates pure bending. When the factor $0 < \zeta < 2.0$, it indicates the combined action of bending and axial compression with different proportions.

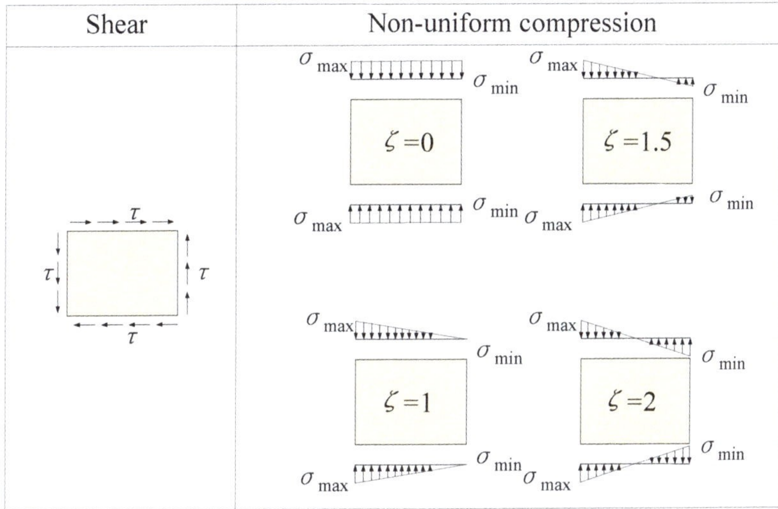

Figure 4. Shear and non-uniform compression.

It should be noted that the value range of ζ is [0.0, 2.0], which can represent any combination of bending and axial compression. However, when the stiffened plate is subjected to pure shear action, the bending and axial compression do not exist, so ζ does not exist. In order to ensure the integrity and effectiveness of the dataset in machine learning, when zeta does not exist, it is assumed to be represented by -1.0.

The relationship between the bending and the axial compression can be greatly expressed by the factor ζ. Here, a factor ψ is introduced to express the relationship between shear and axial compression, as indicated in Equation (24). The compression-to-shear ratio $\psi = -1.0$ indicates compression and $\psi = 1.0$ indicates the pure shear action. The factor $-1.0 < \psi < 1.0$ indicates the combined action of shear and compression with different proportions.

$$\psi = \frac{\tau - \sigma_{max}}{\tau + \sigma_{max}} \tag{24}$$

4. Database

4.1. FE Setting and Verification

Compared with the stiffeners arranged on the tension diagonal, stiffeners arranged on the compression diagonal have a higher buckling load and have a more significant effect on the elastic buckling coefficient [14,29]. Therefore, the subsequent study mainly focused on the stiffened plate over the compression diagonal. Two types of models were used to study the four-edge simply supported plate with compression diagonal stiffener. All of the models were built in the finite element software ABAQUS. The corners of the plate were fixed to prevent movement and the out-of-plane displacements of the four edges were constrained to simulate simple support. Shear and non-uniform compression loads were applied at the edges of the plate.

The first type of stiffened plate model (called Type-1), which is simply supported diagonally, is shown in Figure 5a. Yonezawa [12] assumed that the stiffeners were rigid and could completely restrain the out-of-plane displacement of the diagonals. When the simply supported edges are used to simulate the rigid stiffeners, the results are called critical buckling coefficient k_{cr0}. The plate is used shell element S4R, and S4R is suitable for analyzing the in-plane force of the thin plate and the out-of-plane deformation of the structure. Table 2 shows the comparison between the finite element results and Equations (1)–(9). For example, for a diagonally stiffened plate with dimensions of 3000 mm $\times$ 3000 mm $\times$ 10 mm, the shear buckling stress calculated by finite element analysis is 68 MPa and the shear buckling coefficient is 32.9. This result is consistent with the shear buckling coefficient of 32.9 calculated by Yonezawa's Equation (8). The finite element results agree with the equation results, which shows the feasibility of the FE model.

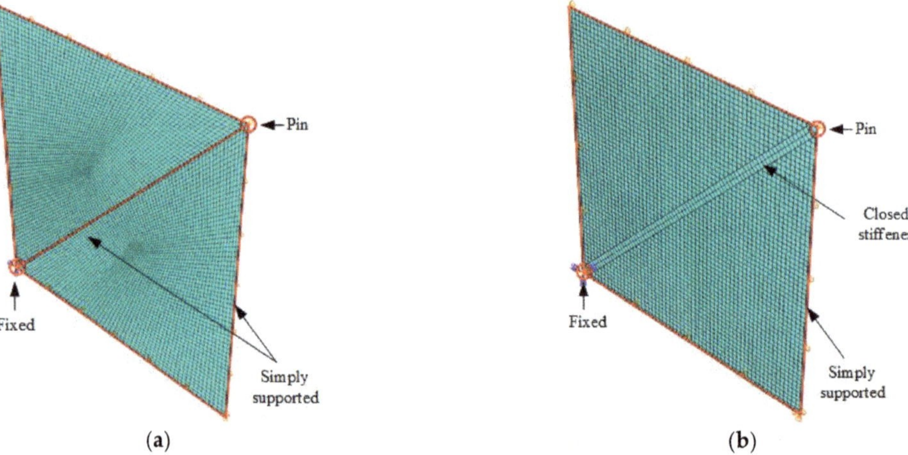

Figure 5. FE model. (**a**) Type-1 model; (**b**) Type-2 model.

Table 2. Verification of finite element results.

Force Conditions	Type	ζ	ψ	σ_{FEA} /MPa	σ_{TH} /MPa	σ_{TH}/σ_{FEA}
Shear	Unstiffened	−1.0	1.0	19.30	19.32	1.001
Axial compression	Unstiffened	0.0	−1.0	8.12	8.27	1.018
Non-uniform compression	Unstiffened	1.0	−1.0	16.05	16.12	1.004
Bending	Unstiffened	2.0	−1.0	52.83	52.90	1.001
Combined shear and non-uniform compression	Unstiffened	1.0	−1.0	15.99	16.16	1.011
Combined shear and non-uniform compression	Unstiffened	1.0	−0.33	14.37	14.43	1.004
Shear (compression)	Diagonally stiffened	−1.0	1.0	63.6	68.0	1.069
Shear (tension)	Diagonally stiffened	−1.0	1.0	23.5	23.8	1.013
Bending	Diagonally stiffened	2.0	−1.0	64.5	63.1	0.978

The second type of model (called Type-2) is shown in Figure 5b, and it has closed cross-section stiffeners that are located diagonally on the plate. The dimensions of C-shaped stiffeners are 100 mm $\times$ 75 mm $\times$ 6 mm. The stiffeners were simulated by the S4R element. It is more elaborate than the Type-1 model. In this way, the flexural rigidity and torsional rigidity of the stiffeners in relation to the buckling coefficient k_{cr} of the stiffened plate can be considered as well.

4.2. Datasets

Based on the model of Section 4.1, the finite element parameter analysis was carried out by changing the geometric size and other variables of the plate. The first type of model included 2136 sets of samples. The statistical characteristics of the dataset are displayed in Table 3. The aspect ratio γ, the non-uniform compression distribution factor ζ, and the compression-to-shear ratio ψ were the three input features of each sample. The second type of model contained 4198 sets of samples, as shown in Table 4. In addition to the above three features of the Type-1 model, the Type-2 model also includes the stiffener-to-plate flexural stiffness ratio η and the stiffener's torsional-to-flexural stiffness ratio K, which has a total of five input features. Figure 6 displays the statistical summary of the input features contained in the database of Type-1 and Type-2 models.

Table 3. Statistical characteristics of features for the Type-1 dataset.

Input Feature	Min	Max	Mean	Standard Deviation
Aspect ratio γ	0.300	4.000	2.160	1.088
Non-uniform compression distribution factor ζ	−1.000	2.000	0.960	0.749
Compression-to-shear ratio ψ	−1.000	1.000	−0.076	0.552
Elastic buckling coefficient k_{cr0}	3.422	285.257	17.766	24.839

Table 4. Statistical characteristics of features for the Type-2 dataset.

Input Feature	Min	Max	Mean	Standard Deviation
Aspect ratio γ	0.200	5.000	0.995	0.730
Non-uniform compression distribution factor ζ	−1.000	2.000	0.416	1.078
Compression-to-shear ratio ψ	−1.000	1.000	0.047	0.793
Stiffener-to-plate flexural stiffness ratio η	1.000	70.000	27.588	20.542
Stiffener's torsional-to-flexural stiffness ratio K	0.192	0.879	0.451	0.247
Elastic buckling coefficient k_{cr}	2.013	676.313	58.551	80.428

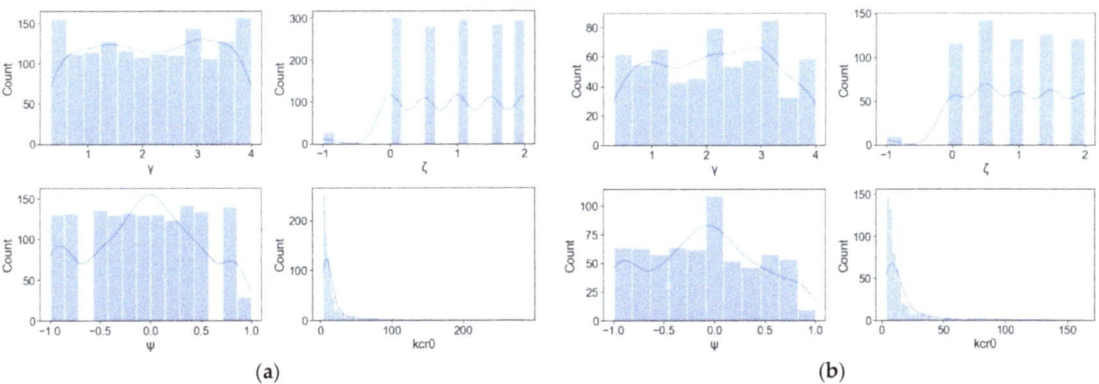

(**a**) (**b**)

Figure 6. *Cont.*

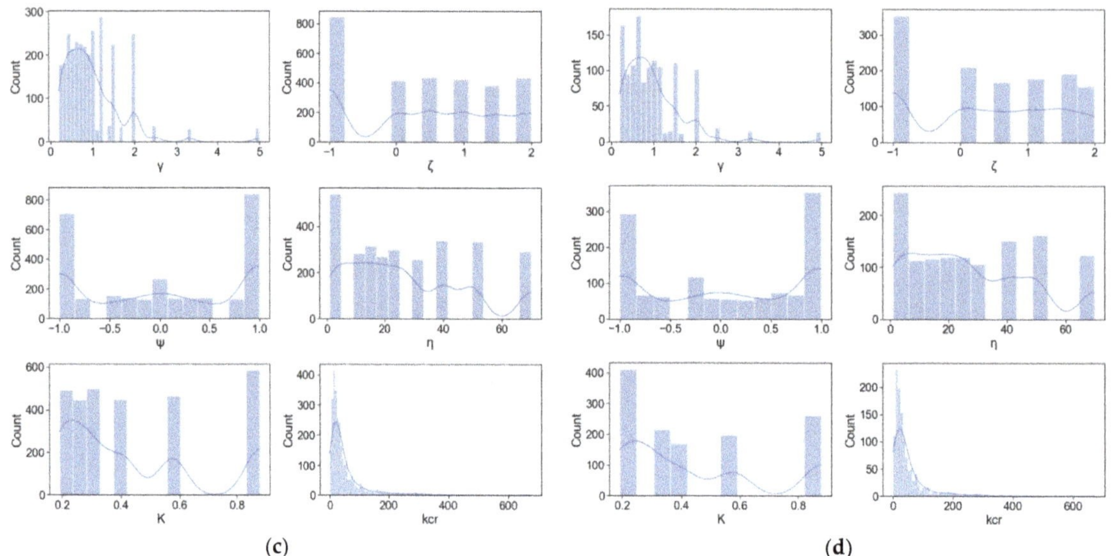

(c)   **(d)**

Figure 6. Distributions of variables of the dataset. (**a**) Distributions of variables of the Type-1 training set; (**b**) Distributions of variables of the Type-1 test set; (**c**) Distributions of variables of the Type-2 training set; (**d**) Distributions of variables of the Type-2 test set.

Figure 7 shows the correlation matrix between the features and the target. A correlation value closer to 1.0 indicates a linearly positive correlation between the two features, closer to −1.0 indicates a linear negative correlation, and closer to 0.0 indicates either no linear correlation or a non-linear correlation. For the Type-1 model in Figure 7a, the correlation among these three input features was weak, and the correlation between aspect ratio γ and target k_{cr0} was strong with a negative correlation. For the Type-2 model in Figure 7b, the correlation coefficients of ζ and ψ were −0.63, indicating a strong correlation due to a large sample of pure shear or uniform compression in the dataset, where ζ or ψ are constant values −1.0 or 1.0. The aspect ratio γ correlates with k_{cr} close to −0.5, indicating that the aspect ratio is negatively correlated with the buckling coefficient k_{cr}.

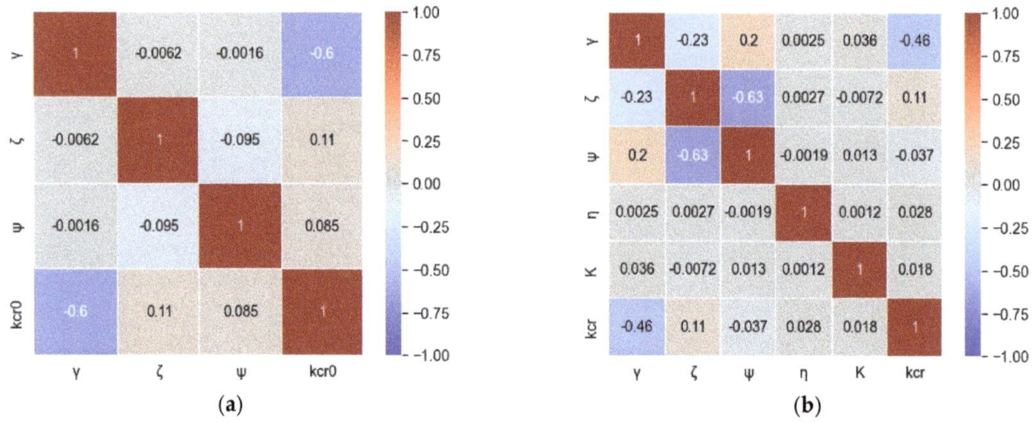

(a)   **(b)**

Figure 7. Correlation matrix heat map of features for the buckling coefficient dataset. (**a**) Correlation matrix of the Type-1 dataset; (**b**) Correlation matrix of the Type-2 dataset.

The dataset was divided into 70% training sets and 30% test sets. All the features were standardized using Equation (25) to make the scales of features uniform. Then, ML algorithms were used to train the model, and the ten-fold cross-validation method (CV_{10}) and random search hyperparameters were used to optimize the predicted model. The 10-fold cross-validation is based on the original training set and divided into new training and validation sets. The ML model was trained using multiple cycles of replacing the training and validation sets, and the results were averaged ten times. The optimal hyperparameters were selected to obtain better performance and generalization. There are two main methods for hyperparameter optimization, including grid search and random search. Among them, grid search aims to search all possible parameter ranges. It is an exhaustive search method, with search time increasing exponentially related to the number of parameters. The random search method takes random sampling in the appropriate parameter space and gradually approaches the optimal parameters.

$$x' = \frac{x - \mu}{\sigma} \tag{25}$$

5. Prediction Results from ML Models

All established ML models were implemented in Python. Various machine learning models were trained in the same training set. The developed ML models were obtained by using a 10-fold cross-validation method and random hyperparameter search. Table 5 contains the hyperparameters of each ML algorithm. Figures 8 and 9 demonstrate the ML model's predictions for Type-1 and Type-2, and compare the prediction values with the finite element results on the training and test datasets.

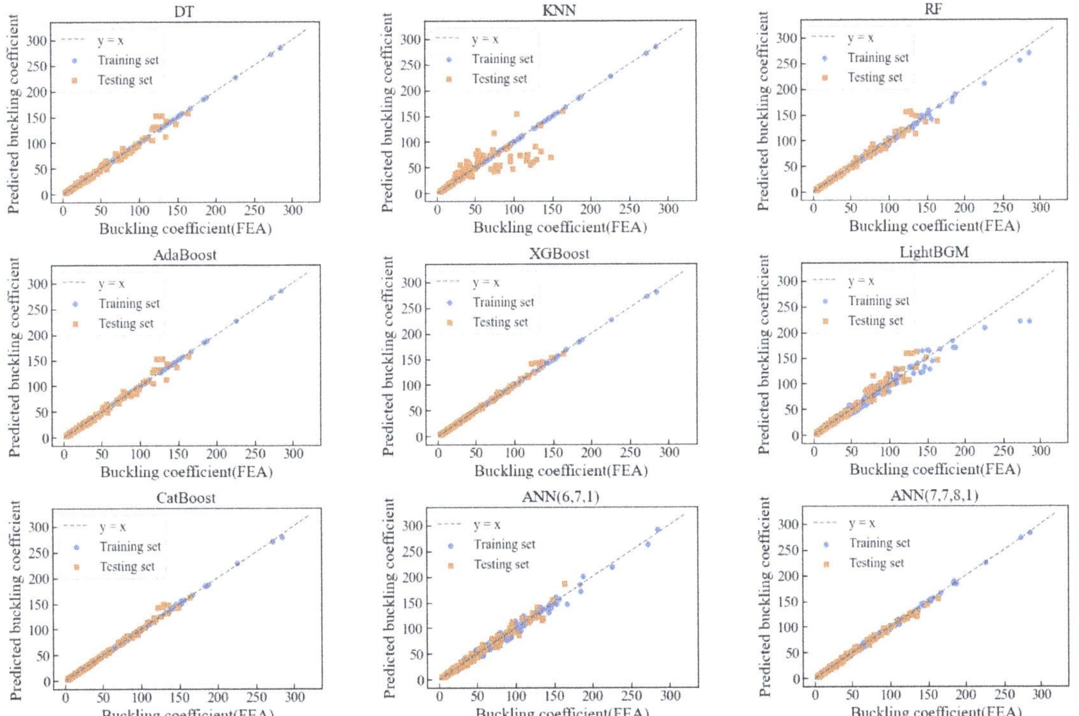

Figure 8. Performance of ML models for predicting k_{cr0}.

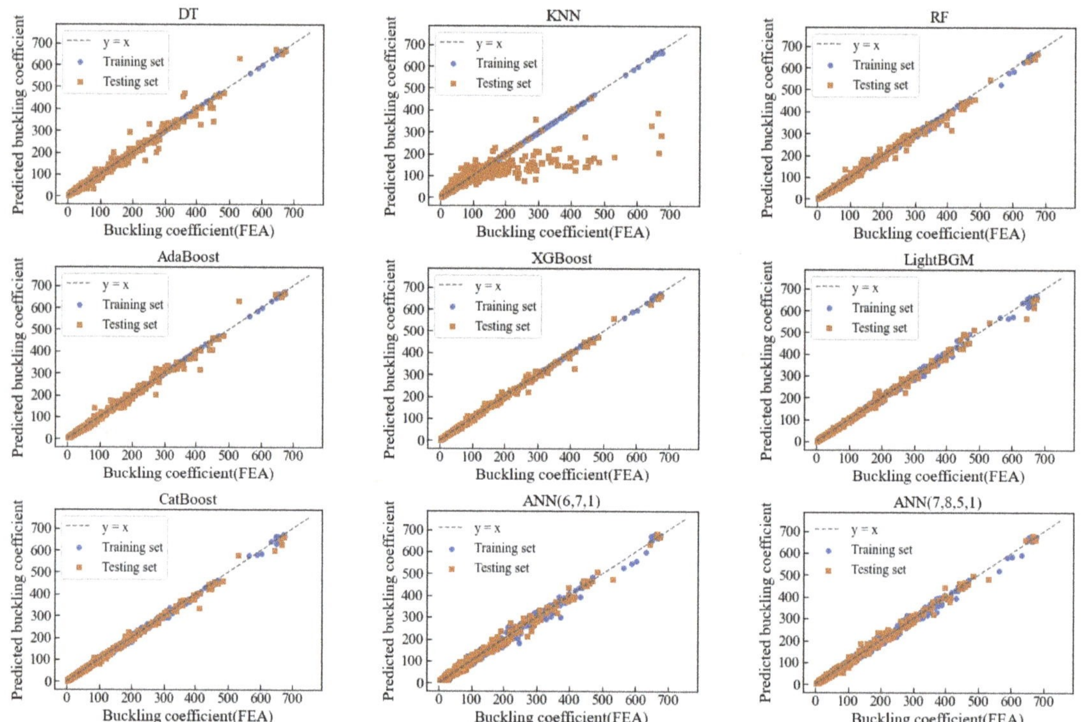

Figure 9. Performance of ML models for predicting k_{cr}.

Table 5. Hyperparameter optimization of machine learning models.

ML Algorithm	Hyperparameter	Value	
		Type-1	**Type-2**
DT	min_samples_split	2	2
	max_depth	18	18
K-NN	n_neighbors	4	16
	leaf_size	37	31
	algorithm	brute	ball_tree
	p	2	1
RF	n_estimators	136	136
	max_depth	100	100
AdaBoost	n_estimators	110	110
	learning_rate	0.6	0.6
	loss	square	square
XGBoost	n_estimators	131	131
	max_depth	7	7
	learning_rate	0.1	0.1
	min_child_weight	2	2
	subsample	0.6	0.6
LightGBM	n_estimators	167	110
	max_depth	4	6
	learning_rate	0.15	0.19
	subsample	0.352	0.226

Table 5. *Cont.*

ML Algorithm	Hyperparameter	Value	
		Type-1	**Type-2**
CatBoost	n_estimators	139	139
	max_depth	5	5
	learning_rate	0.16	0.16
	subsample	0.226	0.226
ANN	hidden_layer_sizes	6,7,1/7,7,8,1	6,7,1/7,8,5,1
	solver	lbfgs	lbfgs
	activation	relu	relu

In Figures 8 and 9, if the relationship between predicted values and actual values is close to the line y = x, and this indicates that the error between the predicted results and the actual values is small. The majority of developed ML models can produce accurate predictions for all data sets. However, when compared to other ML models, K-NN cannot predict the results accurately. K-NN performed well in the training set but poorly in the test set, suggesting that it is overfitted.

Figures 10–13 show the error frequency diagrams and error histogram distribution for each machine-learning model. Different machine learning algorithms have different distributions of target predictions. DT and KNN suffered from overfitting in the Type-1 training set and had large errors in the prediction results in the test set. ANN had larger errors in the Type-2 dataset. In contrast, integrated learning algorithms such as XGBoost had smaller and more uniform errors in both the training and test sets.

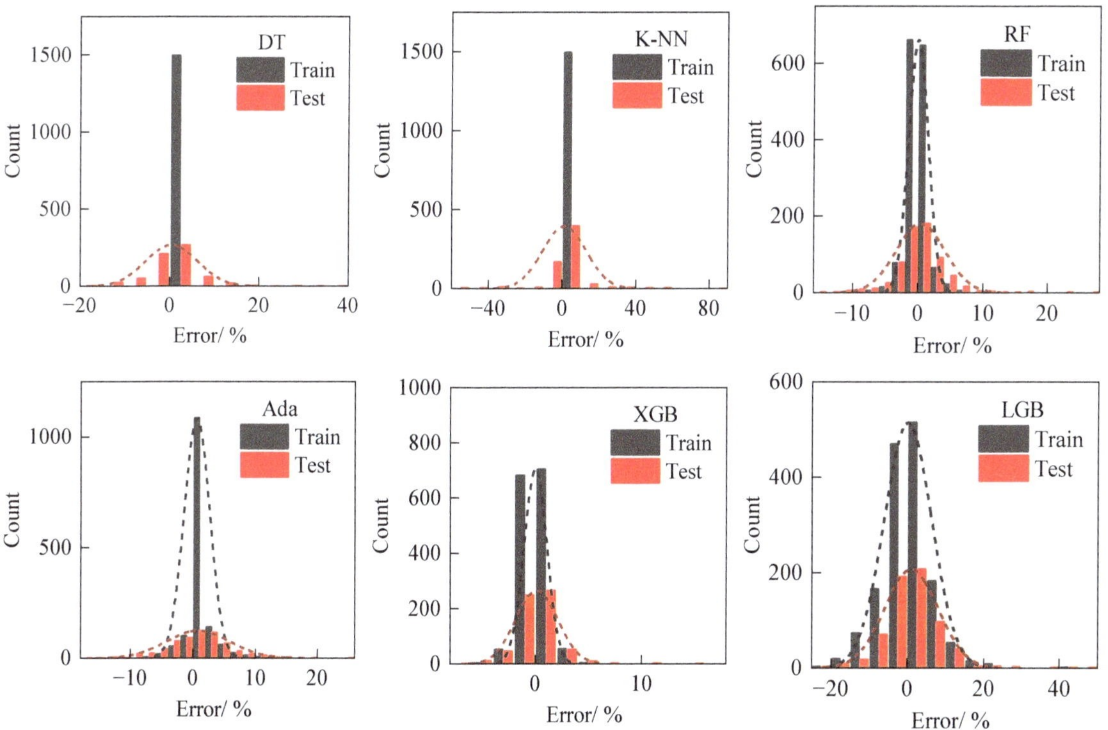

Figure 10. *Cont.*

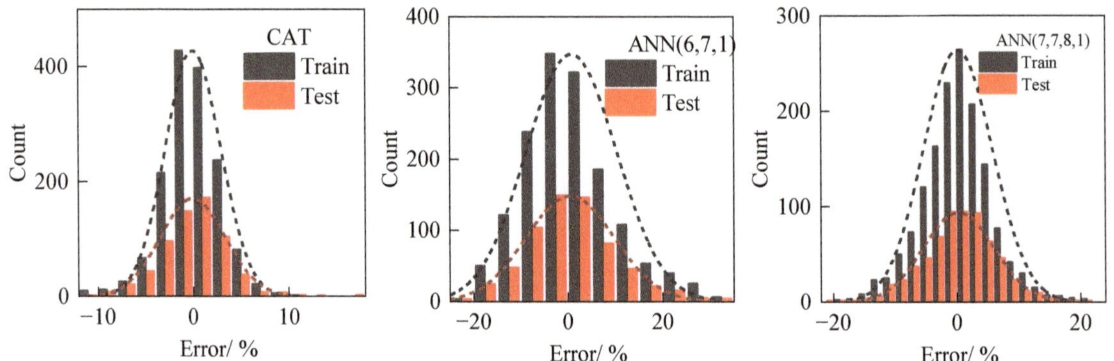

Figure 10. Error histogram distribution for each ML model for k_{cr0}.

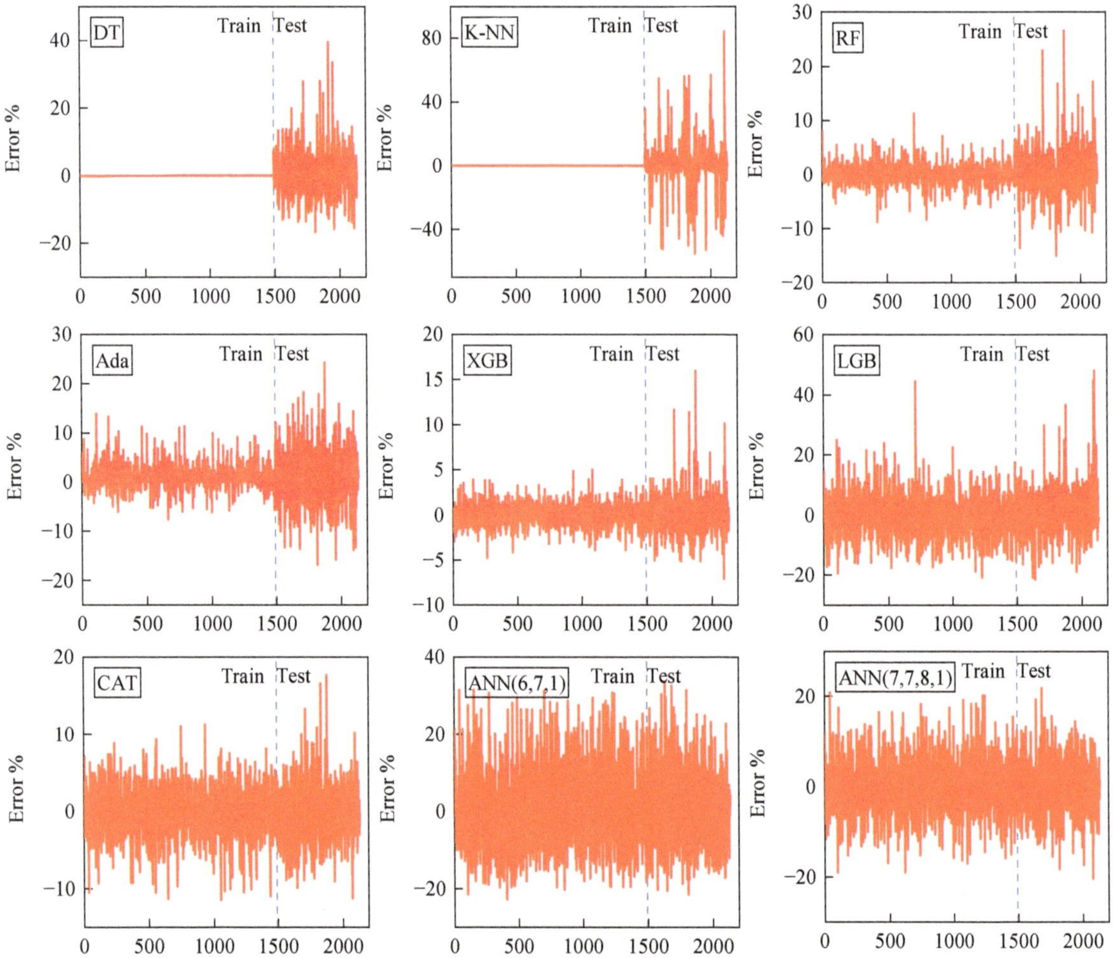

Figure 11. Frequency diagram for each ML model for k_{cr0}.

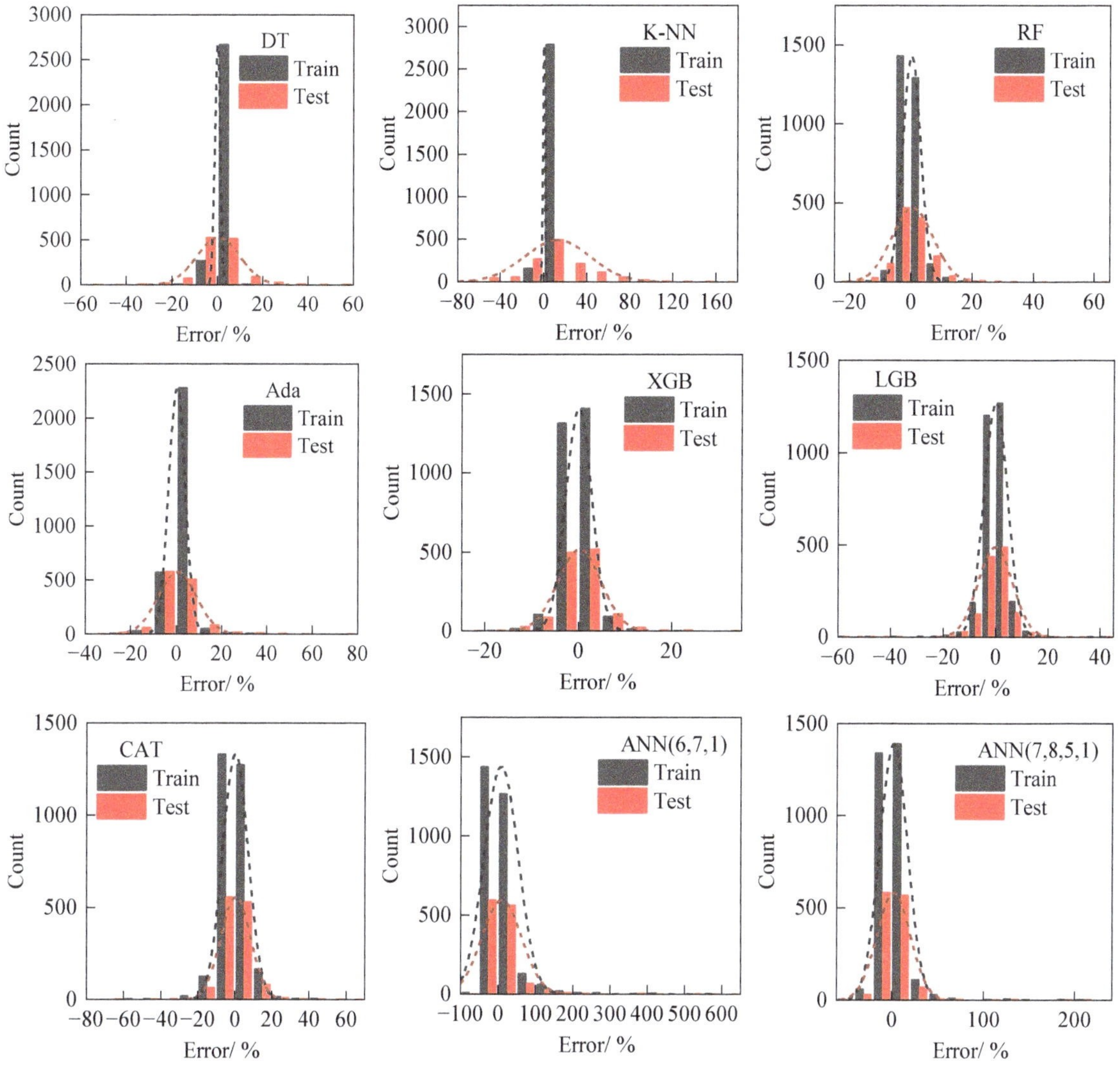

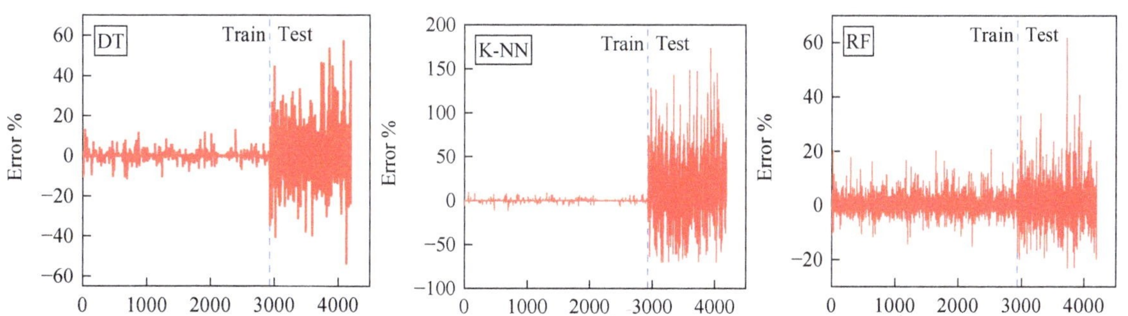

Figure 12. Error histogram distribution for each ML model for k_{cr}.

Figure 13. *Cont.*

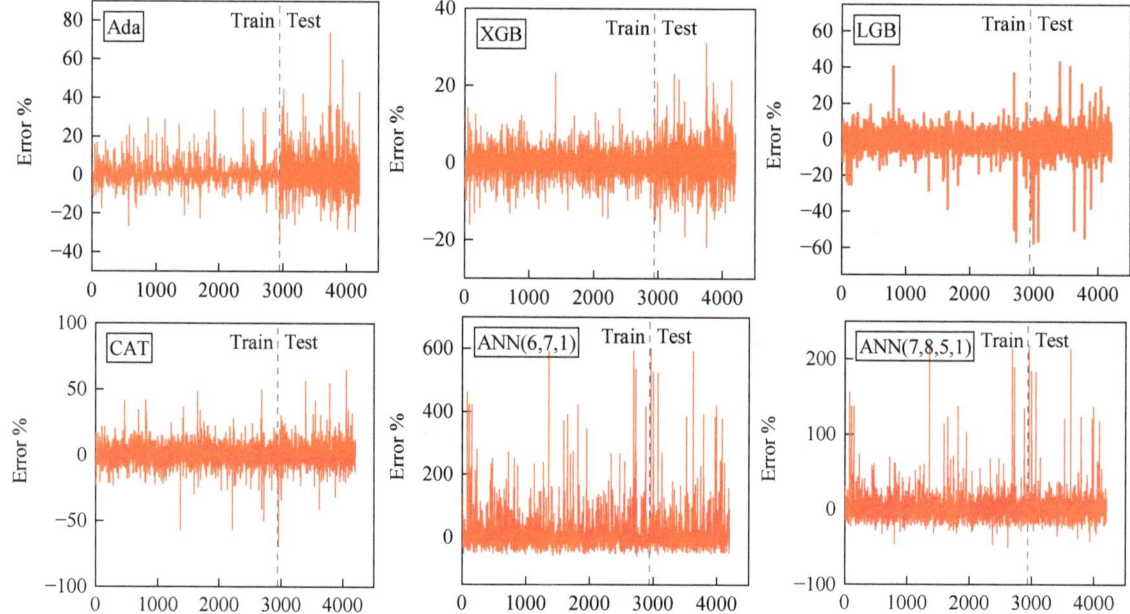

Figure 13. Frequency diagram for each ML model for k_{cr}.

Figure 14 illustrates the U_{95} for the different models, a parameter of the uncertainty analysis, indicating a 95 percent confidence level [21]. In Figure 14a, DT and K-NN model's $U_{95\text{-train}} = 0$, but $U_{95\text{-test}} = 5.88$ and $U_{95\text{-test}} = 20.05$, respectively. The U_{95} of the other models ranged from 0 to 10, among which the $U_{95\text{-train}} = 0.46$ of AdaBoost and $U_{95\text{-test}} = 2.86$ of XGBoost are the lowest. In Figure 14b, $U_{95\text{-test}} = 108.98$ of the K-NN model is significantly larger than others. The U_{95} of the other models was under 20, among which the $U_{95\text{-train}} = 2.55$ of AdaBoost and $U_{95\text{-test}} = 11.18$ of XGBoost were the lowest.

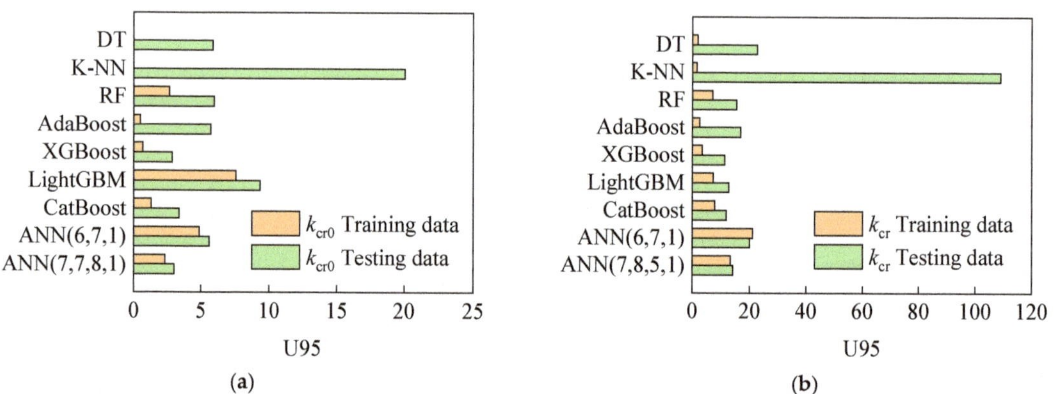

Figure 14. The U_{95} of machine learning models for k_{cr0} and k_{cr}. (**a**) U95 of models for the Type-1 dataset; (**b**) U95 of models for the Type-2 dataset.

Tables 6 and 7 provide evaluation metrics for different machine learning models on the training set and test set. For k_{cr0} in Table 6, most of the ML models showed very high accuracy, with a CV_{10} greater than 0.97. However, the performance of K-NN was worse than other models, and its CV_{10} was only 0.817. XGBoost and ANN (with hidden_layer_sizes 7,7,8,1) were the two models with the best performance metrics and generalization, with

CV_{10} of 0.997 and 0.996 on the training set, respectively. Especially, in the predicted values of the XGBoost model, RMSE was 1.323, MAE was 0.343, MAPE was 0.013%, and the adjusted R^2 was 0.997.

Table 6. Performance metrics of ML models for k_{cr0}.

ML Algorithms	RMSE (Rank)	MAE (Rank)	MAPE/% (Rank)	R^2 (Rank)	Adj. R^2 (Rank)	PI (Rank)	VAF (Rank)	SI (Rank)	CV_{10} (Rank)	Total Rank
				Training data						
DT	0.000 (1)	0.000 (1)	0.000 (1)	1.000 (1)	1.000 (1)	0.000 (1)	100.000 (1)	0.000 (1)	0.976 (7)	15
K-NN	0.000 (1)	0.000 (1)	0.000 (1)	1.000 (1)	1.000 (1)	0.000 (1)	100.000 (1)	0.000 (1)	0.817 (9)	17
RF	1.104 (7)	0.315 (6)	0.010 (4)	0.998 (7)	0.998 (7)	0.043 (7)	99.816 (7)	0.061 (7)	0.989 (4)	56
AdaBoost	0.196 (3)	0.097 (3)	0.012 (5)	1.000 (1)	1.000 (1)	0.007 (3)	99.995 (3)	0.010 (3)	0.982 (6)	28
XGBoost	0.269 (4)	0.122 (4)	0.008 (3)	1.000 (1)	1.000 (1)	0.011 (4)	99.989 (4)	0.015 (4)	0.997 (1)	26
LightGBM	3.183 (9)	0.980 (8)	0.044 (8)	0.985 (9)	0.985 (9)	0.130 (9)	98.368 (9)	0.182 (9)	0.971 (8)	78
CatBoost	0.520 (5)	0.302 (5)	0.021 (6)	1.000 (1)	1.000 (1)	0.020 (5)	99.959 (5)	0.029 (5)	0.997 (1)	34
ANN (6,7,1)	2.026 (8)	1.051 (9)	0.069 (9)	0.994 (8)	0.994 (8)	0.080 (8)	99.377 (8)	0.112 (8)	0.987 (5)	71
ANN (7,7,8,1)	0.955 (6)	0.602 (7)	0.043 (7)	0.999 (6)	0.999 (6)	0.038 (6)	99.862 (6)	0.053 (6)	0.996 (3)	53
				Testing data						
DT	2.733 (6)	1.001 (6)	0.040 (4)	0.986 (5)	0.985 (6)	0.110 (6)	98.631 (6)	0.155 (6)	–	45
K-NN	9.272 (9)	2.584 (9)	0.055 (8)	0.834 (9)	0.833 (9)	0.414 (9)	83.419 (9)	0.540 (9)	–	71
RF	2.757 (7)	0.803 (4)	0.026 (3)	0.985 (7)	0.985 (6)	0.114 (7)	98.529 (7)	0.160 (7)	–	48
AdaBoost	2.665 (5)	0.951 (5)	0.041 (5)	0.986 (5)	0.986 (5)	0.109 (5)	98.671 (5)	0.153 (5)	–	40
XGBoost	1.323 (1)	0.343 (1)	0.013 (1)	0.997 (1)	0.997 (1)	0.055 (1)	99.664 (1)	0.077 (1)	–	8
LightGBM	4.332 (8)	1.197 (8)	0.048 (7)	0.964 (8)	0.964 (8)	0.149 (8)	97.540 (8)	0.208 (8)	–	63
CatBoost	1.573 (3)	0.468 (2)	0.024 (2)	0.995 (3)	0.995 (3)	0.065 (3)	99.525 (3)	0.092 (3)	–	22
ANN (6,7,1)	2.603 (4)	1.176 (7)	0.067 (9)	0.987 (4)	0.987 (4)	0.108 (4)	98.689 (4)	0.151 (4)	–	40
ANN (7,7,8,1)	1.394 (2)	0.725 (3)	0.045 (6)	0.996 (2)	0.996 (2)	0.057 (2)	99.624 (2)	0.081 (2)	–	21

Table 7. Performance metrics of ML models for k_{cr}.

ML Algorithm	RMSE (Rank)	MAE (Rank)	MAPE/% (Rank)	R^2 (Rank)	Adj. R^2 (Rank)	PI (Rank)	VAF (Rank)	SI (Rank)	CV_{10} (Rank)	Total Rank
				Training data						
DT	0.656 (2)	0.141 (2)	0.003 (2)	1.000 (1)	1.000 (1)	0.008 (2)	99.993 (2)	0.011 (2)	0.984 (8)	22
K-NN	0.531 (1)	0.089 (1)	0.002 (1)	1.000 (1)	1.000 (1)	0.006 (1)	99.995 (1)	0.009 (1)	0.721 (9)	17
RF	2.580 (5)	1.103 (5)	0.019 (5)	0.999 (5)	0.999 (5)	0.031 (5)	99.893 (5)	0.044 (5)	0.993 (5)	45
AdaBoost	0.922 (3)	0.369 (3)	0.016 (3)	1.000 (1)	1.000 (1)	0.011 (3)	99.987 (3)	0.015 (3)	0.991 (6)	26
XGBoost	1.204 (4)	0.722 (4)	0.018 (4)	1.000 (1)	1.000 (1)	0.015 (4)	99.977 (4)	0.021 (4)	0.998 (1)	27
LightGBM	2.730 (6)	1.524 (6)	0.035 (6)	0.999 (5)	0.999 (5)	0.039 (7)	99.839 (7)	0.055 (7)	0.996 (3)	52
CatBoost	2.797 (7)	1.776 (7)	0.044 (7)	0.999 (5)	0.999 (5)	0.035 (6)	99.869 (6)	0.049 (6)	0.997 (2)	51
ANN (6,7,1)	7.636 (9)	5.33 (9)	0.204 (9)	0.991 (9)	0.991 (9)	0.093 (9)	99.061 (9)	0.132 (9)	0.990 (7)	79
ANN (7,8,5,1)	4.753 (8)	3.007 (8)	0.087 (8)	0.996 (8)	0.996 (8)	0.058 (8)	99.636 (8)	0.082 (8)	0.994 (4)	68
				Testing data						
DT	9.422 (8)	3.668 (7)	0.061 (6)	0.987 (8)	0.987 (8)	0.111 (8)	98.757 (8)	0.157 (8)	–	61
K-NN	46.564 (9)	16.779 (9)	0.224 (9)	0.693 (9)	0.692 (9)	0.64 (9)	69.783 (9)	0.779 (9)	–	72
RF	6.536 (5)	2.714 (4)	0.044 (2)	0.994 (5)	0.994 (5)	0.077 (5)	99.401 (5)	0.109 (5)	–	36
AdaBoost	7.164 (6)	2.880 (5)	0.055 (5)	0.993 (6)	0.993 (6)	0.086 (6)	99.263 (6)	0.121 (6)	–	46
XGBoost	4.780 (1)	1.866 (1)	0.031 (1)	0.997 (1)	0.997 (1)	0.056 (1)	99.679 (1)	0.080 (1)	–	8
LightGBM	5.508 (3)	2.273 (2)	0.044 (2)	0.996 (2)	0.996 (2)	0.068 (3)	99.53 (3)	0.096 (3)	–	20
CatBoost	4.992 (2)	2.359 (3)	0.050 (4)	0.996 (2)	0.996 (2)	0.059 (2)	99.646 (2)	0.084 (2)	–	19
ANN (6,7,1)	8.530 (7)	5.791 (8)	0.222 (8)	0.990 (7)	0.99 (7)	0.101 (7)	98.971 (7)	0.143 (7)	–	58
ANN (7,8,5,1)	5.997 (4)	3.459 (6)	0.097 (7)	0.995 (4)	0.995 (4)	0.071 (4)	99.491 (4)	0.100 (4)	–	37

For k_{cr} in Table 7, K-NN was still the worst-performing model, while other ML models had a CV_{10} of more than 0.99. The XGBoost model also showed the best performance, with a score of 0.998 for CV_{10} and 0.997 for adjusted R^2. The performance of different ML algorithms varied on different datasets. The k-NN algorithm depends more on data and is more sensitive to feature information than other algorithms because it mainly depends on a limited number of surrounding samples. Additionally, the imbalance of samples can have a significant effect on the performance of K-NN. In the K-NN regression model, the mean method or weighted mean method is mainly used. For example, if the dataset was divided into two datasets greater than 1.0 and less than 1.0 according to the feature of aspect ratio, the accuracy of K-NN on these two datasets improved significantly, and the score of CV_{10} was 0.91, but it was still lower than other ML models.

The performance metrics of the ensemble learning models with Bagging (such as RF) and Boosting (such as AdaBoost, LightGBM, XGBoost, and CatBoost) were better than that of single learners (such as DT and ANN). This is also the advantage of comprehensive learning. When base learners or weak learners are combined correctly, it is possible to get a more accurate and robust learner. On the other hand, in terms of the sample size of the dataset, the total sample sizes of type 1 and type 2 models were 2136 and 4198, respectively. The performance metrics of data sets with larger sample sizes are usually better, which indicates that a large amount of appropriate sample information is conducive to the higher prediction accuracy of the ML model. As can be seen from the ranking, XGBoost had the highest combined ranking (lower is best) in the training and testing sets of Type-1 and Type-2. In particular, the XGBoost model showed the best generalization performance in the test set.

The Taylor diagrams are shown in Figure 15. Where the scatter points represent the different machine learning models, the horizontal and vertical axes represent the standard deviations, the radial lines represent the correlation coefficients, and the green dashed lines represent the root mean square differences. For the training set of Type-1 and Type-2, all ML models were more concentrated and showed good accuracy with correlation coefficients ranging from 0.99 to 1.0. Except for K-NN, the other models also had good performance in the test set. Although K-NN had very high prediction accuracy in the training set, it had poor performance in the test set, and the generalization ability of this model was significantly different from the other models.

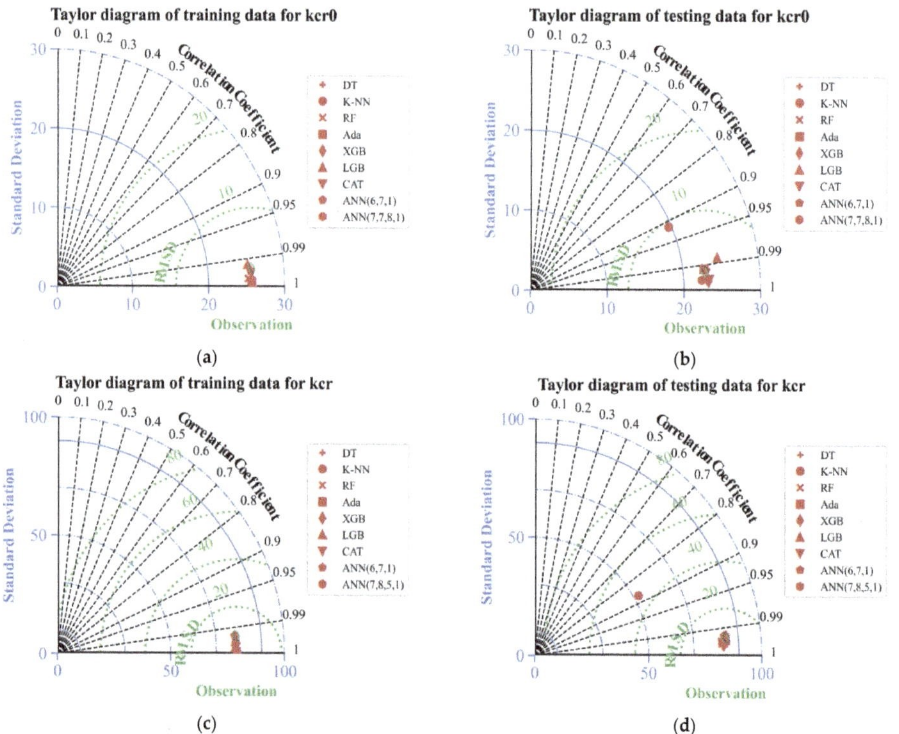

Figure 15. Taylor diagram of training and testing data for k_{cr0} and k_{cr}. (**a**) Taylor diagram of training data for k_{cr0}; (**b**) Taylor diagram of testing data for k_{cr0}; (**c**) Taylor diagram of training data for k_{cr}; (**d**) Taylor diagram of testing data for k_{cr}.

6. ML Model Explanations

Machine learning has high accuracy and precision in regression and classification tasks. However, the machine learning process is often difficult to explain and does not provide a prediction mechanism or reason, which is referred to as a "black box". This makes it difficult for researchers to analyze how the features of a certain sample affect the final output. If a relationship between features and predicted values can be established, it would improve the ML model's credibility.

The SHAP method, derived from coalitional game theory, is a way to allocate expenses to participants based on players' contributions to overall expenditures. SHAP is an additive interpretation model inspired by the Shapley value. The SHAP value has the advantage of not only reflecting the influence of the features in each sample but also showing the positive and negative influence of the features. When used in machine learning, the SHAP value represents each feature's contribution to the predicted value.

According to the performance evaluation results of developed machine learning in Section 5, XGBoost is the ML model with the best R^2 and CV_{10} among the eight machine learning algorithms and showed the best learning and prediction ability on the data sets of the two types of models. Based on the results of the XGBoost model, this section uses the SHAP method to explain the correlation between input features and output values.

6.1. Global Explanations

Figure 16 illustrates the SHAP value of global feature importance, where the importance of each feature is regarded as the mean absolute value of that feature on all given samples. Among the three features in Figure 16a, the effect of aspect ratio γ on the critical buckling coefficient of the diagonally stiffened plate was the most important, which was expected. Its SHAP value was significantly higher than the non-uniform compression distribution factor ζ and the compression-to-shear ratio ψ. The Type-2 model has two additional features than the Type-1 model, and the two features are flexural rigidity and torsional rigidity of stiffeners. Among these five features of the Type-2 model, the aspect ratio γ is still the most important feature affecting the critical buckling coefficient of the structure. Its SHAP value was significantly higher than that of the other four features. The stiffener's torsional-to-flexural stiffness ratio K was the lowest SHAP importance value among the five features.

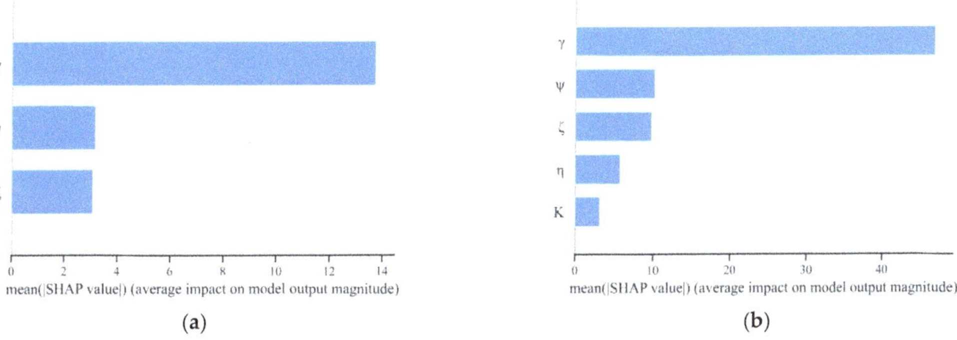

Figure 16. SHAP feature importance. (**a**) Type-1 model; (**b**) Type-2 model.

Each point on the SHAP summary plots represents the SHAP value of the sample in Figure 17. The color of each dot displayed from low (blue) to high (red) represents the corresponding feature value from small to large. The SHAP value represents the positive and negative correlation between the feature and the output. For example, a low γ in blue has a high positive SHAP value in Figure 17a, whereas a high γ in red has a negative SHAP value. The trend of ζ is opposite to that of γ, indicating that axial compression will reduce the buckling coefficient and bending will increase the buckling coefficient. In

Figure 17b, the higher the η and K values are, the greater the positive SHAP values. The increased stiffness of the stiffener will indeed increase the buckling load of the diagonally stiffened plate.

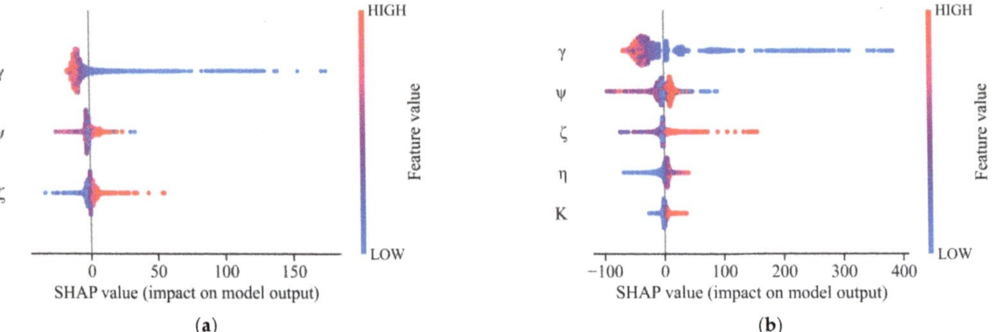

Figure 17. Distribution of Shap values for XGBoost models. (**a**) Type-1 model; (**b**) Type-2 model.

6.2. Feature Dependence

Figures 18 and 19 are the SHAP dependence plots. These figures make it easy to see the interaction of the two features on the SHAP values. The primary feature's values are given on the X-axis and the corresponding SHAP values are displayed on the left Y-axis. In addition, the interaction between the primary feature and the secondary feature is displayed in color on the right Y-axis. The blue dot denotes the low secondary feature value and the red dot denotes the high feature value. It is worth noting that the SHAP values of the samples with the same feature value on the X-axis are not the same. The reason is that this feature interacts with other features. Moreover, the feature dependence plot just explains how the ML model works, not necessarily how its reality works.

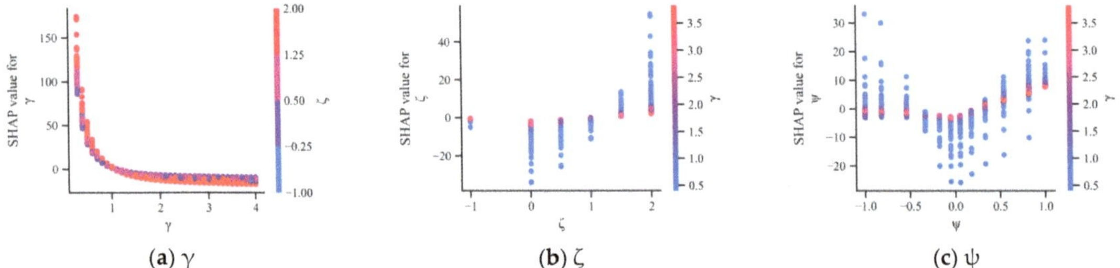

Figure 18. Feature dependence for the Type-1 model.

In Figure 18a, there is a negative correlation between γ and the SHAP value. The SHAP value was positive when γ less than 1.0, and the smaller the γ, the higher the SHAP value, which means that the positive contribution is greater. On the contrary, the SHAP value was negative when γ exceeded 1.0. The SHAP value decreased with the increase in γ, indicating that the negative contribution is larger. Increasing the aspect ratio will reduce the buckling coefficient of the diagonally stiffened plate.

It was assumed that the value of ζ is -1.0 when it does not exist in Section 3.3. Therefore, in the case of excluding $\zeta = -1.0$, the relationship between ζ and the SHAP value is a positive correlation in Figure 18b. It means that the SHAP value increases with the increase of ζ. It can be seen that the SHAP value was lower than 0 when ζ is less than 1.0 indicating a negative contribution. In other words, when the uniform compressive stress was greater than the bending, the buckling coefficient of the stiffened plate decreased. This means that axial compression is disadvantageous to the buckling coefficient, while bending is beneficial to the buckling coefficient.

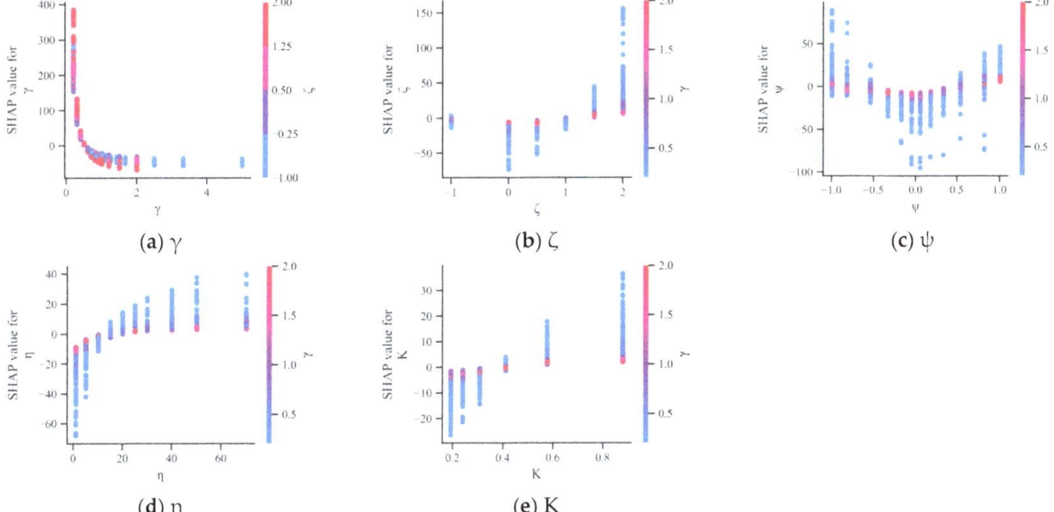

Figure 19. Feature dependence for the Type-2 model.

In Figure 18c, the relationship between ψ and the SHAP value is U-shaped. The SHAP value decreased at first and then increased with an increase of ψ. It changed from positive values to negative values when ψ exceeded -0.5, and it had a negative maximum contribution when ψ was 0.0. Moreover, the color of the right Y-axis indicates that the SHAP value corresponding to ψ is low when the aspect ratio γ is high values (red), and has a large positive or negative SHAP value when the aspect ratio γ is low values (blue).

The first three features' (γ, ζ, and ψ) dependence graphs in Figure 19 are the same as those in Figure 18. In addition, the trend of the feature dependence graphs of η was the same as that of K, and their SHAP value increased with the increase of feature values. When η was greater than 20, its SHAP value was positive, as illustrated in Figure 19d. Similarly, the SHAP value of K was positive when K was greater than 0.4, in Figure 19e. These two features show the contribution of flexural rigidity and torsional rigidity of stiffeners to the buckling coefficient. To improve the buckling coefficient of a compression diagonally stiffened plate under the combined action of compression–bending–shear, it is recommended that K not be lower than 0.4, and η not be lower than 20.

6.3. Individual Explanation

Figure 20 displays the explanation of individual sample prediction. Two samples are used as examples: one from Type-1 and the other from Type-2. Figure 20a shows a sample with the input features $\gamma = 1.0$, $\zeta = -1.0$, and $\psi = 1.0$. Figure 20b shows a sample with the input features $\gamma = 0.6$, $\zeta = 0.0$, $\psi = 0.818$, $\eta = 40.0$, and $K = 0.24$. Features with red arrows will enhance prediction value, whereas blue arrows will decrease prediction value. The base value is the mean prediction value obtained from the ML model on the training set. If there is no information available for the input features, the base value is the output value. For the first sample in Figure 20a, the base value was 18.17 and the predicted buckling coefficient was 30.65. Feature $\psi = 1.0$ had the most impact on increasing the buckling coefficient with a SHAP value of 10.40, followed by $\gamma = 1.0$ with a SHAP value of 2.96. However, $\zeta = -1.0$ had a negative contribution with a SHAP value of -0.88. For this sample, the final prediction result k_{cr0} was derived from the base value and the sum of the SHAP values of the three input features. For the second sample in Figure 20b, the basic value was 58.40, $\psi = 0.818$, and $\eta = 40.0$ increased by 13.73 and 6.00, respectively. While $\gamma = 0.6$, $\zeta = 0.0$, and $K = 0.24$ decreased -8.56, -7.59, and -4.51, respectively. The predicted buckling coefficient k_{cr} of this sample is 57.47.

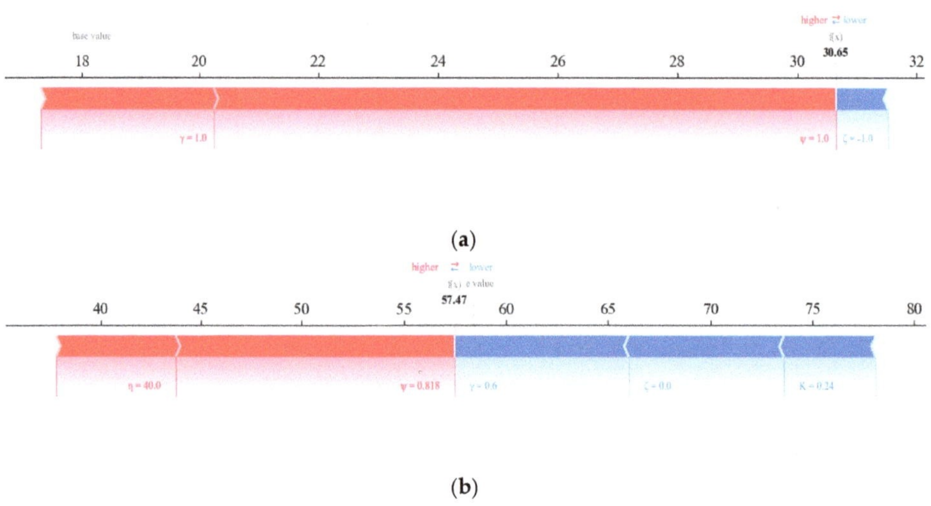

(a)

(b)

Figure 20. Individual explanation by SHAP. (**a**) one simple Type-1 model; (**b**) one simple Type-2 model.

7. Conclusions

Under the combined compression, bending, and shear loads, the buckling mechanism of diagonally stiffened plates is complex. To solve the problem of calculating the elastic buckling coefficient of the diagonally stiffened plate, data-driven machine-learning models were proposed. With and without considering the flexure and torsional rigidity of stiffeners, prediction models for predicting the diagonally stiffened plates' elastic buckling coefficient under combined compression–bending–shear action were developed by using the ML algorithms in Python, including the decision tree (DT), random forest (RF), k-nearest neighbor (K-NN), artificial neural network (ANN), adaptive boosting (AdaBoost), light gradient boosting machine (LightGBM), extreme gradient boosting (XGBoost), and categorical boosting (CatBoost). The main conclusions are as follows:

(1) The ML models were developed using a 10-fold cross-validation method and random search for hyperparameters selection. These models demonstrated excellent accuracy on both the training and test sets. In particular, an ensemble learning algorithm (bagging and boosting) performed significantly better than a single learner (DT). ANN also performed well after selecting an appropriate hidden layer and the number of neurons. However, due to its algorithmic characteristics, the performance of the K-NN algorithm on the test set was poorer than that of other machine learning algorithms, resulting in overfitting. Among all models, XGBoost was found to be the best, with CV_{10} as high as 0.996 and 0.997 on two different datasets;

(2) The XGBoost model was further analyzed using the SHAP method to explain the correlation between the features and the prediction target. The effects of the main features on the buckling coefficient of diagonally stiffened plates under combined action were ranked as follows: the plate's aspect ratio γ, the compression-to-shear ratio ψ, the non-uniform compression distribution factor ζ, the stiffener-to-plate flexural stiffness ratio η and the stiffener's torsional-to-flexural stiffness ratio K. The correlation between these features and the prediction results was found to be inconsistent. There was a negative correlation between γ and the buckling coefficient, indicating that as γ increases, the buckling coefficient decreases. In contrast, ζ, η, and K were positively correlated with the buckling coefficient, meaning that increasing these values can increase the buckling coefficient. The correlation between ψ and the buckling coefficient was found to be U-shaped;

(3) Furthermore, when the plate's aspect ratio γ is small, the other features can reach the maximum and minimum SHAP values under the interaction of γ. This indicates that

their positive or negative contribution to the buckling coefficient of the diagonally stiffened plate is the largest. However, when γ was larger, the SHAP values of other features did not obviously change, meaning that their contribution to the buckling coefficient is not obvious;

(4) The flexural and torsional rigidity of stiffeners have a certain influence on the buckling coefficient of diagonally stiffened plates. Increasing these values can improve the buckling coefficient. Based on machine learning and SHAP interpretation analysis, suggested values for the flexural and torsional rigidity of stiffeners can be provided. It is recommended that the stiffener-to-plate flexural stiffness ratio η be no less than 20 and that the stiffener's torsional-to-flexural stiffness ratio K be no less than 0.4. This will be beneficial for improving the buckling capacity of diagonally stiffened plates.

Author Contributions: Conceptualization, Y.Y.; data curation, Y.Y.; formal analysis, Y.Y.; funding acquisition, Y.Y. and X.G.; methodology, Y.Y. and X.G.; project administration, Y.Y. and Z.M.; software, Y.Y.; supervision, Z.M.; writing—original draft, Y.Y.; writing—review editing, Z.M. and X.G. All authors have read and agreed to the published version of the manuscript.

Funding: This research was funded by FUNDAMENTAL RESEARCH FUNDS FOR THE CENTRAL UNIVERSITIES, grant number FRF-TP-22-117A1; NATURAL SCIENCE FOUNDATION OF HEBEI PROVINCE, grant number E2021202111.

Data Availability Statement: All the data supporting the results were provided within the article.

Conflicts of Interest: The authors declare no conflict of interest.

References

1. Lima, J.P.S.; Cunha, M.L.; dos Santos, E.D.; Rocha, L.A.O.; Real, M.D.V.; Isoldi, L.A. Constructal Design for the ultimate buckling stress improvement of stiffened plates submitted to uniaxial compressive load. *Eng. Struct.* **2020**, *203*, 109883. [CrossRef]
2. Zhou, W.; Li, Y.; Shi, Z.; Lin, J. An analytical solution for elastic buckling analysis of stiffened panel subjected to pure bending. *Int. J. Mech. Sci.* **2019**, *161–162*, 105024. [CrossRef]
3. Deng, J.; Wang, X.; Yuan, Z.; Zhou, G. Novel quadrature element formulation for simultaneous local and global buckling analysis of eccentrically stiffened plates. *Aerosp. Sci. Technol.* **2019**, *87*, 154–166. [CrossRef]
4. Kimura, Y.; Fujak, S.M.; Suzuki, A. Elastic local buckling strength of I-beam cantilevers subjected to bending moment and shear force based on flange–web interaction. *Thin-Walled Struct.* **2021**, *162*, 107633. [CrossRef]
5. Biscaya, A.; Pedro, J.J.O.; Kuhlmann, U. Experimental behaviour of longitudinally stiffened steel plate girders under combined bending, shear and compression. *Eng. Struct.* **2021**, *238*, 112139. [CrossRef]
6. Chen, X.; Real, E.; Yuan, H.; Du, X. Design of welded stainless steel I-shaped members subjected to shear. *Thin-Walled Struct.* **2020**, *146*, 106465. [CrossRef]
7. Jáger, B.; Kövesdi, B.; Dunai, L. Bending and shear buckling interaction behaviour of I-girders with longitudinally stiffened webs. *J. Constr. Steel Res.* **2018**, *145*, 504–517. [CrossRef]
8. Prato, A.; Al-Saymaree, M.; Featherston, C.; Kennedy, D. Buckling and post-buckling of thin-walled stiffened panels: Modelling imperfections and joints. *Thin-Walled Struct.* **2022**, *172*, 108938. [CrossRef]
9. Wu, Y.; Fan, S.; Zhou, H.; Guo, Y.; Wu, Q. Cyclic behaviour of diagonally stiffened stainless steel plate shear walls with two-side connections: Experiment, simulation and design. *Eng. Struct.* **2022**, *268*, 114756. [CrossRef]
10. Timoshenko, S.P.; Gere, J.M. *Theory of Elastic Stability*, 2nd ed.; McGraw-Hill Book Company, Inc.: New York, NY, USA, 1961.
11. Mikami, I.; Matsushita, S.; Nakahara, H.; Yonezawa, H. Buckling of plate girder webs with diagonal stiffener. *Proc. Jpn. Soc. Civ. Eng.* **1971**, *1971*, 45–54. [CrossRef]
12. Yonezawa, H.; Mikami, I.; Dogaki, M.; Uno, H. Shear strength of plate girders with diagonally stiffened webs. *Proc. Jpn. Soc. Civ. Eng.* **1978**, *269*, 17–27. [CrossRef]
13. Yuan, H.; Chen, X.; Theofanous, M.; Wu, Y.; Cao, T.; Du, X. Shear behaviour and design of diagonally stiffened stainless steel plate girders. *J. Constr. Steel Res.* **2019**, *153*, 588–602. [CrossRef]
14. Martins, J.; Cardoso, H. Elastic shear buckling coefficients for diagonally stiffened webs. *Thin-Walled Struct.* **2022**, *171*, 108657. [CrossRef]
15. Zhang, H.; Cheng, X.; Li, Y.; Du, X. Prediction of failure modes, strength, and deformation capacity of RC shear walls through machine learning. *J. Build. Eng.* **2022**, *50*, 104145. [CrossRef]
16. Bin Kabir, A.; Hasan, A.S.; Billah, A.M. Failure mode identification of column base plate connection using data-driven machine learning techniques. *Eng. Struct.* **2021**, *240*, 112389. [CrossRef]
17. Thai, H.-T. Machine learning for structural engineering: A state-of-the-art review. *Structures* **2022**, *38*, 448–491. [CrossRef]
18. Truong, G.T.; Hwang, H.-J.; Kim, C.-S. Assessment of punching shear strength of FRP-RC slab-column connections using machine learning algorithms. *Eng. Struct.* **2022**, *255*, 113898. [CrossRef]

19. Vu, Q.-V.; Truong, V.-H.; Thai, H.-T. Machine learning-based prediction of CFST columns using gradient tree boosting algorithm. *Compos. Struct.* **2020**, *259*, 113505. [CrossRef]
20. Degtyarev, V.V.; Tsavdaridis, K.D. Buckling and ultimate load prediction models for perforated steel beams using machine learning algorithms. *J. Build. Eng.* **2022**, *51*, 104316. [CrossRef]
21. Shi, X.; Yu, X.; Esmaeili-Falak, M. Improved arithmetic optimization algorithm and its application to carbon fiber reinforced polymer-steel bond strength estimation. *Compos. Struct.* **2023**, *306*, 116599. [CrossRef]
22. Zhu, Y.R.; Huang, L.H.; Zhang, Z.J.; Behzad, B. Estimation of splitting tensile strength of modified recycled aggregate concrete using hybrid algorithms. *Steel Compos. Struct.* **2022**, *44*, 389–406.
23. Esmaeili-Falak, M.; Benemaran, R.S. Ensemble deep learning-based models to predict the resilient modulus of modified base materials subjected to wet-dry cycles. *Geomech. Eng.* **2023**, *32*, 583–600.
24. Wang, C.; Song, L.-H.; Fan, J.-S. End-to-End Structural analysis in civil engineering based on deep learning. *Autom. Constr.* **2022**, *138*, 104255. [CrossRef]
25. Lundberg, S.; Lee, S. A unified approach to interpreting model predictions. In Proceedings of the Advances in Neural Information Processing Systems, Long Beach, CA, USA, 4–9 December 2017; pp. 4765–4774.
26. Mangalathu, S.; Hwang, S.-H.; Jeon, J.-S. Failure mode and effects analysis of RC members based on machine-learning-based SHapley Additive exPlanations (SHAP) approach. *Eng. Struct.* **2020**, *219*, 110927. [CrossRef]
27. Bleich, F. *Buckling Strength of Metal Structures*; Engineering Societies Monographs; MacGraw-Hill: New York, NY, USA, 1952.
28. Tong, G. *Out-of-Plane Stability of Steel Structures*; Architecture & Building Press: Beijing, China, 2007. (In Chinese)
29. Yang, Y.; Mu, Z.; Zhu, B. Numerical Study on Elastic Buckling Behavior of Diagonally Stiffened Steel Plate Walls under Combined Shear and Non-Uniform Compression. *Metals* **2022**, *12*, 600. [CrossRef]
30. Adeli, H. Neural Networks in Civil Engineering: 1989–2000. *Comput.-Aided Civ. Infrastruct. Eng.* **2001**, *16*, 126–142. [CrossRef]
31. Breiman, L. Random forests. *Mach. Learn.* **2001**, *45*, 5–32. [CrossRef]
32. Freund, Y.; Schapire, R.E. A Decision-Theoretic Generalization of On-Line Learning and an Application to Boosting. *J. Comput. Syst. Sci.* **1997**, *55*, 119–139. [CrossRef]
33. Ke, G.; Meng, Q.; Finley, T.; Wang, T.; Chen, W.; Ma, W.; Ye, Q.; Liu, T.-Y. LightGBM: A Highly Efficient Gradient Boosting Decision Tree. In Proceedings of the Advances in Neural Information Processing Systems, Long Beach, CA, USA, 4–9 December 2017; pp. 3149–3157.
34. Chen, T.; Guestrin, C. XGBoost: A Scalable Tree Boosting System. In Proceedings of the 22nd ACM SIGKDD International Conference on Knowledge Discovery and Data Mining, San Francisco, CA, USA, 13–17 August 2016; pp. 785–794.
35. Prokhorenkova, L.; Gusev, G.; Vorobev, A.; Dorogush, A.V.; Gulin, A. CatBoost: Unbiased boosting with categorical features. In Proceedings of the 32nd International Conference on Neural Information Processing Systems, Montréal, QC, Canada, 3–8 December 2018; pp. 6639–6649.
36. Feng, D.-C.; Wang, W.-J.; Mangalathu, S.; Taciroglu, E. Interpretable XGBoost-SHAP Machine-Learning Model for Shear Strength Prediction of Squat RC Walls. *J. Struct. Eng.* **2021**, *147*, 4021173. [CrossRef]

sustainability

Article

The Multi-Scale Model Method for U-Ribs Temperature-Induced Stress Analysis in Long-Span Cable-Stayed Bridges through Monitoring Data

Fengqi Zhu [1,2,3], **Yinquan Yu** [1], **Panjie Li** [3,4] **and Jian Zhang** [3,*]

[1] China Institute of Building Standard Design & Research Co., Ltd., Beijing 100048, China; fengqizhu_cbs@outlook.com (F.Z.)
[2] Department of Civil Engineering, Tsinghua University, Beijing 100084, China
[3] School of Civil Engineering, Southeast University, Nanjing 210096, China
[4] School of Civil Engineering, Zhengzhou University, Zhengzhou 450001, China
[*] Correspondence: jian@seu.edu.cn

Abstract: Temperature is one of the important factors that affect the fatigue failure of the welds in orthotropic steel desks (OSD) between U-ribs and bridge decks. In this study, a new analysis method for temperature-induced stress in U-ribs is proposed based on multi-scale finite element (FE) models and monitoring data First, the long-term temperature data of a long-span cable-stayed bridge is processed. This research reveals that a vertical temperature gradient is observed rather than a transverse temperature gradient on the long-span steel box girder bridge with tuyere components. There is a linear relationship between temperature and temperature-induced displacement, taking into account the time delay effect (approximately one hour). Then, a multi-scale FE model is established using the substructure method to condense each segment of the steel girder into a super-element, and the overall bridge temperature-induced displacement and temperature-induced stress of the local U-rib on the OSD are analyzed. The agreement between the calculated temperature-induced stresses and measured values demonstrates the effectiveness of the multi-scale modeling strategy. This approach provides a valuable reference for the evaluation and management of bridge safety. Finally, based on the multi-scale FE model, the temperature-induced strain distribution of components on the OSD is studied. This research reveals that the deflection of the girder continually changes with the temperature variation, and the temperature-induced strain of the girder exhibits a variation range of approximately 100 $\mu\varepsilon$.

Keywords: long-term monitoring data; temperature-induced stress; multi-scale model; substructure; state evaluation

Citation: Zhu, F.; Yu, Y.; Li, P.; Zhang, J. The Multi-Scale Model Method for U-Ribs Temperature-Induced Stress Analysis in Long-Span Cable-Stayed Bridges through Monitoring Data. *Sustainability* **2023**, *15*, 9149. https://doi.org/10.3390/su15129149

Academic Editor: Erika Sabella

Received: 31 March 2023
Revised: 29 May 2023
Accepted: 31 May 2023
Published: 6 June 2023

1. Introduction

Orthotropic steel desks (OSD) are widely used in long-span steel box girder bridges due to their lightweight and excellent bearing capacity [1]. The evaluation of OSD's fatigue performance, considering the coupling effect between vehicle loads and environmental temperature effects, has become an important research topic [2,3]. The asphalt concrete pavement on the OSD, as a temperature-sensitive material, gradually reduces its elastic modulus and the overall stiffness of the bridge deck as the temperature rises. Consequently, the reduced stiffness of the road will inevitably weaken the top plate ability to constrain out-of-plane deformation on the OSD [4]. Researchers have observed that numerous bridges, especially box girder bridges, are subject to temperature-induced stress due to the nonlinear temperature gradient of the structure caused by solar radiation, atmospheric temperature and other external factors [5–10]. However, there are limited studies on the temperature effect evaluation of U-ribs on the OSD at present. Therefore, it is necessary to analyze temperature-induced stress and evaluate the performance of the OSD on long-span steel bridges.

Research on bridge temperature-induced responses has a long history, dating back to the 1960s. Zuk [11] studied the effect of ambient temperature on concrete bridge structures. Over the years, numerous of these influencing factors (e.g., atmospheric temperature and solar radiation) have been extensively investigated to determine their impact on temperature-induced stress. Structural health monitoring (SHM) systems can conduct real-time monitoring and condition assessment through sensors [12–15], unmanned aerial vehicles (UAVs) [16], computer vision [17], mobile robots [18], and other techniques. The first-hand field temperature and strain data sampling by the SHM system is employed to obtain the structural performance under temperature load. Therefore, the SHM system can be implemented for the safety evaluation of bridge structures based on temperature and temperature-induced displacement and stress [8,9,19,20]. However, there are many components of bridges that cannot be directly monitored due to the limited number of sensors in the SHM system [21]. As a result, it is necessary to establish a finite element (FE) model to evaluate the structural performance of long-span steel bridges under temperature loads. Xia et al. [22,23] developed the FE model for structural damage identification using temperature-induced strain data. Yu et al. [24,25] proposed a digital twin-based structure health hybrid monitoring, which included a multi-scale FE model to analyze the impact of pavement temperature vibration on the OSD. Research has shown that the effect of temperature on the fatigue performance of the OSD is mainly manifested in two aspects: the vehicle load dispersion effect and the composite stiffness effect between asphalt concrete pavement and bridge decks. In order to achieve more accurate response predictions and performance evaluations, FE model updating can be carried out on large-span bridges [26,27].

The U-ribs are widely applied and are crucial components of the OSD on long-span steel bridges. However, the in-depth investigation of U-ribs is proved to be difficult. Especially in the conventional FE model method, bridge girders are commonly simplified using beam elements based on the principle of equivalent cross-section, which can not be analyzed for the temperature-induced stress of U-ribs. If a global fine FE model is used, the element size of the FE model should be determined by U-ribs. However, the global fine FE model requires significant computational resources, and high degrees of freedom (DOFs) hinder an effective dynamic analysis. In order to solve the above problems, a multi-scale FE model can be applied to provide both global and local information for the assessment of bridge safety performance [28–32]. Existing multi-scale FE models include the sub-model, the interface constraint method, and the substructure method.

Chen et al. [33] used the sub-model method to analyze the stress of bridge deck pavements and considered the effect of vehicle-bridge coupling. The segment model was analyzed separately as a sub-model, and its boundary was simulated using the finite mixed element method. Yu et al. [24] also used a multi-scale FE model based on the sub-model method to analyze the temperature-induced stress of U-ribs. Identifying the boundary conditions in the sub-model analysis is key. Compared to traditional FE analysis, the quadratic (sub-model) analysis demonstrated improved feasibility in solving the linear problem, while errors accumulated for the nonlinear problem, may distort the results [34].

A local model is built into the global model through the interface constraint method to directly address the distortion issues occurring in the sub-model method. In the interface constraint method, key areas can be simulated by shell or solid elements, while other areas can be simulated by beam or truss elements. The beam elements are connected to the shell/solid elements using constraint equations or constraining elements. The constraint equation or the constraint element is employed to standardize the DOFs of the global model boundary, which are equal to the DOFs of the local model boundary [35]. Nie et al. [36] used the constraint equation method to build a multi-scale model of a cable anchorage system for a suspension bridge. The results demonstrated an improved reliability compared to the traditional modeling method. Based on the multi-point constraint element MPC184 in ANSYS, Wang et al. [37] established a multi-scale FE model for a cable-stayed steel bridge. Despite its success, the interface constraint method is limited by the number of general local models. If all local models need to be concerned, the multi-scale FE model can have similar DOFs to the global fine FE model, which reduces the potential advantage.

The aforementioned issue can be solved by implementing the substructure method. A group of elements is condensed into a super-element based on the linear response in the substructure method [38]. Li et al. [39] established a substructure model considering U-ribs for the mid-span segment on Runyang Bridge. The model can be used for the global dynamic analysis and hot spot stress analysis of local welds. Zhu et al. [40] built a multi-scale FE model using the substructure method to simulate the concrete box and steel box girders in a cable-stayed bridge in ANSYS. The substructures were connected by the main DOF coupling. Comparing the other two modeling methods, the substructure method has significant advantages when applied to structures with many repetitive geometries. For such geometries, the super-element is generated and copied to different locations. This advantage enables the substructure method to effectively solve the large DOFs problem of the long-span bridge. Connected by standard segments, the girder of long-span cable-stayed bridges contains a large number of repetitive geometric structures. Therefore, in this paper, a multi-scale FE model based on the substructure method is used to analyze the temperature-induced stress of U-ribs on a long-span cable-stayed bridge.

When evaluating the structural performance of the U-ribs on the Sutong Yangtze River Bridge (STB) under the influence of temperature, the length of the girder on the STB is 2088 m while the size of a U-rib on the girder is 300~400 mm, resulting in a scale difference of approximately 5000 times. If the global fine FE method is used to analyze the temperature-induced strain of the U-ribs, the model would involve a large number of DOFs, leading to significant hardware resource consumption, long computation time and even infeasibility of calculation. Therefore, the multi-scale FE model method is used to address these challenges. It is the first time a multi-scale FE modeling method based on substructure technology is used to analyze the temperature-induced stress of U-ribs on long-span bridges through monitoring data.

The framework of U-ribs state evaluation through the multi-scale FE model and monitoring data are shown in Figure 1. It can be seen from Figure 1 that the global response temperature-induced displacement and the local response temperature-induced stress can be obtained by the SHM system and multi-scale FE model. The rest of the paper is organized as follows. The long-term monitoring data of STB are analyzed in Section 2. In Section 3, the substructure method is used to establish the multi-scale FE model and the thermal field and temperature-induced effects are analyzed and discussed. Section 4 provides the concluding remarks of the paper.

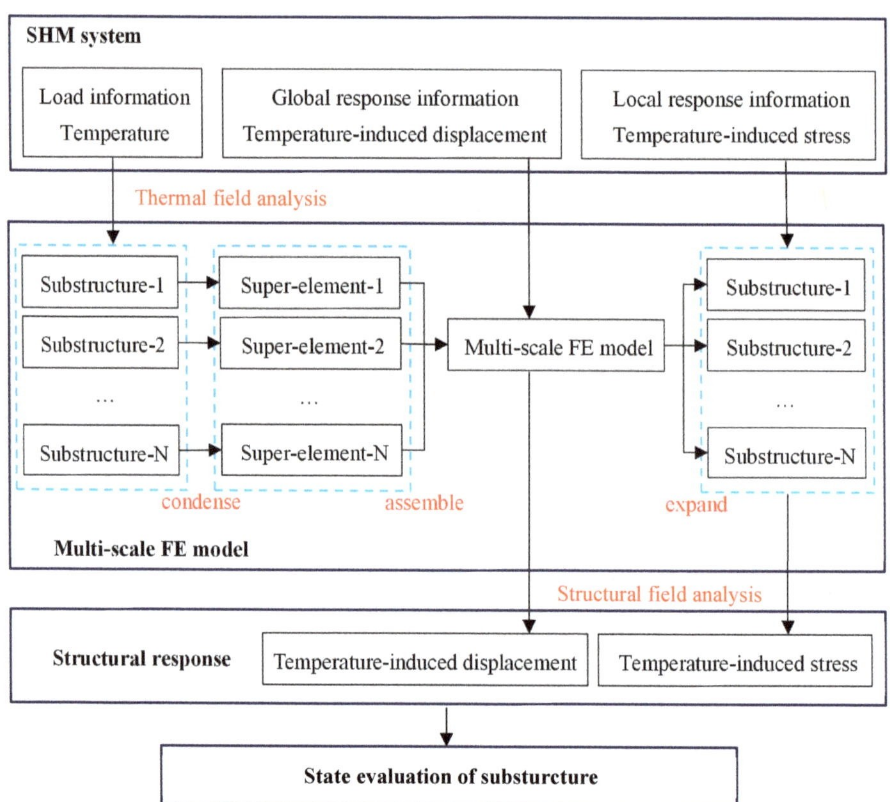

Figure 1. Framework of U-ribs state evaluation through multi-scale FE model and monitoring data.

2. STB Monitoring Data and Statistical Analysis

2.1. STB and Its SHM System

The STB is located downstream of the Yangtze River of China and connects the cities of Nantong and Suzhou. It is a double towers cable-stayed bridge with three piers arranged on the side spans (Figure 2a). Initially, the STB held the record for the world's largest cable-stayed bridge with a span of 1088 m, and the total length of the girder is 2088 m. The STB's steel box girder is connected to different components, including the top plate, bottom plate, diaphragms, longitudinal plates (or longitudinal trusses), top plate U-ribs, bottom plate U-ribs, tuyeres (Figure 2b). The STB towers and piers are constructed using reinforced concrete structures. The bridge is equipped with a total of 272 stay cables. Figure 2c illustrates the general layout of thermometers, displacement gauges, strain gauges and cable accelerometers of the STB SHM system. Structural thermometers are installed in the mid-span section of the girder (section 5-5). Displacement gauges are placed at the end of the girder (sections 1-1 and 9-9) to measure the longitudinal deformation of expansion joints. Strain gauges are installed in three sections of the girder: the mid-span section, the 1/4 section of the main span (section 4-4), and the girder section fitted at the north tower (section 3-3).

The schematic diagram in Table 1 depicts the sensor layout and SHM data for a day of the mid-span section (section 5-5). The red and blue lines represent the test results of the temperature sensors (T-5-#) and strain sensors (S-5-#), respectively.

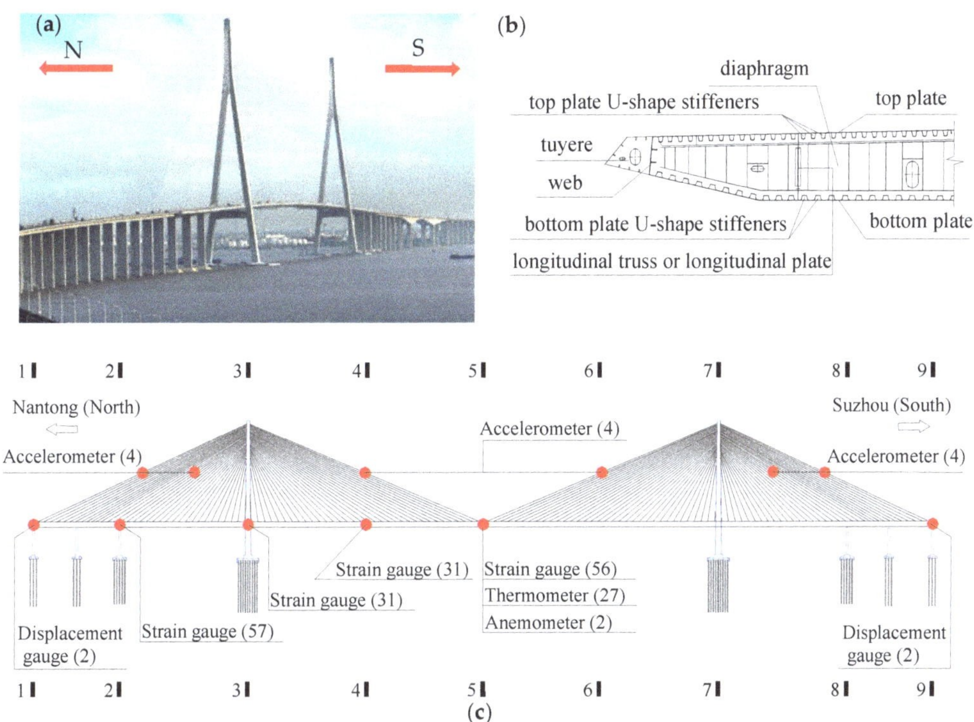

Figure 2. The cable-stayed bridge and its SHM system. (**a**) bridge image (**b**) components of steel girder (**c**) sensor layout.

Table 1. Temperature and strain time histories for a day of the mid-span section.

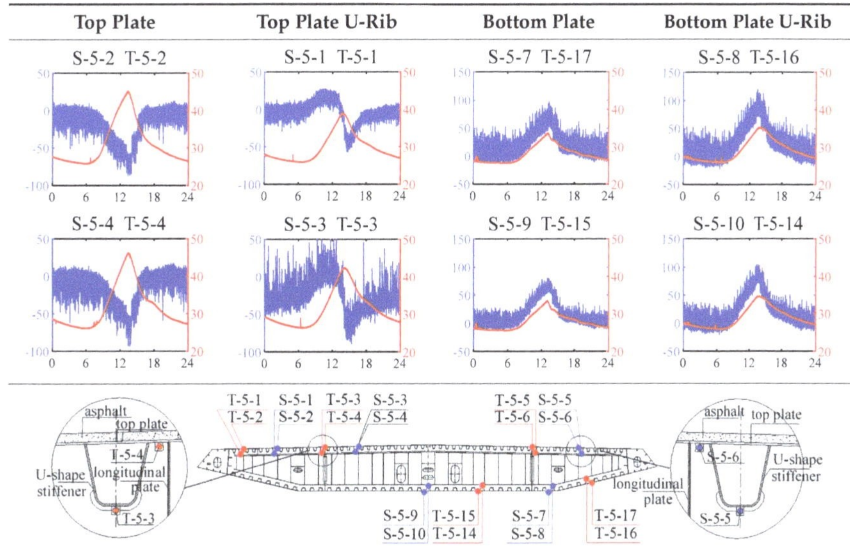

In the data graph, X label is hour, blue Y label is strain (unit: microstrain), and red Y label is temperature (unit: centigrade).

2.2. Monitoring Data and Statistical Analysis of the Temperature and Structure Response

The steel box girder of the investigated bridge is longer in the longitudinal direction compared to the lateral and vertical directions. Additionally, the longitudinal curvature is small, and the cross-section along the longitudinal direction changes uniformly. Therefore, it is assumed that the thermal field along the longitudinal direction of the bridge is consistently exposed to solar radiation and atmospheric temperature. The lateral and vertical temperature distribution of the STB steel box girder is focused. Previous studies have found that the temperatures at the top and bottom plates of the bridge remain consistent along the longitudinal direction, while the temperatures at the web, which is the vertical component connecting the top and bottom plates, show variations between the upstream and downstream of the bridge [41]. However, the analysis of the STB temperature monitoring data does not reveal any temperature gradient along the lateral direction of the top plate. The same observation holds for the bottom plate, top plate U-ribs and bottom plate U-ribs. Moreover, the web temperatures at both sides are the same as shown in Figure 3a, which is different from previous research results. The temperatures of diaphragms are consistent with those of the webs. This is because the tuyere, as a non-structural component, is directly exposed to solar radiation, while the diaphragm and the web are located inside of the tuyere and are not directly affected by solar radiation. The girder without the tuyere component generally exhibits a temperature gradient in the lateral direction. However, after the tuyere is installed, the web is not directly exposed to the sun, resulting in the disappearance of the lateral temperature gradient. Figure 3b presents the vertical temperature distribution of the mid-span cross-section. The components are arranged sequentially from high to low temperatures, namely, the top plate, top plate U-rib, diaphragm, bottom plate U-rib and bottom plate, respectively. Figure 3c illustrates the cross-section temperature gradient data, which are compared with the BS 5400 code. According to the data collected by the temperature sensor, the measured temperatures of the top plate and the top plate U-ribs comply with BS 5400, while the measured temperatures of the diaphragm and U-ribs of the bottom plate exceed the values specified by BS 5400. This raises potential concerns for the structural design of the bridge and requires more attention. Based on the above statistical analysis, the thermal field of the STB can be analyzed using a 1D vertical thermal field approach.

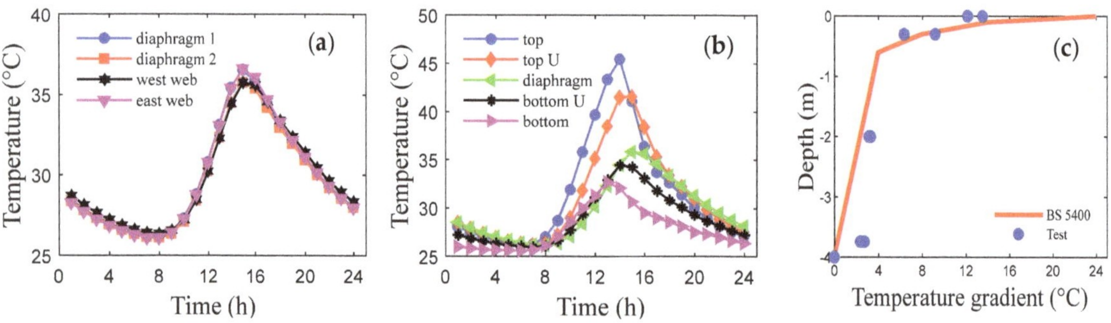

Figure 3. Temperature distribution. (**a**) lateral direction (**b**) vertical direction (**c**) temperature gradient.

The temperature-induced displacement of the girder on the STB is typically related to the average temperature [42]. Figure 4a shows a significant correlation between the average displacement and average temperature. As the girder temperature increases, the displacement also increases, and vice versa. However, this trend is not consistently observed. There is a time delay effect in which the temperature data lag behind the displacement data for a certain period. Perhaps this is because the temperature sensors are installed inside the steel box girder, and it takes a certain time for the heat to transfer to the thermometers. The time delay effect between the structural displacement response and the temperature load on the STB is approximately one hour. Figure 4b presents the correlation between the average temperature difference ΔT and the average displacement difference δ_u. Both values are calculated as the differences from their initial values. The correlation does not follow a linear relationship. The nonlinear relationship is mainly attributed to the time delay effect. By considering the time delay effect and shifting the temperature axis in the negative direction, a linear relationship between displacement and temperature can be observed (Figure 4c).

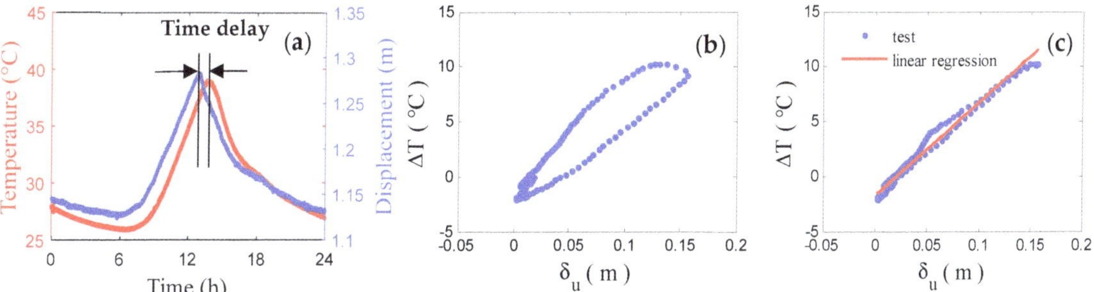

Figure 4. Temperature-Displacement correlation analysis. (**a**) time-delay effect (**b**) without considering time-delay (**c**) considering time-delay.

It is shown in Table 1 that there is a significant correlation between temperature and temperature-induced strain for a single day. Owing to the temperature of the top plates being higher than the bottom plates, the top plates (S-5-2, 4 and 6) represent compression, while the bottom plates (S-5-7 and 9) exhibit tension. The compressive strain of top plates increases continuously with increasing temperature, and vice versa. For instance, if the temperature rises by 20 °C, the compressive strain of the top plate increases to 60 $\mu\varepsilon$. On the other hand, the tensile strain of bottom plates increases continuously with the rising temperature, and vice versa. When the temperature increment of the bottom plate reaches 7 °C, the tensile strain of the bottom plate increases to 80 $\mu\varepsilon$. However, the temperature-induced strain of the U-ribs is influenced by the combined temperature of the decks and U-ribs. Between 10:00 and 15:00, the top plate U-ribs are under tension. In other time periods, the top plate U-ribs are under compression. This phenomenon occurs because the temperature of the top plates is higher than the top plate U-ribs between 10:00 and 15:00. Therefore, the top plates exhibit compression and the top plate U-ribs exhibit tension. However, the bottom plate U-ribs are under tension throughout.

Figure 5 shows the correlation between temperature and longitudinal temperature-induced strain of the girder under the influence of seasonal temperature variations in one year. The temperature and longitudinal strain were obtained from the temperature sensor T-5-5 and strain gauge S-5-5 on the top plate of the mid-span section on the STB. Firstly, the temperature and longitudinal strain of the first three days of each month were selected as representative data, and the data for each hour were averaged. Then, the monthly temperature and longitudinal strain data were averaged, respectively, resulting in a total of 12 monthly average temperature and 12 monthly average strain data points. The results indicate a sinusoidal relationship between seasonal temperature and longitudinal strain within a year. The top plate temperature reached its maximum value in August and its minimum value in February. On the other hand, the top plate strain reached its maximum value in September and its minimum value in February.

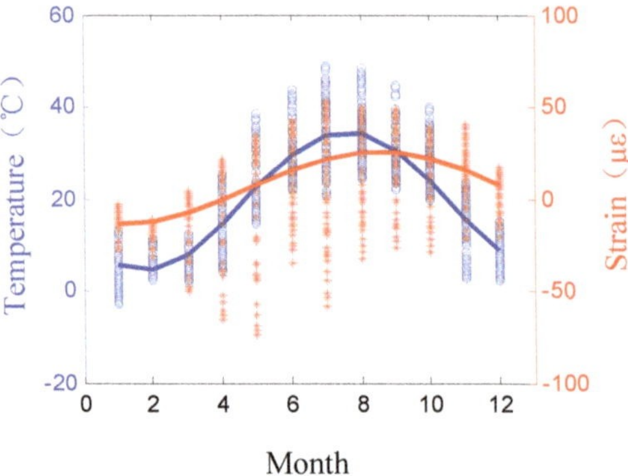

Figure 5. Temperature-induced strain correlation analysis in a year.

3. Analysis of Temperature-Induced Stress through Multi-Scale Modelling

3.1. STB Multi-Scale Modelling Using the Substructure Method

By providing both global and local information to assess the structural safety performance, the multi-scale FE model can not only be used to perform static and dynamic analysis at the global level but also can analyze the stress of local components such as local strain, fatigue and the fracture of cracks.

The STB multi-scale FE model was established through ANSYS. Figure 6a shows the division of the girder into segments on the STB. The segment among the stay cables is selected as a segment model considering the connections between the girder, cables, towers and piers. The diaphragm enhances the load-bearing capacity of the segment at the end of the girder and the segment at the tower. Each segment is considered a substructure. Accordingly, the segment with a longitudinal length of 6.7 m is divided into a substructure GEN3, located at the end of the girder; the segment between stay cable No.2, with a longitudinal length of 48 m, is divided into a substructure GEN2, located at the tower; the standard segment between the adjacent stay cables, with a longitudinal length of 16 m, is divided into a substructure GEN0 and the standard segment between the contiguous stay cable with a longitudinal length of 12 m, is divided into a substructure GEN01. As the longitudinal truss of the mid-span segment is designed differently from the standard segment GEN0, it was divided into a single substructure GEN1. Overall, there are 53 standard segment substructures GEN0, 11 standard segment substructures GEN01, 1 mid-span segment substructure GEN1, 1 segment substructure GEN2 under the tower, and 1 segment substructure GEN3 at the end of the girder, all positioned on the side of the symmetrical axis on the STB. Figure 6d presents the FE model of the segment substructure GEN1 on the STB, modeled according to the design drawing. The segment, including top plates, bottom plates, webs, diaphragms and U-ribs, is simulated through the SHELL63 shell element in ANSYS. The top and bottom flanges of the longitudinal truss are welded with the top and bottom plates by steel plates, and thus the SHELL63 element is used to simulate the top and bottom flanges. The diagonal braces of the longitudinal truss adopt angle steel, and therefore, the BEAM4 element is used to simulate the longitudinal braces.

In the STB multi-scale FE model, each segment substructure model (Figure 6c) should be condensed into a super-element (Figure 6b), which is the condensation of a group of elements into one element using the substructure method. The purpose of this condensation is to simplify the model and improve computational efficiency. Once the super-element matrices have been formed, they are stored in a file and can be used as normal finite elements in subsequent analyses. The master nodes (purple nodes in Figure 6c) are selected to connect segment models and the other components in the multi-scale FE model. In the substructure model, all nodes (with the exception of the master nodes) are eliminated through the substructure technique. More specifically, a group of elements is condensed into one super-element, with representative nodes of the segment model interface selected as the master nodes. In the case of segment substructure model GEN2, three additional rows of master nodes are additionally selected based on the connection of the beam between the substructure and tower, as well as the connection between the substructure and stay cable No.1. The DOFs of the master nodes on the interface of each segment substructure were then coupled.

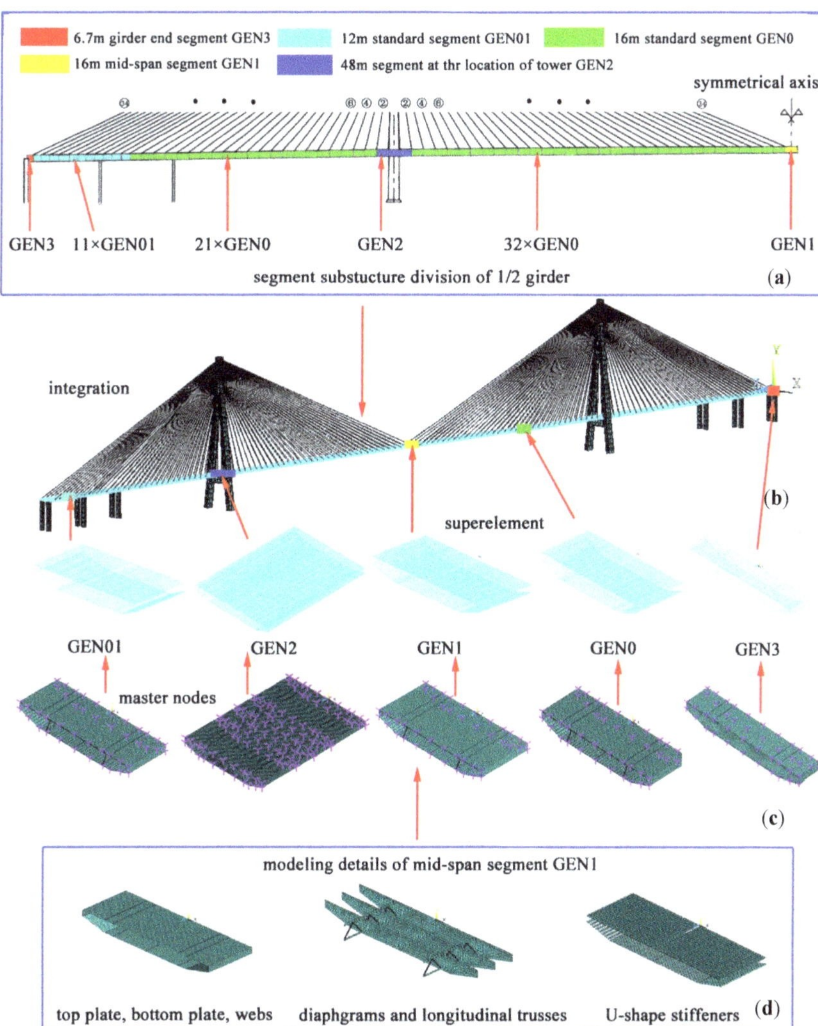

Figure 6. Multi−scale modeling. (**a**) segment substructure division (**b**) multi−scale FE model and super-elements (**c**) segment substructure models (**d**) modeling details of GEN1.

Owing to the length of the girder being long, the mass block was used inside the steel box girder at the piers to endure the warping deformation of the girder at these locations. The mass block was simulated by structural mass element MASS21 at the corresponding master nodes. In addition, the secondary dead load exerts a great impact on the static and dynamic performance of the bridge structures and is non-negligible. The secondary dead load of the STB deck was also simulated via MASS21. Based on the design drawing, the secondary dead loads of the bridge deck were converted into the mass of the MASS21 elements. The girder model was built following the completion of all the above steps.

The tower and pier of the STB were modeled by the 3D beam element BEAM188, which could accurately simulate the hollow cross-section characteristics of the concrete structure. Nonlinear truss element LINK10 was employed to simulate the stay cable, which was set to tension-only. The stay cable force of the bridge design was converted into the initial strain to simulate the initial tension of the stay cables. The stay cable experiences sag due to its self-weight, and cannot be calculated using the tensile truss element. In this case, the sag effect needs to be considered by the equivalent elasticity modulus. The Ernst formula (Equation (1)), initially proposed by German scholar Ernst in 1965, is widely used for this purpose. The formula assumes that the cable weight is uniformly distributed along chord length (not arc length) and neglects the cable's ability to bear bending moments. As a result, the shape of the stay cable follows a parabolic curve.

$$E_l = \frac{E_e}{1 + \frac{E_e \gamma^2 l_x^2}{12\sigma^3}} \tag{1}$$

where E_l is the corrected elastic modulus of the stay cable; E_e is the initial elastic modulus of the stay cable; σ is the stress of the stay cable; γ is the unit weight of the stay cable; l_x is the horizontal projection length of the stay cable.

As STB bearings in the piers constrain the transverse and vertical displacement of the girder, these constraints were used at the joints between the piers and the girder. With the presence of the support beam between the tower and the girder, constraints were used to connect the support beam and the girder. The foundation at the bottom of the towers and three piers were constrained by fixed constraints. The multi-scale FE model was developed by coupling the master node DOFs on the girder super-elements with the nodes on stay cables, towers, and piers. Figure 5b presents the assembly of the STB multi-scale model by various components.

The proposed model was verified using SHM data prior to its application for the temperature-induce response analysis. The results obtained from the modal analysis of the multi-scale FE model were compared with measured values reported by Zhang and Chen [43]. The first three order modes of the multi-scale FE model were identified as the first-order longitudinal mode, the first-order symmetric lateral bending mode and the first-order symmetric vertical bending mode, with corresponding frequencies of 0.0693, 0.1069, 0.1884 Hz, respectively. By comparing with the frequency in the reference [43], the errors are determined as 8.1%, 5.3%, and 3.5%, respectively. The values calculated with the proposed method are consistent with the test values, indicating the proposed modeling strategy is reliable and the model can be adopted for further analysis.

The above models can be directly applied to stress field analysis. However, it is necessary to investigate the thermal field analysis before analyzing the temperature-induced displacement/stress. Since the longitudinal thermal variation of the bridge is not obvious, only the thermal field analysis of segment substructures with different geometry features was carried out. The thermal field was calculated using the heat-transfer analysis on the segment substructures GEN0, GEN01, GEN1, GEN2 and GEN3, which had not yet been condensed into super-elements. The structural elements should be replaced by thermal elements. The 3D thermal shell element SHELL57 replaced the elastic shell element SHELL63. Similarly, the convection link element LINK34 was used instead of the 3D elastic beam element BEAM4. The bridge deck pavement has a great influence on the results of thermal field analysis. Therefore, the SOLID70 thermal solid element was employed to simulate the bridge deck pavement, which is made up of 5.5 cm thick epoxy asphalt concrete on the bridge deck. During the thermal field analysis, the secondary dead load was deleted. When the stress field model was analyzed, the temperature results calculated by the thermal field were substituted into the segment substructures GEN0, GEN01, GEN1, GEN2 and GEN3 as the element internal force, and then each segment substructure was condensed into a super-element.

The advantages of the multi-scale FE model were illustrated by comparing with the global fine FE model. As shown in Table 2, the DOFs of the global fine FE model are 11,274,550, while the DOFs of the multi-scale FE thermal field model proposed in this paper are only 471,721. Based on the ANSYS run-time stats module, the number of the nodes, elements and DOFs of the global fine thermal field FE model is about 22 times, 25 times and 24 times that of the multi-scale FE model, respectively, and the time required for thermal analysis is about 25 times longer. As shown in Table 3, the DOFs of the global fine stress field FE model are 11,274,550, while the DOFs of the multi-scale FE model proposed in this paper are only 69,272. The number of nodes, elements and DOFs of the global fine stress field FE model is approximately 113 times, 2280 times and 162 times that of the multi-scale FE model, respectively, and the time required for static self-balance analysis (under the action of gravity load and measured cable force, the calculated value of girder deformation is consistent with the measured value) and temperature-induced stress analysis is about 1630 times and 40 times longer. Therefore, if the global fine FE method is used to analyze the temperature-induced strain of U-ribs, the model has a significantly large number of DOFs, leading to the consumption of extensive hardware resources and calculation time, and in some cases, it may not even be feasible to perform the calculation. However, the multi-scale FE model analysis can save a significant amount of computing time on the premise of accuracy, which is convenient for subsequent parameter analysis and structural performance evaluation.

Table 2. Comparison of calculation scale and time between thermal field models.

	Nodes	Elements	DOFs	Thermal Analysis Time (s)
global fine	1,690,000	2,359,862	11,274,550	625,920
multi-scale	78,620	94,395	471,720	25,180

Table 3. Comparison of calculation scale and time between stress field models.

	Nodes	Elements	DOFs	Static Time (s)	Stress Analysis Time (s)
global fine	1,690,000	2,359,862	11,274,550	26,080	625,920
multi-scale	14,955	1035	69,272	16	15,491

3.2. Thermal Field Analysis

The thermal parameters of the steel components and deck pavement asphalt concrete, including thermal conductivity, specific heat capacity and density, were determined as shown in Table 4 [44–46]. Generally, the atmospheric temperature one hour before sunrise is considered the initial temperature of the bridge. The temperature distribution inside and outside the STB box girder at 6:00 a.m. is relatively uniform, so it is appropriate to take the atmospheric temperature at 6:00 a.m. as the initial atmospheric temperature. The time interval for the thermal field analysis is set to one hour.

Table 4. Thermal parameters of material.

	Steel	Asphalt
thermal conductivity k (W/(m·K))	60.5	2
heat capacity $c \left(\frac{J}{kg \cdot K} \right)$	460	900
density ρ (kg/m^3)	7850	2100

Thermal field analysis was performed to determine the boundary conditions, including absolute radiation temperature T_v and overall heat transfer coefficient h_o. The relation between the radiation and reflection balance of the deck surface on the bridge is described as Equation (2) [44]:

$$q_B + q_k = q_J + q_H + q_R + q_{G_a} + q_{UR} \tag{2}$$

where q_B is the structural radiation; q_k is the thermal irradiation; q_J is the solar direct radiation; q_H is the scattered radiation; q_R is the reflected radiation; q_{G_a} is the atmospheric radiation and q_{UR} is the surface effective radiation. The equations for the above parameters can be found in reference [45].

Equation (2) is a fourth-order transcendental equation of T_v and cannot be solved directly. Therefore, the following formula is used to simplify the equation [45]:

$$\varepsilon_{bl} C_s T_v{}^4 = h_r(T_v - T_a) + \varepsilon_{bl} C_s T_a{}^4 \tag{3}$$

where ε_{bl} is the structural radiation coefficient; C_s is the Stefan–Boltzmann constant, $C_s = 5.677 \times 10^{-4} \text{ W/m}^2\text{K}^4$; T_a is the atmospheric temperature; $h_o = h_r + h_k$, h_r is the coefficient of radiation heat transfer and h_v is the exchange heat transfer coefficient.

The absolute radiation temperature T_v can be determined as the following formulas [45]:

$$\begin{cases} T_{v,top} = T_a + \dfrac{a_{bk}}{h_r+h_k}(I_J sinh + I_H) - (1 - \varepsilon_a)\dfrac{\varepsilon_{bl}C_s}{h_r+h_k}T_a{}^4 \\ T_{v,bot} = T_a + \dfrac{\gamma_{uk}a_{bk}}{h_r+h_k}(I_J sinh + I_H) \\ T_{v,web} = T_a + \dfrac{a_{bk}}{h_r+h_k}[\gamma_{uk}(I_J sinh + I_H)cos^2\frac{\beta}{2} + I_H cos\left(\frac{\pi}{2} + h - \beta\right) \\ cos(\alpha - \alpha_w)sinh + I_H sin^2\frac{\beta}{2}] - (1 - \varepsilon_a)\dfrac{\varepsilon_{bl}C_s}{h_r+h_k}T_a{}^4 sin^2\frac{\beta}{2} \end{cases} \tag{4}$$

where T_v includes $T_{v,top}$, $T_{v,bot}$ and $T_{v,web}$. $T_{v,top}$ is the absolute radiation temperature of the top plate; $T_{v,bot}$ is the absolute radiation temperature of the bottom plate; $T_{v,web}$ is the absolute radiation temperature of the web; a_{bk} is the shortwave radiation absorptivity; h is solar altitude angle; ε_a is the atmospheric radiation coefficient, $\varepsilon_a = 0.82$; γ_{uk} is the ground reflection coefficient; I_J is the direct solar radiation intensity and I_H is the solar scattered energy; β is included angle between web and horizontal plane; α is the azimuth of the sun, which is the included angle between the projection of the line from the observer to the sun on the ground plane and the due south direction; α_w is the azimuth of the external normal of the web. The steel and asphalt shortwave radiation absorptivity were considered to be 0.685 and 0.9, and the steel and asphalt surface radiation coefficients were assumed to be 0.8 and 0.92, respectively [44]. For more detailed information on these expressions, please refer to [45].

Based on the measured meteorological data and calculation parameters of the thermal field, the thermal boundary conditions were calculated by Equation (4). Then these boundary conditions were applied to the segment substructures for transient thermal analysis. Figure 7 presents the thermal field of the substructure GEN1 calculated for one day. Through conducting statistical analysis of the errors between calculated and measured values, it was found that the maximum error of temperature for the top plate is 8.2%, and the weighted average error within a day is 3.9%. The maximum error and the weighted average error for the bottom plate are 7.8% and 3.5%, respectively. For the web, corresponding values are 7.1% and 3.5%, respectively. For the diaphragm, they are 5.6% and 4.1%, respectively. For the top plate U-rib, they are 9.9% and 6.2%, respectively. For the bottom plate U-rib, they are 5.0% and 2.1%, respectively. There is an agreement between the predicted and measured temperatures. This verifies the effectiveness of the numerical model and the heat transfer analysis.

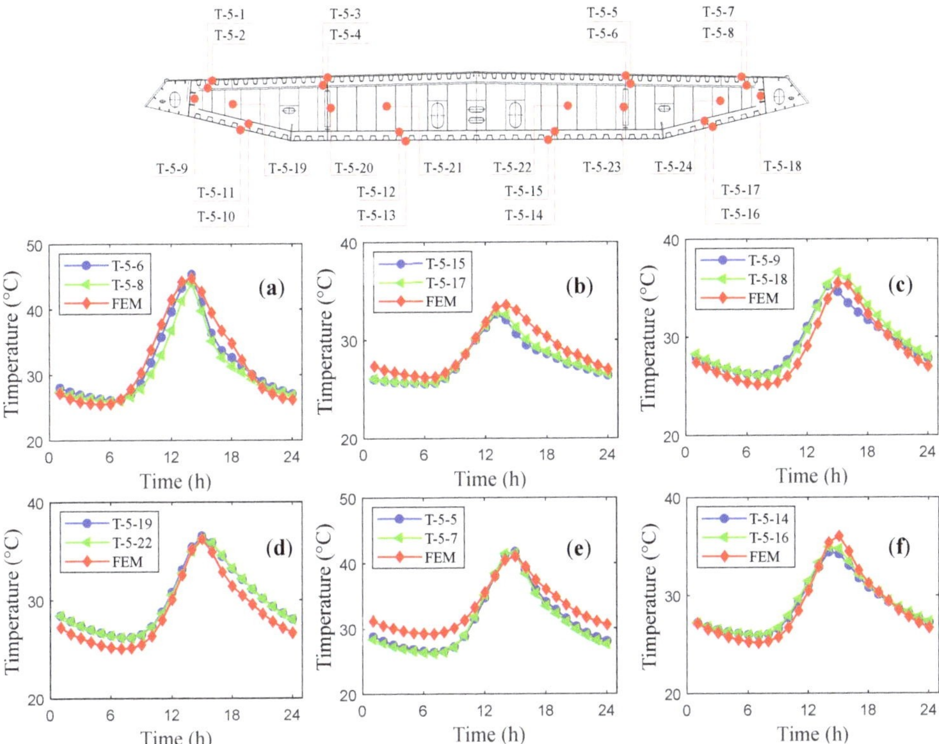

Figure 7. Comparison of thermal field measured and calculated values of each component on substructure GEN1. (**a**) top plate, (**b**) bottom plate, (**c**) web, (**d**) diaphragm, (**e**) top plate U-rib, (**f**) bottom plate U-rib.

3.3. Temperature-Induced Structural Responses

The temperature-induced responses can be investigated by combing the temperature time history with the multi-scale FE stress field model. In this section, the focus is on the analysis of longitudinal temperature-induced displacements at the end of the girder and the longitudinal temperature-induced strains on the sensor layout cross-section. First, the results of thermal field analysis are imported into each segment substructure model (Figure 6c) and then the temperature load is treated as the internal force, each segment substructure model is condensed into a super-element (Figure 6b); at last, these super-elements are applied to the multi-scale FE model for stress field analysis. The isotropic linear constitutive model for thermal expansion is adopted. Temperature-induced displacement can be directly calculated from the master nodes of the girder end in the multi-scale FE model. Figure 8a,b depicts the temperature-induced displacements at the end of the girder. The calculated values of the FE model are in agreement with the measured values of the D-9-1 displacement gauges. However, there is a certain deviation between the temperature-induced displacements of the D-1-1 displacement gauges and the calculated values. It should be noted that the STB multi-scale FE model is assumed symmetry with respect to the mid-span cross-section of the girder, resulting in consistent displacements calculated by the FE model on both sides. However, this assumption does not match the actual situation. Figure 8c,d displays the correlation between the temperature and temperature-induced displacements. The statistical results indicate the accuracy of the correlation between temperature and temperature-induced displacement based on the proposed method.

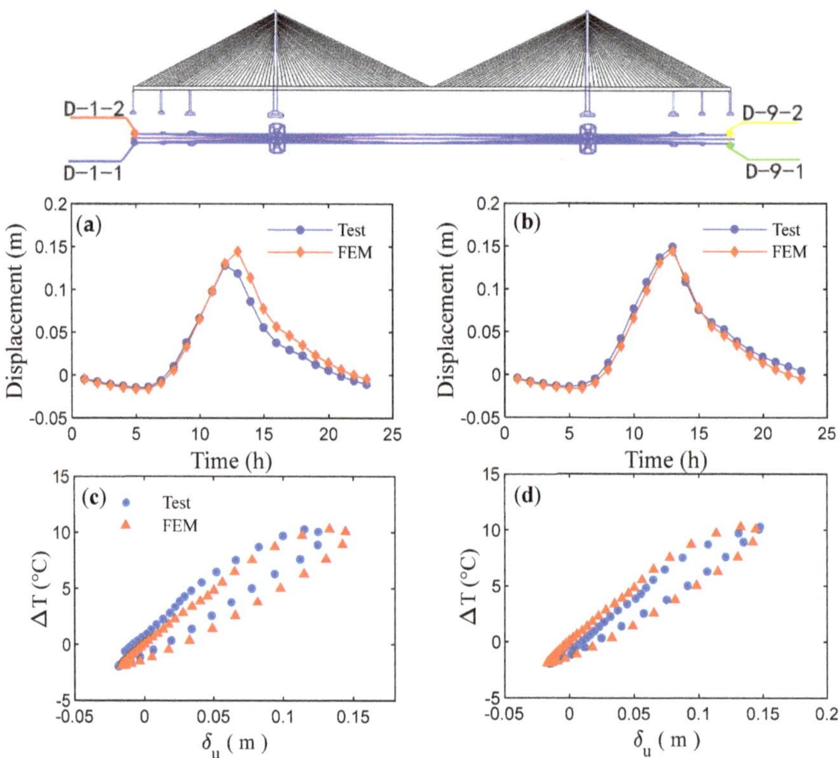

Figure 8. Comparison between calculated displacements and measured values. (**a**) D-1-1 displacement history, (**b**) D-9-1 displacement history, (**c**) D-1-1 temperature-displacement correlation, (**d**) D-9-1 temperature-displacement correlation.

The temperature-induced displacement can be calculated directly. However, in order to obtain the temperature-induced stress and strain, an additional step is required. At first, it is necessary to reverse the displacement of the master nodes back to the segment substructure to calculate the displacement of the slave nodes; then the element temperature-induced stress and strain in the segment substructure are calculated based on the displacement of the slave node. Figure 9 shows that the calculated values of decks and U-ribs are consistent with the measured values. Equation (5) is used as an evaluation indicator to compare the calculated and measured values of temperature-induced strain. The evaluation indicator value ranges between 0 and 1, with a value closer to 1 indicating a better similarity between the two sets of values.

$$I = \frac{\left|\sum_{i=1}^{n} S_{ai}S_{ei}\right|^2}{\left(\sum_{i=1}^{n} S_{ai}S_{ai}\right)\left(\sum_{i=1}^{n} S_{ei}S_{ei}\right)} \tag{5}$$

where I is the evaluation indicator, S_{ai} is the calculated value of temperature-induced strain for i-th sampling point, S_{ei} is the measured value of temperature-induced strain for i-th sampling point, n is the number of samples for one day.

The evaluation indicator values for S-5-1 to S-5-6 in the top plate or top plate U-rib are 0.96, 0.97, 0.94, 0.95, 0.75 and 0.98, respectively. The evaluation indicator values for S-5-8 to S-5-10 in the bottom plate or bottom plate U-rib are 0.90, 0.77 and 0.93, respectively. Therefore, compared with the traditional FE model, the multi-scale FE model based on the substructure method can be adopted to analyze the temperature-induced stress of the local members (i.e., U-ribs) on long-span bridges.

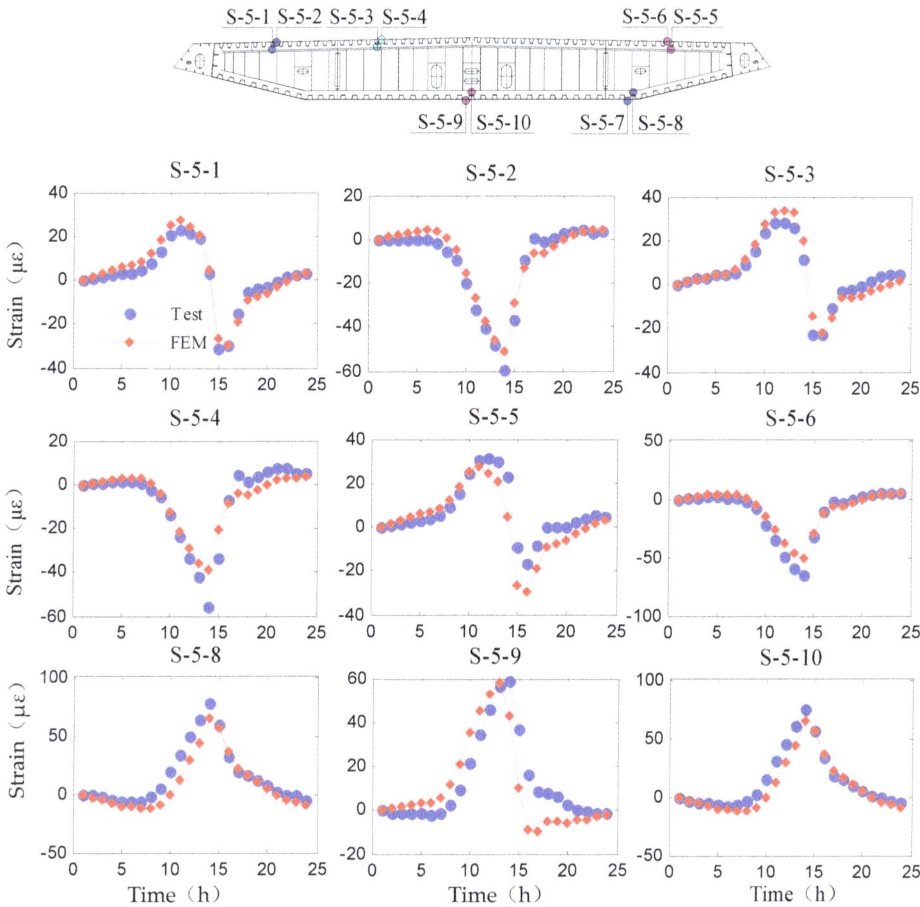

Figure 9. Calculated structural strain response.

In order to evaluate the variations in the strain distribution across the longitudinal and vertical directions of the girder cross-section for one day, strains at different cross-sections and positions were analyzed at 6:00 and 14:00, respectively. Figure 10a shows that the top plates (SA-N-1, 3, 5) are subject to compression, while the bottom plates (SA-N-2, 4, 6) suffer from tension at 14:00. Moreover, the girder exhibits an upward bending behavior. However, the inverse phenomenon is observed in Figure 10b, whereby the top and bottom deck are subject to tension and compression, respectively, at 6:00, and the girder exhibits a downward bending behavior. Similar strain trends are observed at the top and bottom plate U-ribs in Figure 10c,d, respectively. This is attributed to the greater temperature of the top girder compared to the bottom girder at 14:00, while an inverse temperature distribution along the vertical direction is present at 6:00. The maximum variation range of the temperature-induced strain at the top plate and bottom plate is approximately 100 με over the course of one day.

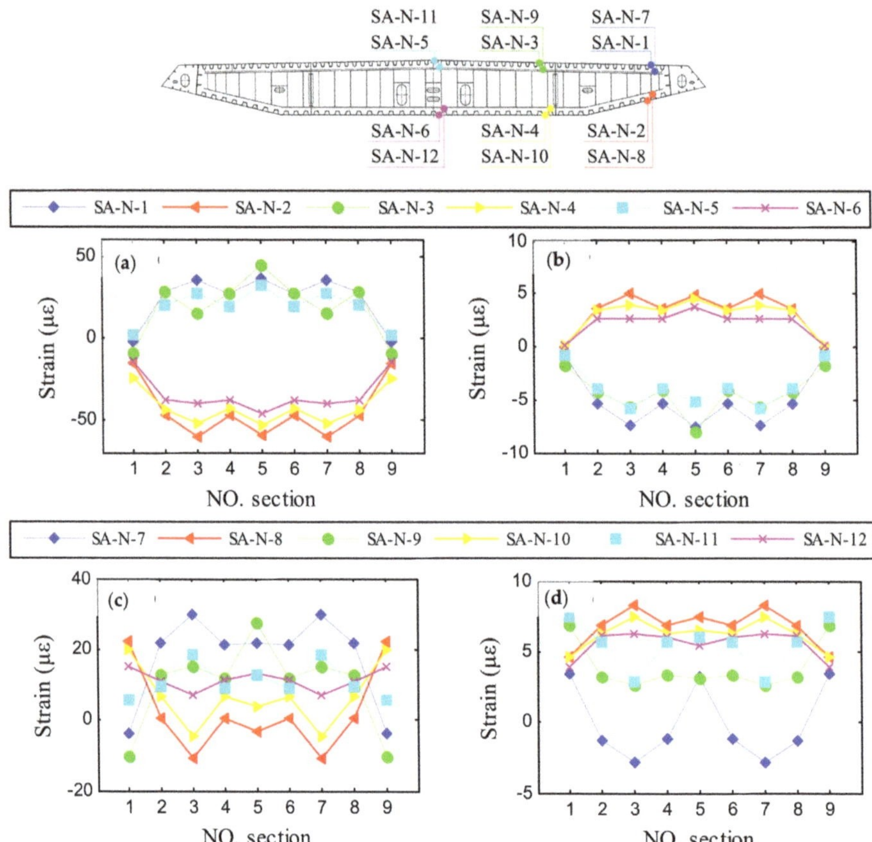

Figure 10. Longitudinal and vertical distributions of strain. (**a**) Strain distribution of plates at 14:00, (**b**) Strain distribution of plates at 6:00, (**c**) Strain distribution of U-ribs at 14:00, (**d**) Strain distribution of U-ribs at 6:00.

4. Conclusions

The performance evaluation of OSD by considering the temperature load is very important. In previous studies, the strain of local components needed to be analyzed in the global fine finite FE model. In this paper, based on substructure technology, the temperature-induced stress of U-ribs is evaluated using a multi-scale FE model on a long-span cable-stayed bridge. The following key conclusions are obtained.

(1) The temperature-induced stress of U-ribs on the STB was analyzed based on monitoring data and the multi-scale FE method. This method can be applied to other long-span bridges to address the issue of low computational efficiency in analyzing U-ribs in the global fine FE model.

(2) Analysis of monitoring data indicates that the long-span steel box bridge with the tuyere components exhibits a vertical temperature gradient rather than a transverse temperature gradient. The correlation between temperature-induced displacement and temperature demonstrates a linear relationship once the time delay effect is considered. The temperature-induced strain of the top plates and bottom plates is influenced by the temperature between them. The temperature-induced strain of U-ribs is influenced by the temperature of the decks and U-ribs. Furthermore, the seasonal temperature and longitudinal strain over time within a year exhibit a sinusoidal relationship.

(3)   A multi-scale FE model, which can effectively reduce the calculation time based on the substructure method, has been established to analyze the temperature-induced stress of U-ribs on long-span bridges. The accuracy of the multi-scale FE model results for the temperature-induced stress of U-ribs has been confirmed through monitoring data.

(4)   By evaluating the temperature-induced strain during the highest and lowest temperatures of one day on the multi-scale FE model, it indicates that the deflection of the girder, a key index for bridge design and SHM assessment, exhibits dynamic changes in response to temperature loads. The temperature-induced strain of the top and bottom plates displays a maximum variation range of approximately 100 $\mu\varepsilon$.

5. Recommendation

Due to the limitations of research time, the evaluation and discussion are focused on the longitudinal temperature-induced strain of specific segments on the bridge, specifically the top plate, bottom plate, and U-ribs. In future research, the temperature-induced strain of web plates, diaphragms, and diaphragms can be studied in depth, as well as more segments of the girder. Additionally, it is recommended to investigate the influence of temperature load on the fatigue behavior of U-ribs under varying thermal conditions.

Author Contributions: Conceptualization, F.Z. and J.Z.; methodology, F.Z.; software, F.Z.; validation, F.Z., Y.Y. and J.Z.; formal analysis, F.Z.; investigation, F.Z.; resources, F.Z.; data curation, F.Z.; writing—original draft preparation, F.Z.; writing—review and editing, F.Z. and P.L.; visualization, F.Z.; supervision, F.Z.; project administration, F.Z.; funding acquisition, Y.Y. All authors have read and agreed to the published version of the manuscript.

Funding: The research presented was financially supported by the National Science Foundation of China (No.: 51578139, 51778134), Key Scientific and Technological Research Projects of Henan Province, China (No.: 222102320006) and Postdoctoral research grant in Henan Province.

Institutional Review Board Statement: Not applicable.

Informed Consent Statement: Not applicable.

Data Availability Statement: The data presented in this study are available on request from the corresponding author. The data are not publicly available due to copyright issues.

Conflicts of Interest: The authors declare no conflict of interest.

References

1.   Zhao, R.; Zheng, K.; Wei, X.; Jia, H.; Liao, H.; Li, X.; Wei, K.; Zhan, Y.; Zhang, Q.; Xiao, L.; et al. State-of-the-art and annual progress of bridge engineering in 2020. *Adv. Bridge Eng.* **2021**, *2*, 1–105.
2.   Zeng, Y.; Qiu, Z.; Yang, C.C.; Haozheng, S.; Xiang, Z.F.; Zhou, J.T. Fatigue experimental study on full-scale large sectional model of orthotropic steel deck of urban rail bridge. *Adv. Mech. Eng.* **2023**, *15*, 16878132231155271. [CrossRef]
3.   Guo, T.; Li, A.Q.; Wang, H. Influence of ambient temperature on the fatigue damage of welded bridge decks. *Int. J. Fatigue* **2008**, *30*, 1092–1102. [CrossRef]
4.   Liu, Y.; Zhang, H.P.; Liu, Y.M.; Deng, Y.; Jiang, N.; Lu, N.W. Fatigue reliability assessment for orthotropic steel deck details under traffic flow and temperature loading. *Eng. Fail. Anal.* **2017**, *71*, 179–194. [CrossRef]
5.   Xia, Q.; Xia, Y.; Wan, H.P.; Zhang, J.; Ren, W.X. Condition analysis of expansion joints of a long-span suspension bridge through metamodel-based model updating considering thermal effect. *Struct. Control. Health Monit.* **2020**, *27*, e2521. [CrossRef]
6.   Xiao, F.; Hulsey, J.L.; Balasubramanian, R. Fiber optic health monitoring and temperature behavior of bridge in cold region. *Struct. Control. Health Monit.* **2017**, *24*, e2020. [CrossRef]
7.   Wang, D.; Liu, Y.M.; Liu, Y. 3D temperature gradient effect on a steel–concrete composite deck in a suspension bridge with field monitoring data. *Struct. Control. Health Monit.* **2018**, *25*, e2179. [CrossRef]
8.   Yarnold, M.T.; Moon, F.L. Temperature-based structural health monitoring baseline for long-span bridges. *Eng. Struct.* **2015**, *86*, 157–167. [CrossRef]
9.   Ni, Y.Q.; Xia, H.W.; Wong, K.Y.; Ko, J.M. In-Service Condition Assessment of Bridge Deck Using Long-Term Monitoring Data of Strain Response. *J. Bridge Eng.* **2012**, *17*, 876–885. [CrossRef]
10.   Xia, Y.Y.; Shi, M.S.; Zhang, C.; Wang, C.X.; Sang, X.X.; Liu, R.; Zhao, P.; An, G.F.; Fang, H.Y. Analysis of flexural failure mechanism of ultraviolet cured-in-place-pipe materials for buried pipelines rehabilitation based on curing temperature monitoring. *Eng. Fail. Anal.* **2022**, *142*, 106763. [CrossRef]

11. Zuk, W. *Simplified Design Check of Thermal Stresses in Composite Highway Bridges*; The National Academies of Sciences, Engineering, and Medicine: Washington, DC, USA, 1965; pp. 10–13.
12. Peiris, A.; Sun, C.; Harik, I. Lessons learned from six different structural health monitoring systems on highway bridges. *J. Low Freq. Noise V. A.* **2018**, *39*, 616–630. [CrossRef]
13. Ni, Y.Q.; Xia, Y.; Liao, W.Y.; Ko, J.M. Technology Innovation in Developing the Structural Health Monitoring System for Guangzhou New TV Tower. *Struct. Control Health Monit.* **2009**, *16*, 73–98. [CrossRef]
14. Deng, Y.; Li, A.Q.; Chen, S.; Feng, D.M. Serviceability assessment for long-span suspension bridge based on deflection measurements. *Struct. Control. Health Monit.* **2018**, *25*, e2254. [CrossRef]
15. Yuan, J.; Gan, Y.X.; Chen, J.; Tan, S.M.; Zhao, J.T. Experimental research on Consolidation creep characteristics and microstructure evolution of soft soil. *Front. Mater.* **2023**, *10*, 1137324. [CrossRef]
16. Yang, Z.Y.; Yu, X.Y.; Dedman, S.; Rosso, M.; Zhu, J.M.; Yang, J.Q.; Xia, Y.X.; Tian, Y.C.; Zhang, G.P.; Wang, J.Z. UAV remote sensing applications in marine monitoring: Knowledge visualization and review. *Sci. Total Envrion.* **2022**, *838*, 155939. [CrossRef]
17. Zhang, B.; Zhou, L.; Zhang, J. A methodology for obtaining spatiotemporal information of the vehicles on bridges based on computer vision. *Comput. Aided Civ. Inf.* **2018**, *34*, 471–487. [CrossRef]
18. Zhao, W.; Guo, S.; Zhou, Y.; Zhang, J. A Quantum-Inspired Genetic Algorithm-Based Optimization Method for Mobile Impact Test Data Integration. *Comput. Aided Civ. Inf.* **2018**, *33*, 411–422. [CrossRef]
19. Wang, G.X.; Ye, J.H. Localization and quantification of partial cable damage in the long-span cable-stayed bridge using the abnormal variation of temperature-induced girder deflection. *Struct. Control Health Monit.* **2019**, *26*, e2281. [CrossRef]
20. Salcher, P.; Pradlwarter, H.; Adam, C. Reliability assessment of railway bridges subjected to high-speed trains considering the effects of seasonal temperature changes. *Eng. Struct.* **2016**, *126*, 712–724. [CrossRef]
21. Siringoringo, D.M.; Fujino, Y. Seismic response of a suspension bridge: Insights from long-term full-scale seismic monitoring system. *Struct. Control Health Monit.* **2018**, *25*, e2252. [CrossRef]
22. Xia, Q.; Cheng, Y.Y.; Zhang, J.; Zhu, F.Q. In-Service Condition Assessment of a Long-Span Suspension Bridge Using Temperature-Induced Strain Data. *J. Bridge Eng.* **2017**, *22*, 04016124. [CrossRef]
23. Xia, Q.; Zhang, J.; Tian, Y.D.; Zhang, Y. Experimental Study of Thermal Effects on a Long-Span Suspension Bridge. *J. Bridge Eng.* **2017**, *22*, 04017034. [CrossRef]
24. Yu, S.; Li, D.S.; Ou, J.P. Digital twin-based structure health hybrid monitoring and fatigue evaluation of orthotropic steel deck in cable-stayed bridge. *Struct. Control Health Monit.* **2022**, *29*, e2976. [CrossRef]
25. Yu, S.; Ou, J.P. Fatigue life prediction for orthotropic steel deck details with a nonlinear accumulative damage model under pavement temperature and traffic loading. *Eng. Fail Anal.* **2021**, *126*, 105366. [CrossRef]
26. Asadollahi, P. Bayesian-Based Finite Element Model Updating, Damage Detection, and Uncertainty Quantification for Cable-Stayed Bridges. Ph.D. Thesis, University of Kansas, Lawrence, KS, USA, 2018.
27. Mensah-Bonsu, P.O. Computer-Aided Engineering Tools for Structural Health Monitoring under Operational Conditions. Master's Thesis, University of Connecticut Graduate School, Storrs, CT, USA, 2012.
28. Wang, F.Y.; Xu, Y.L.; Zhan, S. Multi-scale model updating of a transmission tower structure using Kriging meta-method. *Struct. Control Health Monit.* **2017**, *24*, e1952. [CrossRef]
29. Li, Z.; Chen, Y.; Shi, Y. Numerical failure analysis of a continuous reinforced concrete bridge under strong earthquakes using multi-scale models. *Earthq. Eng. Eng. Vib.* **2017**, *16*, 397–413. [CrossRef]
30. Kojic, M.; Milosevic, M.; Kojic, N.; Starosolski, Z.; Ghaghada, K.; Serda, R.; Annapragada, A.; Ferrari, M.; Ziemys, A. A multi-scale FE model for convective-diffusive drug transport within tumor and large vascular networks. *Comput. Method Appl. M.* **2015**, *294*, 100–122. [CrossRef]
31. Pletz, M.; Daves, W.; Yao, W.; Kubin, W.; Scheriau, S. Multi-scale FE modeling to describe rolling contact fatigue in a wheel-rail test rig. *Tribol. Int.* **2014**, *80*, 147–155. [CrossRef]
32. Trofimov, A.; Le-Pavic, J.; Ravey, C.; Albouy, W.; Therriault, D.; Levesque, M. Multi-scale modeling of distortion in the non-flat 3D woven composite part manufactured using resin transfer molding. *Compos. Part A.* **2021**, *140*, 106145. [CrossRef]
33. Chen, L.; Qian, Z.; Wang, J. Multiscale Numerical Modeling of Steel Bridge Deck Pavements Considering Vehicle-Pavement Interaction. *Int. J. Geomech.* **2016**, *16*, B4015002. [CrossRef]
34. Chan, T.H.T.; Li, Z.X.; Ko, J.M. Fatigue analysis and life prediction of bridges with structural health monitoring data—Part II: Application. *Int. J. Fatigue* **2001**, *23*, 55–64. [CrossRef]
35. Li, Z.X.; Jiang, F.F.; Tang, Y.Q. Multi-scale analyses on seismic damage and progressive failure of steel structures. *Finite Elem. Anal. Des.* **2012**, *48*, 1358–1369. [CrossRef]
36. Nie, J.; Zhou, M.; Wang, Y.; Fan, J.; Tao, M. Cable Anchorage System Modeling Methods for Self-Anchored Suspension Bridges with Steel Box Girders. *J. Bridge Eng.* **2014**, *19*, 172–185. [CrossRef]
37. Wang, F.; Xu, Y.; Sun, B.; Zhu, Q. Updating Multiscale Model of a Long-Span Cable-Stayed Bridge. *J. Bridge Eng.* **2018**, *23*, 04017148. [CrossRef]
38. Jensen, H.A.; Munoz, A.; Papadimitriou, C.; Vergara, C. An enhanced substructure coupling technique for dynamic re-analyses: Application to simulation-based problems. *Comput. Method Appl. M.* **2016**, *307*, 215–234. [CrossRef]
39. Li, Z.X.; Chan, T.H.T.; Yu, Y.; Sun, Z.H. Concurrent multi-scale modeling of civil infrastructures for analyses on structural deterioration-Part I: Modeling methodology and strategy. *Finite Elem. Anal. Des.* **2009**, *45*, 782–794. [CrossRef]

40. Zhu, Q.; Xu, Y.L.; Xiao, X. Multiscale Modeling and Model Updating of a Cable-Stayed Bridge. I: Modeling and Influence Line Analysis. *J. Bridge Eng.* **2015**, *20*, 04014112. [CrossRef]
41. Roberts-Wollman, C.L.; Breen, J.E.; Cawrse, J. Measurements of Thermal Gradients and their Effects on Segmental Concrete Bridge. *J. Bridge Eng.* **2002**, *7*, 166–174. [CrossRef]
42. Specifications ALBD. *Final Report of the AASHTO Standing Committee on Highways Task Force on Commercialization of Interstate Highway Rest Areas*; American Association of State Highway and Transportation Officials (AASHTO): Washington, DC, USA, 2007.
43. Zhang, X.G.; Chen, A.R. *Design and Structural Performance of Sutong Bridge*; CnDao: Beijing, China, 2010.
44. Xia, Y.; Chen, B.; Zhou, X.Q.; Xu, Y.L. Field monitoring and numerical analysis of Tsing Ma Suspension Bridge temperature behavior. *Struct. Control Health Monit.* **2013**, *20*, 560–575. [CrossRef]
45. Kehlbeck, F. *Influence of Solar Radiation on Bridge Structure*; China Communication Press: Beijing, China, 1981.
46. Zhou, L.R.; Xia, Y.; Brownjohn, J.M.W.; Koo, K.Y. Temperature Analysis of a Long-Span Suspension Bridge Based on Field Monitoring and Numerical Simulation. *J. Bridge Eng.* **2016**, *21*, 04015027. [CrossRef]

sustainability

MDPI

Article

The Influence Depth of Pile Base Resistance in Sand-Layered Clay

Dianfu Fu [1], Shuzhao Li [1], Hui Zhang [1], Yu Jiang [2,*], Run Liu [2] and Chengfeng Li [2]

[1] Research Institute Ltd., CNOOC, Beijing 100028, China
[2] State Key Laboratory of Hydraulic Engineering Simulation and Safety, Tianjin University, Tianjin 300072, China
* Correspondence: jy_td@tju.edu.cn

Abstract: Pile base resistance is an important part of the ultimate bearing capacity, and the soil within a certain range above and below the pile end contributes to the pile base resistance. In general, pile base resistance is calculated according to the average value of soil strength within a certain range of the pile end in the current calculating method, so it is very important to determine the influence range of pile base resistance. Based on the soil parameters and the results of the cone penetration test of the LiuHua 11-1 site in the South China Sea, the difference of pile base resistance calculated by different methods, the regularity of pile base resistance affected by calculation depth range is revealed. Additionally, the numerical simulation method is used to analyze the distribution of a plastic zone around the pile end in homogeneous soil and stratified soil, the results show the influence depth range of pile base resistance is 0.12 D above the pile end to 0.83 D below the pile end in clay, and the influence depth range is 0.9 D above the pile end to 1.3 D below the pile end in stratified soil.

Keywords: clay; pile foundation; base resistance; influence depth; cone penetration test

1. Introduction

Pile foundation is the main foundation form of offshore platform and offshore wind power [1–4]. The ultimate bearing capacity is an important content in the design of pile foundation, and there are two main methods to calculate the bearing capacity of pile foundation in the current relevant specifications for marine engineering; one is based on the undrained shear strength of soil [5], and the other is based on the results of an in situ cone penetration test (CPT/CPTU). Regardless of the calculation method, the bearing capacity is divided into skin friction and base resistance.

The method, which is based on undrained shear strength to calculate pile base resistance, can be divided into theoretical method and empirical method based on experimental data. Terzaghi [6], Meyerhof [7–10], Berezantzev [11], and Hu [12] proposed theoretical calculation methods of unit pile base resistance based on limit analysis theory according to the failure mode of soil in the pile end. Vesic [13] proposed a calculation method of pile base resistance based on the theory of cavity expansion, and the formula for calculating pile base resistance by all theoretical methods is $N_c S_u$, where N_c is the coefficient of pile base resistance. Wilson [14] suggested that N_c should be taken as 8.0 by the pile base resistance tests in clay. To measure pile base resistance, six tests were conducted by Meigh [15] and Yassin [16] in undisturbed soil and remolded soil in Imperial College London, and Skempton [17] suggested that N_c should be taken as 9.0 based on six test results and the theoretical analysis of Meyerhof and Mott–Gibson. Randoplh et al. [18], who combined with the empirical coefficient of deep foundation base resistance of Skempton and the theoretical solution of base resistance coefficient established by Vesic based on cavity expansion theory, recommended N_c as 9.0. This coefficient was introduced into the American Petroleum Institute (API), which is a standard method widely used in offshore pile foundation engineering for determining unit pile base resistance based on undrained

Citation: Fu, D.; Li, S.; Zhang, H.; Jiang, Y.; Liu, R.; Li, C. The Influence Depth of Pile Base Resistance in Sand-Layered Clay. *Sustainability* **2023**, *15*, 7221. https://doi.org/10.3390/su15097221

Academic Editor: Marco Lezzerini

Received: 7 March 2023
Revised: 19 April 2023
Accepted: 21 April 2023
Published: 26 April 2023

shear strength. The end is at least three diameters above the bottom of the layer to preclude punch-through in layered soils, and the unit pile base resistance should be corrected if this distance cannot be reached.

The penetration mechanism of the cone penetration test and driven pile is similar, and with the improvement of in situ test technology and data acquisition, the measurement results are more accurate and reliable, so the method of calculating pile base resistance based on in situ cone penetration test results gradually emerged in recent years [19–22]. Scholars proposed a variety of methods for calculating pile base resistance based on the cone penetration test. These methods all adopt the method of multiplying cone penetration resistance by empirical coefficient. Each method considers the influence of soil in a certain range above and below the pile end on the pile base resistance because the soil in a certain range above and below the pile end contributes to the pile base resistance. However, due to the complexity of determining the depth range of pile base resistance, the depth range of cone penetration resistance is different when calculating pile base resistance by various methods, which leads to the difference of calculation results. For example, the average cone penetration resistance depth proposed by Schmertmann [23] ranges from 8 D above the pile end to 0.7 D–4 D below the pile end. The average cone penetration resistance depth range of European [24] is 0.7 D–4 D below the pile end. ICP [25] and LCPC [26] adopt the depth range of ±1.5 D at the pile end. Based on the test results of the unified database, Lehane [27] proposed to use the range of 20 times the pile wall thickness below the pile end. Although the depth range of the Fugro [28] method also uses ± 1.5 D at the pile end, it takes the average value of the net cone penetration resistance.

It can be seen that the influence range of pile base resistance has a great influence on the calculation of pile base resistance, especially when the pile end is in the clay layer with the thin sand layer. Based on the soil layer data and cone penetration test data of the Liuhua site in the South China Sea, this paper analyzes the differences of the calculation methods of ultimate unit base resistance, analyzes the distribution of the plastic zone of the soil around the pile end by numerical simulation, and gives the recommended range of influence of the pile base resistance. The research results can provide a reference for the pile base resistance in the South China Sea.

2. Difference in Calculation Formula of Pile Base Resistance

2.1. Engineering Example

The design parameters of the pile foundation of a deep water jacket platform in Liuhua are shown in Table 1. The sampling depth of the borehole in the site is 170 m, and the geological data exposed are shown in Table 2.

Table 1. Design parameters of pile foundation.

Pile Length (m)	Length of Embedded Pile (m)	Diameter (m)	Thickness (mm)
171	134	2.743	70

Table 2. Soil parameters.

Stratum	Soil Description	Thickness (m)	Submerged Unit Weight γ' (kN/m³)	Internal Friction Angle φ (°)	Design Shear Strength S_u (kPa)
1	Very soft sandy clay	0.8	8.2	—	11
2	Soft calcareous silty clay	5.8	7.7	—	12–24
3	Slightly hard calcareous sandy clay	2.5	7.0	—	28–38
4	Slightly hard to hard calcareous silty clay	36.4	7.6	—	30–75
5	Hard to very hard calcareous sandy clay	3.9	9.8	—	90–120
6	Loose to medium dense clayey sand	2.2	9.3	25	—
7	Hard to very hard calcareous silty clay	21.1	8.6	—	80–130
8	Medium dense carbonate silty fine sand	6.1	6.5	30	—
9	Very hard to hard calcareous silty clay	91.4	9.6	—	160–400

The penetration depth of an in situ cone penetration test is 170 m, and the test results are shown in Figure 1.

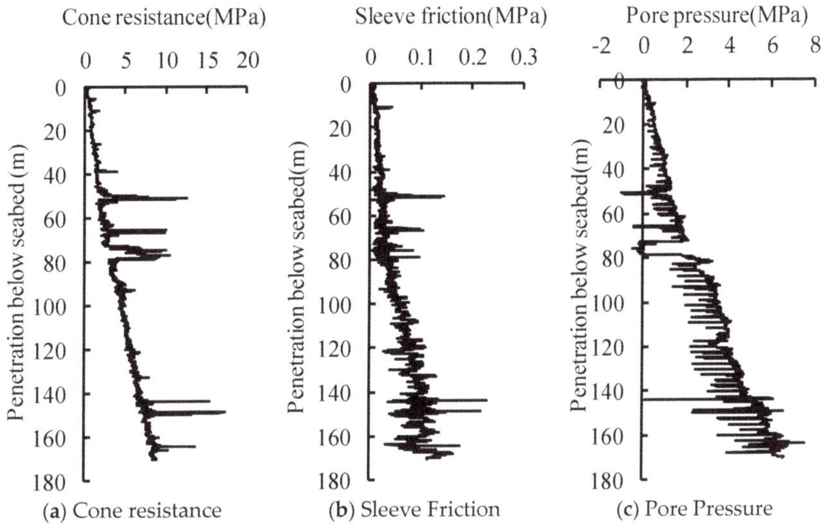

Figure 1. Data of cone penetration test.

2.2. Results of Pile Base Resistance

The common methods for calculating pile base resistance include API/ISO, NGI, ICP, LCPC, and Fugro, etc. The soil parameter types (soil shear strength S_u or cone resistance) and the empirical coefficients are different in these methods. Because the pile base resistance has a certain range of influence above and below the pile end, the depth range of the soil at the pile end is different when calculating the pile base resistance by various methods, and the comparison is shown in Table 3.

Table 3. Methods for calculating unit pile base resistance.

Method	Formula	Kernel Parameter	Range of Soil
API/ISO	$q = 9S_u$	Design Shear Strength, S_u	Pile end
NGI	$q = 9S_u^{UU}$	UU Shear Strength, S_u^{UU}	Pile end
Lehane	$q = q_t[0.2 + 0.6D^*/D]$	q_t	20 times the pile wall thickness below the pile end
Schmertmann	$q = \frac{q_{c1}+q_{c2}}{2}$	q_c	8 D above the pile end, 0.7 D–4 D below the pile end
European	$q = 9C_u \le 15\text{MPa } C_u = q_c/N_k$	q_c	0.7 D–4 D below the pile end
ICP	Closed: $q = 0.8\,q_c$ (undrained) $q = 0.8\,q_c$ (drained) Open (fully plugged): $q = 0.4\,q_c$ (undrained) $q = 0.65\,q_c$ (drained) Open (unplugged): $q = 1.0\,q_c$ (undrained) $q = 1.6\,q_c$ (drained)	q_c	$\pm 1.5\,D$ at the pile end
LCPC	$q = kq_c$	q_c	$\pm 1.5\,D$ at the pile end
Penpile [29]	$q = 0.25q_c$	q_c	Pile end
Takesue [30]	$q = q_t - u_2$	q_t	Pile end
Fugro	$q = 0.7q_{n,av}$	q_n	$\pm 1.5\,D$ at the pile end

Notes: S_u is design shear strength; S_u^{UU} is shear strength determined by UU test; D is the out diameter of pile, $D^* = \left(D^2 - D_i^2\right)^{0.5}$, D_i is the inner diameter of pile; q_{c1} is average cone resistance within the range of 8 D above the pile end, q_{c2} is average cone resistance within the range of 0.7–4 D below the pile end; C_u is the shear strength of soil, N_k is empirical coefficient; q_c is cone resistance; q_t is corrected cone resistance, $q_t = q_c + (1 - a)u_2$; q_n is net cone resistance, $q_n = q_c + (1 - a)u_2 - \sigma_{v0}$; u_2 is pore pressure; k is empirical coefficient; and $q_{n,av}$ is the average net cone resistance within the 1.5 D above and below the pile end.

The methods for calculating pile base resistance can be divided into three categories as shown in Table 3. The first class of methods is based on the shear strength at the pile end, including API/ISO, NGI, Penpile, and Takesue; the second class of methods is based on the average of cone resistance within a certain range below the pile end, including Lehane and European; and the last class of methods is based on the average of the cone resistance, corrected cone resistance, and other parameters of the soil within a certain range above and below the pile end. The calculation range of soil at the pile end is different, and the specific calculation range is shown in Table 3.

The bearing capacity of pile foundation in the Liuhua site is calculated using the method in Table 3, and the distribution of pile base resistance along the depth is shown in Figure 2.

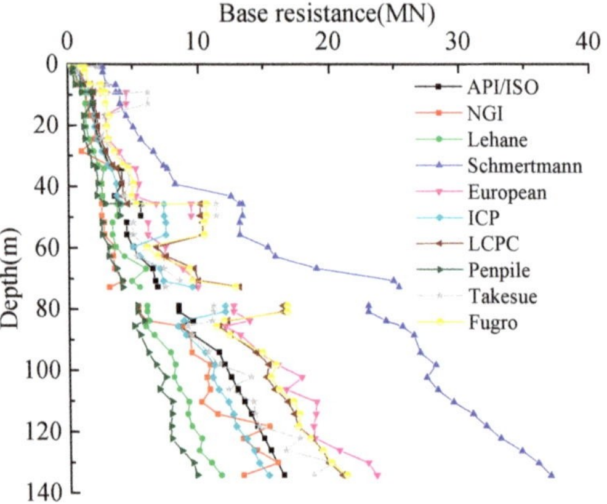

Figure 2. Distribution of pile base resistance along the depth.

The difference of pile base resistance calculated by each method is obvious as shown in Figure 2; the value of Schmertmann is the largest, and the value of Penpile is the minimum. The soil within the depth range of 49.4–51.6 m and 72.7–78.8 m is sand, so the cone resistance increases rapidly, which results in a sudden change in the pile base resistance calculated near this depth. In order to verify the effect of thin sand on pile base resistance, the pile base resistance at the depth of z = 33 m, 49 m, 80 m, and 134 m of each method is taken as shown in Figure 3.

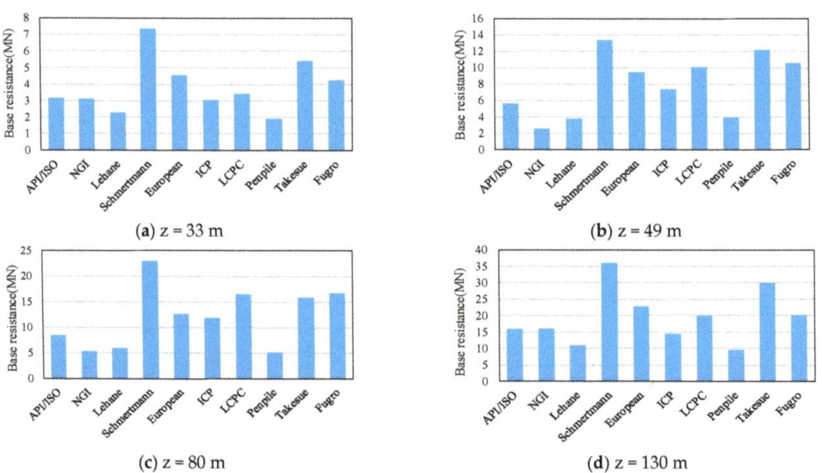

Figure 3. Base resistance calculated by methods at different depths.

It can be seen from Figure 3 that the base resistance calculated by each method has large dispersion and difference. The maximum, minimum, mean, variance, and coefficient of variation in pile base resistance at different depths are summarized in Table 4.

Table 4. Difference of base resistance at different depths.

Depth (m)	Maximum (MN)	Minimum (MN)	Mean (MN)	Variance	COV
33	7.4	1.9	3.9	1.6	0.41
49	13.4	2.5	7.9	3.8	0.48
80	22.9	5.3	12.2	5.9	0.49
130	36.0	9.5	19.6	8.3	0.42

It can be seen from Table 4 that the maximum base resistance calculated by various methods at different depths is about 4–6 times the minimum. The differences are mainly caused by two aspects: one is the difference of soil parameter and empirical coefficient, and the other is the difference of calculation depth range of soil at the pile end. At the depth of 33 m and 130 m, the calculation results of each method are similar because the soil distribution is uniform and because of the absence of the sand interlayer. While at the depth of 49 m and 80 m, due to the presence of the sand layer, the soil strength is affected by the calculation depth range of the pile end, which leads to significant differences in the calculation results of each method. It can be seen that the difference of pile base resistance caused by the different calculation depth range of the pile end is far greater than the difference caused by the calculation method itself.

The soil within the depth of 49.4–51.6 m of the Liuhua site is sand, and within a range above and below the sand, the pile base resistance is calculated by each method at the depth of 44 m, 49 m, 52 m, 57 m, and 62 m. The results are shown in Figure 4.

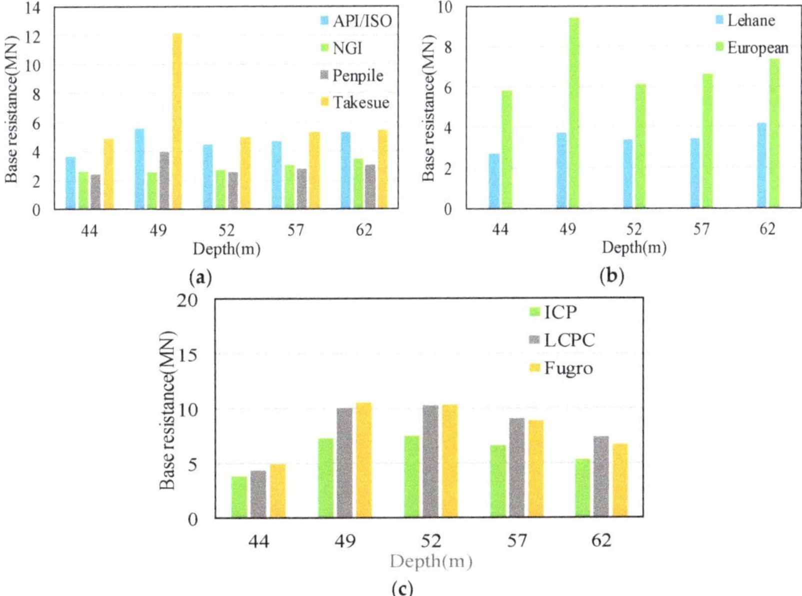

Figure 4. Base resistance at representative depth. (**a**) Based on the parameters at the pile end, (**b**) based on the average of parameters below the pile end, (**c**) and based on the average of parameters above and below the pile end.

The method of calculating the pile base resistance uses the cone resistance at the pile end in Figure 4a. In these methods, the pile base resistance in the upper and lower clay is not affected by sand, so it does not have a sudden change. However, the influence of pore pressure cannot be ignored in the Takesue method, and the pore pressure suddenly dissipates when close to the sand in Figure 1c. As a result, the pile base resistance increases sharply when it is close to the sand, which also indirectly proves that the sand has an impact on the pile base resistance in the overlying clay layer. The two methods in Figure 4b are to calculate the pile base resistance by using the average of the soil parameters below the pile end. When only considering the influence of the soil below the pile end, the base resistance increases at a depth of 49 m near the sand layer, while the European method increases more significantly at the position close to the sand because the large range of soil depth is calculated downward.

The three methods in Figure 4c are the methods of calculating the pile base resistance using the average of the soil above and below the pile end. When considering the influence of the soil above and below the pile end, the cone resistance of the sand is much larger than that of the clay, resulting in the pile base resistance calculated at the depth of 49 m and 52 m being the same, with a sharp increase trend compared with the pile base resistance at the depth of 44 m.

The difference of pile base resistance calculated by different methods is analyzed, and the conclusions are obtained: when calculating the pile base resistance, the range of the soil near the pile end has a significant impact on the results, and the distance between the sand and the clay and the thickness of the sand have an important impact on the pile base resistance.

3. Numerical Simulation on Influence Range of Pile Base Resistance

3.1. Numerical Model and Verification

The finite element method (FEM) is used to analyze the influence range of pile base resistance in homogeneous clay and sand-layered clay. The linear model is adopted for the pile, with the diameter $D = 2.743$ m, the length $L_0 = 85.29$ m, the length of the embedded $L = 82.29$ m, and the length–diameter ratio $L/D = 30$. The Mohr–Coulomb model is adopted for soil, the effective The relationship between mass and mobility is often expressed with the effective density, which is defined as the mass of the particle divided by its mobility equivalent volume, of soil is taken, Poisson's ratio is taken as 0.49, and the elastic modulus is taken as $500\,S_u$. The parameters of FEM are shown in Table 5.

Table 5. Parameters of FEM.

Model	Effective Density (kg/m^3)	Elastic Modulus (kPa)	Poisson's Ratio	Shear Strength (kPa)	Internal Friction Angle (°)
Pile	6850	210×10^6	0.25	—	—
Clay	600	$500\,S_u$	0.49	20–100	—
Sand	900	50,000	0.3	—	30

Note: The relationship between mass and mobility is often expressed with the effective density, which is defined as the mass of the particle divided by its mobility equivalent volume.

The diameter of soil is $20\,D$ and the vertical height is $L + 20\,D$ in the numerical model to avoid boundary effect. In order to improve the calculation speed, the axisymmetric model is used in the numerical simulation. In the model, the vertical displacement U3 is limited at the bottom, horizontal displacement U1 is limited at the right, and symmetric boundary conditions are set on the left side of the model. The numerical model of a steel pipe pile is shown in Figure 5.

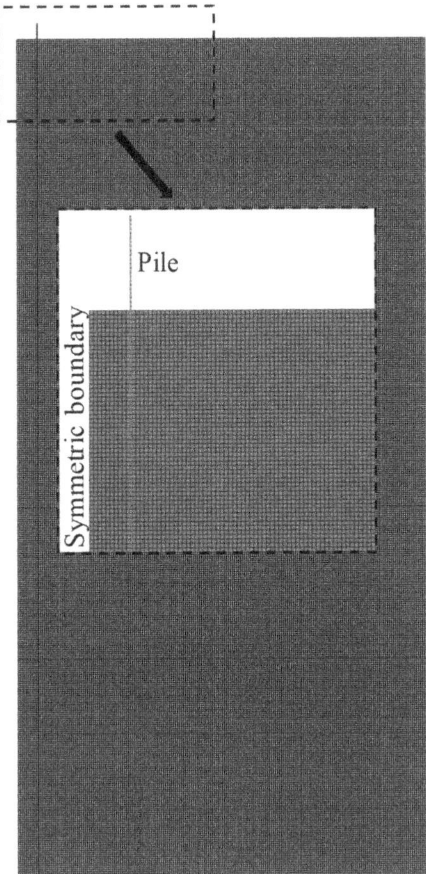

Figure 5. Axisymmetric model of a steel pipe pile.

The contact pair is set between pile and soil, the pile surface is regarded as the main surface, and the soil surface is treated as the slave surface. The 'Penalty Function' is established in the contact property. The friction coefficient between pile and clay is set to 0.4, and the friction coefficient between piles and sand is set to 0.3.

The lateral boundary of the soil is subject to horizontal constraints, the bottom boundary is subject to horizontal and vertical constraints, and the symmetric boundary is subject to axisymmetric constraints. The soil unit type is C3 D8R, and the soil grid is 0.05 D. The reference point is coupled in the middle of the pile top surface, and vertical displacement of 0.05 D is applied on the reference point to calculate the base resistance.

In order to verify the rationality of the calculation results of FEM, the undrained shear strength of the soil is set to 40 kPa, and the load–displacement curve calculated by FEM and API is compared, as shown in Figure 6. The trend of the load–displacement curve calculated by FEM is basically the same as that calculated by the API, and the ultimate bearing capacity of the pile foundation calculated by FEM is 28.1 MN, and the ultimate bearing capacity calculated by the API is 26.5 MN, the difference between the two is 5.7%, and the error is small, so the rationality of FEM is verified.

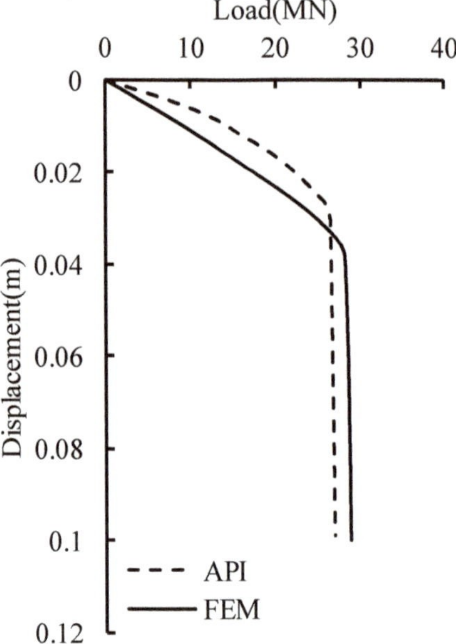

Figure 6. Verification of FEM.

3.2. Results in Homogeneous Clay

The influence range of pile base resistance in homogeneous clay is studied. In the numerical model, the undrained shear strength of homogeneous clay is set as 20 kPa, 40 kPa, 60 kPa, 80 kPa, and 100 kPa, respectively, to analyze the distribution of the plastic zone of the pile end and determine the influence range of pile base resistance. The results are shown in Figure 7.

It can be seen from Figure 7 that with the increase in undrained shear strength of soil, the distribution range of the plastic zone near the pile end did not change significantly, that is, the undrained shear strength in homogeneous clay has little effect on the development of plastic zone. The influence range of pile base resistance is about 0.12 D at the upper part of the pile end and about 0.83 D at the lower part of the pile end.

The undrained shear strength of homogeneous clay is taken as 40 kPa, and the distance between the pile end and the soil bottom boundary is set as 0.5 D, 1.0 D, 1.5 D, 2.0 D, and 3.0 D. The distribution of the plastic zone at the pile end, load–displacement curve, and relationship between ultimate bearing capacity and distance are shown in Figures 8 and 9.

It can be seen from Figure 8 that when the distance between the pile end and the soil bottom boundary reaches 1.0 D, the plastic zone at the pile end does not reach the soil bottom boundary. With the increase in the distance, the development depth of the plastic zone is unchanged. Therefore, the influence range of the pile base resistance is still 0.12 D above the pile end and 0.83 D below the pile end.

It can be seen from Figure 9 that when the distance between the pile end and the soil bottom boundary reaches 1.0 D, the load–displacement curve is nearly coincident. The ultimate bearing capacity of the pile foundation is reduced with the distance between the pile end and the soil bottom boundary, but the change is not obvious, which indicates that the distance 1.0 D can meet the development of the plastic zone. Based on the results in Figures 7 and 8, it can be considered that the distribution range of the plastic zone and the range of the influence depth of the pile base resistance reflected by the load–displacement curve have good consistency. From the results in Figures 7–9, it can be considered that the influence range of the pile base resistance in the homogeneous clay is 0.12 D above the pile end and 0.83 D below the pile end.

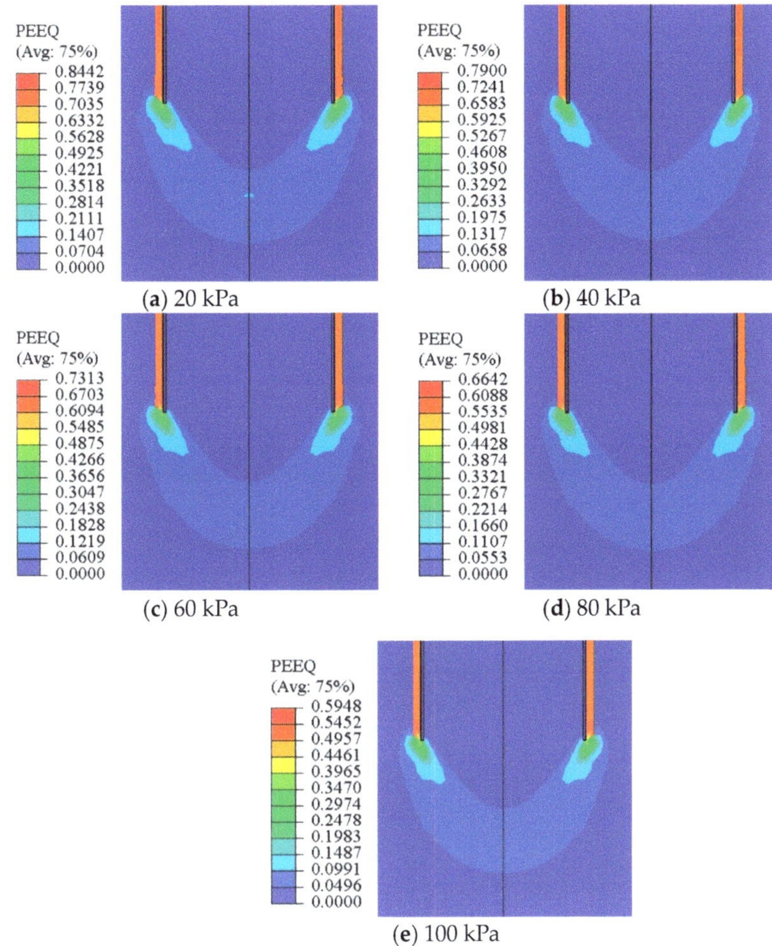

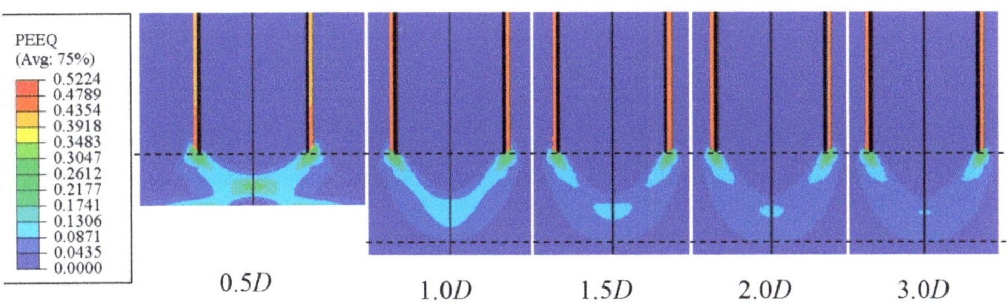

Figure 7. Distribution of plastic zone at pile end under the different undrained shear strength.

Figure 8. Distribution of plastic zone at pile end under the different distance between the pile end and the soil bottom boundary.

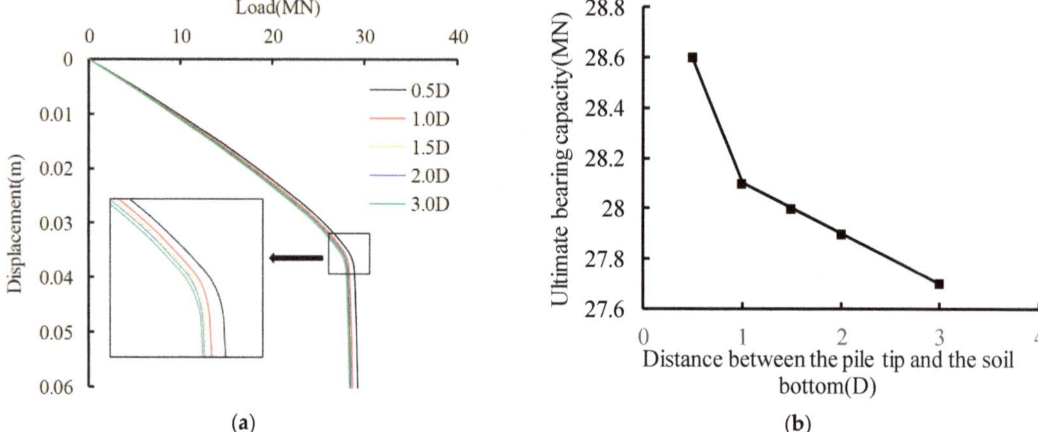

Figure 9. Ultimate bearing capacity. (**a**) Load–displacement curve, (**b**) relationship between ultimate bearing capacity and distance.

3.3. Results when the Clay Is above the Sand

3.3.1. Influence of Distance between Sand and the Pile End

A sand layer is set under the pile end to analyze the influence of the sand under the clay on the pile base resistance. Additionally, the influence of the distance between the top of the sand and the pile end on the pile base resistance is analyzed. This distance H_1 is taken as 0.5 *D*, 1.0 *D*, 1.5 *D*, 1.7 *D*, and 2.0 *D*, respectively. The results are shown in Figure 10b.

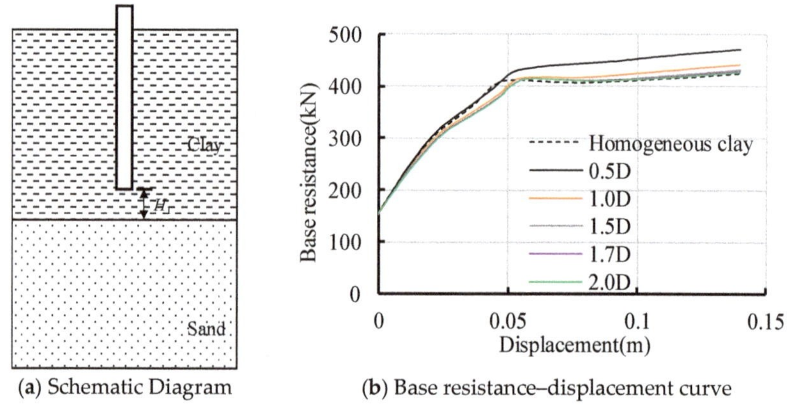

(**a**) Schematic Diagram   (**b**) Base resistance–displacement curve

Figure 10. Influence of the distance between the pile end and top of the sand on pile base resistance.

It can be seen from Figure 10 that the pile base resistance is enhanced when the distance between the top of the sand and the pile end is 0.5 *D*, increasing by about 10.8%. With the increase in the distance, the pile base resistance decreases gradually. When the distance exceeds 1.5 *D*, the sand under the clay has little effect on the pile base resistance.

3.3.2. Influence of Sand Thickness

The influence of sand on pile base resistance is mainly within the range of the distance between the top of the sand and the pile end and is less than 1.5 *D* in the above analysis. Therefore, the distance is set to 0.5 *D*, and the thickness of the sand is changed to 0.3 *D*, 0.5 *D*, 0.8 *D*, 1.0 *D*, and 1.5 *D*, respectively, to analyze the influence of sand thickness on pile base resistance; the thickness of sand T_1 is shown in Figure 11a, and the result is shown in Figure 11b.

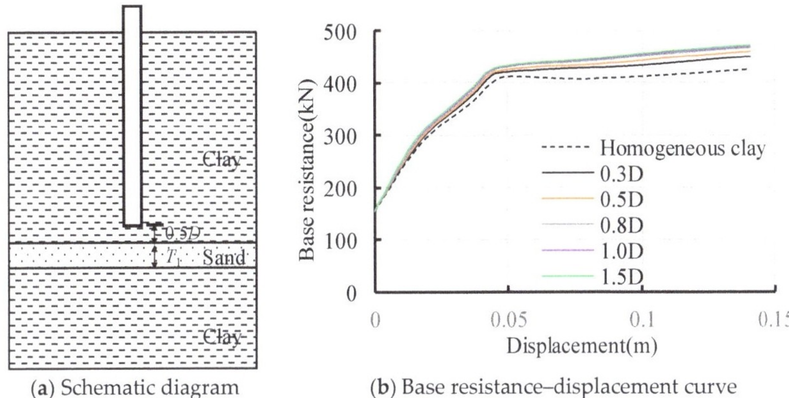

(**a**) Schematic diagram   (**b**) Base resistance–displacement curve

Figure 11. Influence of sand thickness under the pile end on pile base resistance.

With the increase in the thickness of the sand, the pile base resistance gradually increases. When the thickness of the sand exceeds 0.8 D, the continuous increase in the thickness of sand has no effect on the pile base resistance. Compared with the pile end in homogeneous clay, when the sand thickness is 0.3 D and 0.5 D, respectively, the pile base resistance increases by 5.7% and 8.4%. Therefore, when there is a sandy layer under the clay, and the distance between the pile end and the top surface of the sand is not more than 1.5 D, it is recommended that the influence range of the pile base resistance is 1.3 D.

3.4. Results when the Clay Is under the Sand

3.4.1. Influence of Distance between Sand and the Pile End

A sand layer is set above the pile end to analyze the influence of the sand above the clay on the pile base resistance. Additionally, the influence of the distance between the bottom of the sand and the pile end on the pile base resistance is analyzed. This distance H_2 is taken as 0.1 D, 0.2 D, 0.5 D, 1.0 D, and 1.5 D, respectively. The results are shown in Figure 12.

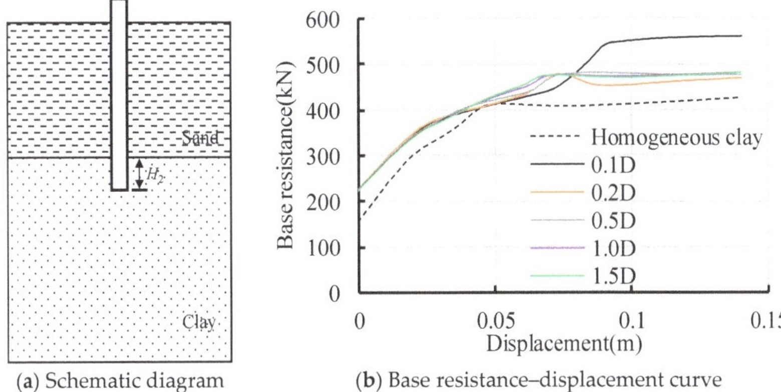

(**a**) Schematic diagram   (**b**) Base resistance–displacement curve

Figure 12. Influence of distance between the pile end and bottom of sand on pile base resistance.

The pile base resistance increases when the sand is above the pile end compared with the pile end in homogeneous clay, and the influence is very significant. When the distance between the bottom of the sand and the pile end is 0.1 D and 0.2 D, the pile base resistance increases by 32.1% and 10.0%. When the distance exceeds 0.5 D, pile base resistance is not affected by distance.

3.4.2. Influence of Sand Thickness

The distance between the bottom of the sand and the pile end is 0.1 D, and the change in the thickness of sand to 0.3 D, 0.5 D, 0.8 D, 1.0, D and 1.5 D, respectively, to analyze the influence of sand thickness on pile base resistance, the thickness of sand T_2 is shown in Figure 13a, and the result is shown in Figure 13b.

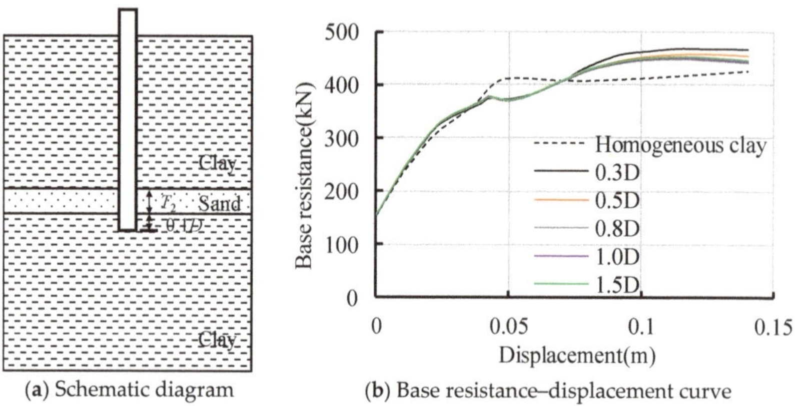

(**a**) Schematic diagram   (**b**) Base resistance–displacement curve

Figure 13. Influence of sand thickness above the pile end on pile base resistance.

As shown in Figure 13, with the increase in the thickness of sand, the pile base resistance decreases gradually. When the thickness of sand exceeds 0.8 D, the thickness of sand has no effect on the pile base resistance. Therefore, in the case of sand above the clay, when the distance between the bottom of sand above the clay and the pile end is less than 0.5 D, it is suggested that the influence range of pile base resistance is 0.9 D.

4. Conclusions

In the present study, the influence range of pile base resistance is studied in homogeneous clay and sand-layered clay. The influence of the strength and thickness of the soil near the pile end on the pile base resistance is revealed, and the recommended influence range of pile base resistance is given. The following conclusions can be achieved:

The pile base resistance has a certain influence range above and below the pile end, and the range of soil used to calculate the average strength at the pile end is different in the current calculation methods, resulting in different pile base resistance calculated by different methods; especially when the there is sand-layered clay, t he difference is more significant.

The influence range of pile base resistance is 0.12 D above the pile end and 0.83 D below the pile end when the pile end is located in homogeneous clay.

When there is sand above or below the clay where the pile end is located, the distance between the sand and the pile end and the thickness of the sand has an impact on the pile base resistance. When the sand is above the clay, the influence range of the pile base resistance is 0.9 D above the pile end, and when the sand is below the clay, the influence range of pile base resistance is 1.3 D below the pile end.

Author Contributions: Conceptualization, D.F. and H.Z.; methodology, S.L.; software, Y.J.; validation, H.Z., R.L. and C.L.; formal analysis, D.F.; investigation, S.L.; resources, D.F.; data curation, R.L.; writing—original draft preparation, D.F.; writing—review and editing, Y.J.; visualization, C.L.; supervision, C.L.; All authors have read and agreed to the published version of the manuscript.

Funding: This research received no external funding.

Institutional Review Board Statement: Not applicable.

Informed Consent Statement: Not applicable.

Data Availability Statement: Not applicable.

Conflicts of Interest: The authors declare no conflict of interest.

References

1. Wu, X.; Hu, Y.; Li, Y.; Yang, J. Foundations of offshore wind turbines: A review. *Renew. Sustain. Energy Rev.* **2019**, *104*, 379–393. [CrossRef]
2. Li, L.; Zheng, M.; Liu, X.; Wu, W. Numerical analysis of the cyclic loading behavior of monopile and hybrid pile foundation. *Comput. Geotech.* **2022**, *144*, 104635. [CrossRef]
3. Basack, S.; Goswami, G.; Dai, Z.H.; Baruah, P. Failure-mechanism and design techniques of offshore wind turbine pile foundation: Review and research directions. *Sustainability* **2022**, *14*, 12666. [CrossRef]
4. Zhang, X.; Liu, C.; Wang, P.; Zhang, G. Experimental simulation study on dynamic response of offshore wind power pile foundation under complex marine loadings. *Soil Dyn. Earthq. Eng.* **2022**, *156*, 107232.
5. American Petroleum Institute. Geotechnical and Foundation Design Considerations. In *ANSI/API Recommended Practice 2GEO*, 1st ed.; API Publishing Services: Washington, DC, USA, 2014.
6. Terzaghi, K.T. *Theoretical Soil Mechanics*; John Wiley and Sons Inc.: Hoboken, NJ, USA, 1943.
7. Meyerhof, G.G. The ultimate bearing capacity of foundations. *J. Geotech.* **1951**, *2*, 301–332. [CrossRef]
8. Meyerhof, G.G. The ultimate bearing capacity of foundations under eccentric and inclined loads. In Proceedings of the third International Conference on Soil Mechanics and Foundation Engineering, Zurich, Switzerland, 16–27 August 1953; Volume 1.
9. Meyerhof, G.G. Influence of roughness of base and ground-water conditions on the ultimate bearing capacity of foundations. *J. Geotech.* **1955**, *5*, 227–242. [CrossRef]
10. Meyerhof, G.G. Bearing capacity and settlement of pile foundations. *J. Geotech. Eng. Div.* **1976**, *102*, 197–228. [CrossRef]
11. Berezantzev, V.G. Load bearing capacity and deformation of piled foundations. In Proceedings of the 5th International Conference on Soil Mechanics and Foundation Engineering ISSMFE, Paris, France, 17–22 July 1961; Volume 2, pp. 11–12.
12. Hu, G.C.Y. Variable-factors theory of bearing capacity. *J. Soil Mech. Found. Div.* **1964**, *90*, 85–95. [CrossRef]
13. Vesic, A.S. *Design of Pile Foundations*; Transportation Research Board, National Research Council: Washington, DC, USA, 1977.
14. Wilson, G. The Bearing Capacity of Screw Piles and Screwcrete Cylinders. *J. Inst. Civ. Eng.* **1950**, *34*, 4–73. [CrossRef]
15. Meigh, A.C. Model Footing Tests on Clay. Master's Thesis, Imperial College London, London, UK, 1950.
16. Yassin, A.A. Model Studies on Bearing Capacity of Piles. Ph.D. Thesis, Imperial College London, London, UK, 1950.
17. Skempton, A.W. The Bearing Capacity of Clays. *Build. Res. Congr.* **1951**, *1*, 180–189.
18. Randolph, M.; Gourvenec, S. *Offshore Geotechnical Engineering*; CRC Press: Boca Raton, FL, USA, 2017.
19. Gavin, K.; Kovacevic, M.S.; Igoe, D. A review of CPT based axial pile design in The Netherlands. *Undergr. Space* **2019**, *6*, 85–99. [CrossRef]
20. Amirmojahedi, M.; Abu-Farsakh, M. Evaluation of 18 Direct CPT Methods for Estimating the Ultimate Pile Capacity of Driven Piles. *Transp. Res. Rec. J. Transp. Res. Board* **2019**, *2673*, 127–141. [CrossRef]
21. Leetsaar, L.; Korkiala-Tanttu, L.; Kurnitski, J. CPT, CPTu and DCPT methods for predicting the ultimate bearing capacity of cast in situ displacement piles in silty soils. *Geotech. Geol. Eng.* **2022**, *41*, 631–652. [CrossRef]
22. Heidari, P.; Ghazavi, M. Statistical Evaluation of CPT and CPTu Based Methods for Prediction of Axial Bearing Capacity of Piles. *Geotech. Geol. Eng.* **2021**, *39*, 1259–1287. [CrossRef]
23. Schmertmann, J.H. *Guidelines for Cone Penetration Test: Performance and Design*; Federal Highway Administration: Washington, DC, USA, 1978.
24. De Kuiter, J.; Beringen, F.L. Pile foundations for large North Sea structures. *J. Mar. Georesour. Geotechnol.* **1979**, *3*, 267–314. [CrossRef]
25. Jardine, R.; Chow, F.; Overy, R. *ICP Design Methods for Driven Piles in Sands and Clays*; Thomas Telford: London, UK, 2005.
26. Bustamante, M.; Gianeselli, L. Pile bearing capacity prediction by means of static penetrometer CPT. In Proceedings of the 2nd European Symposium on Penetration testing, Amsterdam, The Netherlands, 24–27 May 1982; Volume 2, pp. 493–500.
27. Lehane, B.M.; Liu, Z.; Bittar, E.J. CPT-based axial capacity design method for driven piles in clay. *J. Geotech. Geoenviron. Eng.* **2022**, *148*, 04022069. [CrossRef]
28. Van Dijk, B.F.J.; Kolk, H.J. CPT-based design method for axial capacity of offshore piles in clays. *J. Front. Offshore Geotech. II* **2011**, 555–560.

29. Price, G.; Wardle, I.F. A comparison between cone penetration test results and the performance of small diameter instrumented piles in stiff clay. In Proceedings of the 2nd European Symposium on Penetration Testing, Amsterdam, The Netherlands, 24–27 May 1982; Volume 2, pp. 775–780.
30. Takesue, K.; Sasao, H.; Matsumoto, T. Correlation between ultimate pile skin friction and CPT data. In Proceedings of the First International Conference on Site Characterization, Atlanta, Georgia, 19–22 April 1998; pp. 1177–1182.

 *sustainability*

Article

Seismic Performance Comparison of Three-Type 800 m Spherical Mega-Latticed Structure City Domes

Zibin Zhao [1,2] and Yu Zhang [1,2,*]

[1] Key Lab of Structures Dynamic Behavior and Control of the Ministry of Education, Harbin Institute of Technology, Harbin 150090, China
[2] Key Lab of Smart Prevention and Mitigation of Civil Engineering Disasters of the Ministry of Industry and Information Technology, Harbin 150090, China
* Correspondence: zhangyuhit@hit.edu.cn

Abstract: With changes in the city environment and advances in engineering technologies, there is an increasing demand for the construction of super-large span city domes that can cover a large area to create a small internal environment within a specific region. However, the structural design must overcome various challenges in order to break the current structural span limitations. Moreover, there is little research on structures achieving such large spans. The seismic performance of the selected Kiewitt-type, Geodesic-type, and Three-dimensional grid-type mega-latticed structures is further investigated upon previous studies of the model selection, static and stability analysis results of the 800 m span mega-latticed structures. Finite element models were established with ANSYS to analyze the modal properties and earthquake response of the structures. The study evaluated the impact of earthquake directionality on the structural response as well as the response pattern of the structure under frequent and rare earthquake actions. It was found that the overall integrity of the structures is good, with strong coupling effects in three directions. The multi-dimensional seismic input method should be applied to solve the structural response. Combining the plastic development of the structure under rare earthquakes, the top and the circumferential trusses of the third and fourth rings are relatively weak parts of the structures. According to this study, given the known static analysis results, the maximum displacement and maximum stress of the structures under frequent and rare earthquake actions can be estimated. Furthermore, the study highlights that Three-dimensional grid-type mega-latticed structures should be prioritized designing structures with spans of 800 m, providing helpful guidance for the practical application of this type of structure.

Keywords: super-large span; mega-latticed structure; dynamic characteristics; seismic performance; plastic development; model selection

Citation: Zhao, Z.; Zhang, Y. Seismic Performance Comparison of Three-Type 800 m Spherical Mega-Latticed Structure City Domes. *Sustainability* **2023**, *15*, 7240. https://doi.org/10.3390/su15097240

Academic Editor: Gianluca Mazzucco

Received: 13 March 2023
Revised: 5 April 2023
Accepted: 24 April 2023
Published: 26 April 2023

1. Introduction

Structural engineers in the field of space structure have been pursuing larger structural spans. The reinforced concrete Centennial Hall in Wroclaw [1], with a span of 65 m that was built in 1913, is still in use today. After more than a century, the structural materials have advanced from concrete to steel [2] and aluminum alloys [3], and the structural span has increased from tens of meters to hundreds of meters. At the same time, structural forms are also becoming increasingly varied and complicated [4,5]. With the continuous breakthrough of the structural span, people's living demands and the pursuit of environmental quality are gradually rising. In recent years, climate change has intensified, and extreme weather has occurred frequently [6], causing many inconveniences to people's normal lives and leading to losses of life and property, thus prompting people to think about protecting existing cities. Engineers are currently exploring the possibility of constructing super-large-span structures, and such a kilometer-level "city dome" could create many new challenges [7]. The "city dome" could reduce the huge costs of air conditioning and snow removal in winter, while creating a comfortable indoor environment. Additionally, because of the considerable interior space, the dome may experience uplift due to the hot air inside, which will offset some of the gravity loads of the structure. Although there are already a few structures that exceed 300 m [8–10], the development of kilometer-level "city domes" requires continuous exploration and innovation of structural systems.

The increase in span leads to more complex structural forms having an increasing number of structural members and a greater weight. As a result, structural designs encounter new challenges to increase the bearing capacity against static and dynamic loads and improve structural stability. The mega-latticed structure proposed by He and Zhou [11] has a dual force transmission system of the main structure and substructure, which is more reasonable when the span reaches more than 200 m. He and Zhou first studied cylindrical mega-latticed structures with a double-layer grid shell as the substructure [12], providing a preliminary understanding of their static performance and buckling modes. Then, they investigated the cylindrical mega-latticed structure with a single-layer-latticed membranous shell substructure [13] and a cylindrical mega-latticed structure with the single-layer intersectional grid cylindrical shell substructure [14] and gave a range of values for the rise/span ratio of the substructure and the main structure. When the rise/span ratio for both the main structure and substructure is 1/6, the ultimate bearing capacity of the structure reaches its peak [14]. It is also noted that the buckling mode of the structure depends on the stiffness ratio between the main structure and the substructure, and for safety considerations, the instability of the main structure should occur after that of the substructure [13]. By studying the stability during the construction process of the spherical mega-latticed structures [15], it was concluded that the geometric nonlinearity of the structure should be considered in the stability analysis during construction. However, their studies on this structural form have primarily been concentrated on spans between 80 m and 160 m, without delving into even larger spans. Zhang [16,17] carried out studies on the optimization and model selection of spherical mega-latticed structures with a span of 800 m, as well as the static and stability performance of the structures. It is shown that at the span of 800 m, the maximum displacement and the structural steel consumption under static forces of the Kiewitt-type, Geodesic-type, and Three-dimensional grid-type mega-latticed structures are smaller than those of the Ribbed-type, Schwedler-type, and Sunflower-type mega-latticed structures, and their stability capacity is higher, making them suitable as structural forms for spans up to 800 m [17]. The mega-latticed structure has better light transmittance and requires less steel than traditional double-layer-reticulated shell structures, which can effectively reduce the carbon emissions of the structure and play an important role in energy saving and emission reduction.

Earthquakes are sudden natural disasters that can have significant impacts on personal safety, the natural environment, and human society. Both high-rise structures and large-span space structures are voluminous architectural structures which often accommodate a high concentration of people. If they suffer damage during an earthquake, it could result in unpredictable and devastating consequences. Thus, it is essential to conduct seismic research on them. Research on the seismic resistance of high-rise buildings is relatively mature, with current efforts focused on energy dissipation and vibration reduction in structures. Wang [18] and Zhang [19] conducted research on the seismic performance of steel-truss-reinforced concrete core walls and steel frames with replaceable energy dissipation elements, respectively. The steel truss reinforcement significantly improved the load-bearing capacity, ductility, and energy dissipation capability of reinforced concrete core walls, while the replaceable energy dissipation elements effectively reduced the structural damage and maintenance cost. Applying the numerical simulation method, Barkhordari compared the performance of different types of passive damping systems under seismic action [20]. The results showed that friction-based and viscous-based damping systems were the most effective type of passive damping systems. In terms of seismic research on large-span space structures, Abdalla [21] analyzed the dynamic characteristics of spherical and parabolic reinforced-concrete domes and obtained the relationship between the structural frequency and dome height as well as thickness. Feizolahbeigi [22] discussed the seismic performance of double-shell spherical domes in Iran during the 16th to 18th centuries. Through the study of many cases, it was found that rational geometric shapes and proportions can improve the overall stability of the structure, while appropriate construction techniques can reduce the structure's seismic response. Research on isolation systems for large-span spatial structures is also being conducted. The use of hybrid three-directional isolation systems can significantly reduce the seismic response of dome structures while possessing a good energy dissipation capacity and control effect [23].

The seismic analysis of super-large-span structures becomes more complex as the span increases. The significant increase in the number of rods increases the number of low-order vibration modes [24], and the relatively dense structural frequencies and the similarity of the periods of the different modes reflect the complexity of the dynamic properties of the structure due to the existence of multiple symmetry axes [25]. In addition, the height of the structure increases at the same time as the span of the structure increases, resulting in the dual characteristics of large-span and high-rise structure. Therefore, it is necessary to conduct research on the dynamic performance of the structures based on the previous static and stability studies [17]. The analysis of the dynamic characteristics of the structures and the discussion of the influence of the earthquake direction on their response can help to understand the overall stiffness. The calculation of the structures' response under frequent and rare earthquakes provides a rough estimate of the response at a specific seismic intensity with known static results. Studying the plastic development of the structure under rare earthquakes contributes to identifying weak parts and enables appropriate design measures. Finally, we propose the optimal structural model for an 800 m span spherical mega-latticed structure based on the structures' dynamic performance, providing guidance for practical applications.

2. Spherical Mega-Latticed Structure Model

2.1. Geometric Model of the Spherical Mega-Latticed Structure

Based on the static calculation results by Zhang [17], a spherical mega-latticed structure can be adopted with three structural forms: Kiewitt-type, Geodesic-type, and Three-dimensional grid-type, as shown in Figure 1a, for spans of up to 800 m. These three structural forms all divide the sphere into relatively uniform triangular grids, exhibiting good and similar mechanical performances. The primary geometric parameters that need to be determined to establish a structural model include structural span L, rise/span ratio H/L (Figure 1b), the number of circumferential divisions n_K, the number of radial divisions n_N, the height of the truss h, the width of the truss b, and the length of the truss internodes m (Figure 1c). At the same time, all members in the structure are classified into eight categories, including upper chord, lower chord, web member, cross rod between upper chords, diagonal rod between upper chords, and members at the pyramid position, as well as pyramid adjacent lower chords and pyramid adjacent webs listed separately in consideration of the stress characteristics of the structure. The positions of the rods are shown in Figure 1d, and the size of the section is identical for the same rod type in the same ring.

2.2. Finite Element Model Parameters

In this study, ANSYS [26] was used to model, calculate, and analyze the structure. The element selection of the finite element model, the number of element mesh divisions, model materials, and load value are described as follows:

a. Element selection: The overall stress state of the mega-latticed structure is between that of the single-layer reticulated shell and the double-layer reticulated shell. To improve the structural safety while considering a simplified calculation model, it is assumed that the connections between the structural members are rigid; meaning that the rods not only bear the axial force but also the bending moment, the torque, the shear force, etc. [27]. In addition, to consider the subsequent analysis of the plastic development of the structure under the rare earthquake excitation, the PIPE20 element that specifically simulates the pipe structure was selected.

b. The number of element mesh divisions and support form: To achieve a balance between computational accuracy and efficiency, it is necessary to determine the number of element mesh divisions. Fan [28] studied the ultimate bearing capacity of the structure with the rods divided into different element numbers. The results show that the bearing capacity remained stable when the rod was divided into more than three elements, as shown in Figure 2. Therefore, the rods of the structures were divided into three elements along the length. The three-direction-hinged supports were set at the lower chord nodes of the outermost ring.

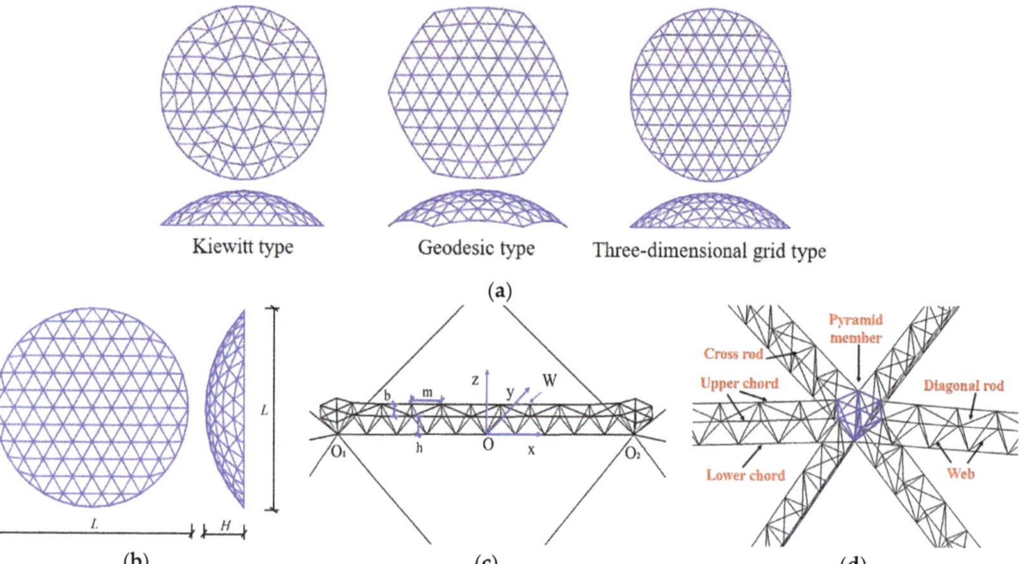

Figure 1. Schematic diagram of the mega-latticed structures (**a**) Three types of spherical mega-latticed structure models; (**b**) Geometric parameters of mega-latticed structure; (**c**) Geometric parameters of truss; and (**d**) Location of mega-latticed structure members.

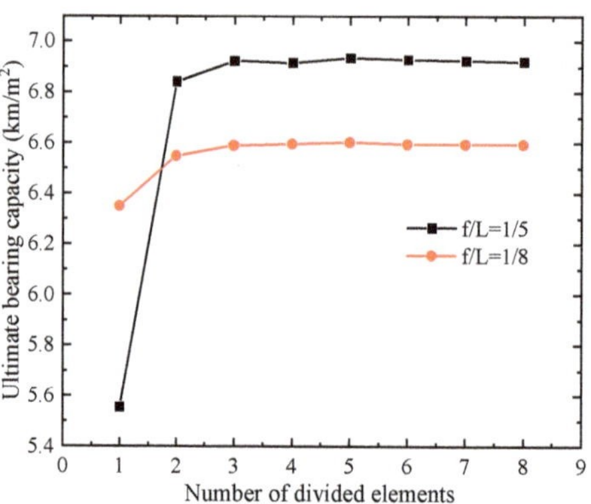

Figure 2. Comparison of the bearing capacity of different number of divided elements [28].

c. Materials: All members in the research structures were made of Steel Q420, which has a yield strength of 420 MPa. Additionally, the ideal elastoplastic model was used as the constitutive model. Considering that the structural members may enter the plastic stage under the rare earthquake excitation, the Bilinear kinematic (BKIN) hardening [29] was adopted to calculate the structural response. The material's elastic modulus, the Poisson's ratio, and the density are 210 GPa, 0.3, and 7850 kg/m^3, respectively.

d. Gravity standard value: In accordance with the Code for Seismic Design of Buildings (GB50011-2010) [30], the representative value of the gravity load is composed of the standard value of the dead load and 0.5 times the standard value of snow load, where the standard value of snow load is 50 kg/m^2, and the combined representative value of the gravity load is 120 kg/m^2. The load is converted into concentrated mass and applied to the node of the finite element model in the form of the MASS21 element.

To obtain the influence of different structural forms on the seismic performance more intuitively, the geometric parameters of the structures in Figure 1 were kept the same. After selecting the member section under the static action, the steel consumption of the Kiewitt-type, the Geodesic-type, and the Three-dimensional grid-type mega-latticed structures was 201.90 kg/m^2, 165.58 kg/m^2, and 157.21 kg/m^2, respectively.

3. Structural Dynamic Characteristic Analysis

This section conducted a modal analysis on three spherical mega-latticed structures of Kiewitt-type, Geodesic-type, and Three-dimensional grid-type. The first 3000 modes of frequencies, periods, and vibration modes of the three structures were calculated. The frequency and period distribution diagrams of the modes that cause the structure's cumulative effective mass coefficient to exceed 90% were plotted, and the vibration modes were analyzed to understand the basic dynamic characteristics of the three spherical mega-latticed structures.

3.1. Analysis of Structural Spectral Characteristics

Spectral analysis of the three spherical mega-latticed structures was performed to obtain their spectral characteristics. Figure 3a shows the frequency and period distributions. The fundamental frequency f_1 and fundamental period T_1 of Kiewitt-type, Geodesic-type, and Three-dimensional grid-type mega-latticed structures are 0.426 Hz and 2.35 s, 0.429 Hz and 2.33 s, and 0.407 Hz and 2.46 s, respectively. The fundamental period exceeds that of the ordinary-span-reticulated shell and is far from the predominant period of the ground. Taking the Three-dimensional grid-type mega-latticed structure as an example, Figure 3b shows the frequency change rate $p_i\%$ ($p_i = 100 \times (f_{i+1} - f_i)/f_i$) of the first 1000 modes. About 71.8% of the values fall in the interval [0.01–1%], and only 3.9% of the values are distributed in [1–10%]. The mega-latticed structure of the Kiewitt-type and the Geodesic-type exhibit similar distribution laws. It shows that the frequency changes of the structure are not significant, and the distribution is uniform, reflecting the complexity of the dynamic characteristics of such super-large-span structures.

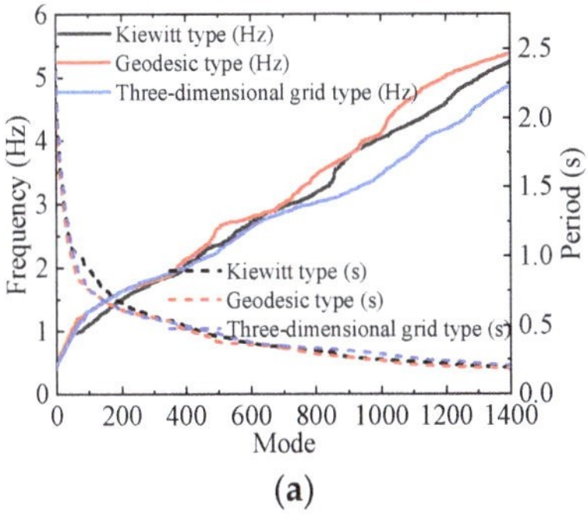

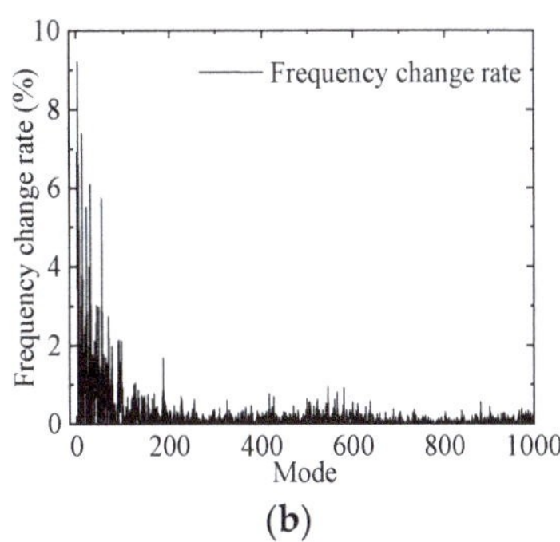

Figure 3. Spectral characteristics of the structures (**a**) Distribution of structural frequency and period; (**b**) Frequency change rate of Three-dimensional grid-type mega-latticed structure.

3.2. Characteristic of Structural Mode

Since the three spherical mega-latticed structures studied in this paper are similar in many respects, the Three-dimensional grid-type structure is taken as an example. Figure 4 shows the first 6 modes of the structure. Due to the presence of multiple axes of symmetry, the frequency and period of two adjacent modes are the same, such as Mode-II and Mode-III, Mode-IV, and Mode-V.

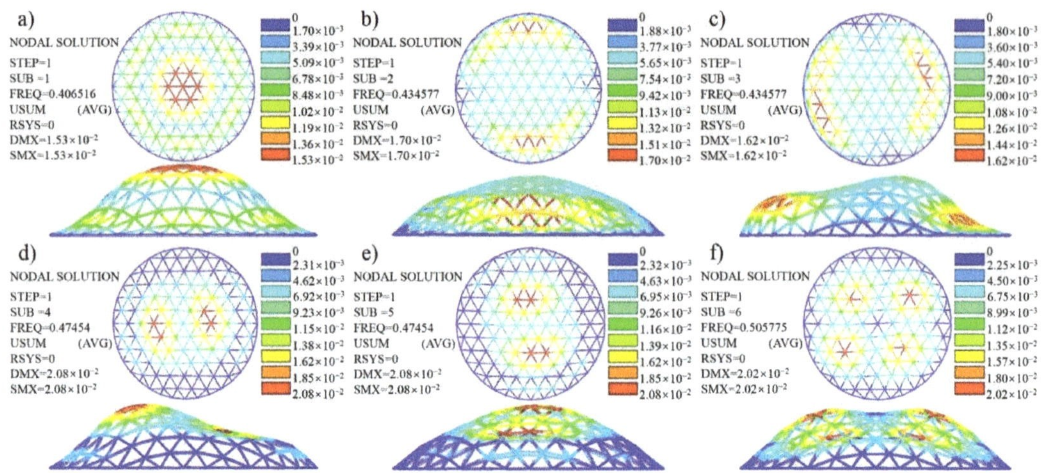

Figure 4. First 6 modes of Three-dimensional grid-type mega-latticed structure (**a**) Mode-I, T_1 = 2.46 s; (**b**) Mode-II, T_2 = 2.30 s; (**c**) Mode-III, T_3 = 2.30 s; (**d**) Mode-III, T_4 = 2.30 s; (**e**) Mode-V, T_5 = 2.11 s; and (**f**) Mode-VI, T_6 = 1.98 s.

The order of the mode occurrence can reflect the stiffness levels of the structure in different directions. In all three types of mega-latticed structures, the vertical mode is the first to occur, suggesting that the horizontal stiffness of these three mega-latticed structures is greater than the vertical stiffness. Considering the relatively small rise/span ratio of the structures, the conclusion of weak vertical stiffness is reasonable. In addition, the first several modes are all global modes, indicating that there are no obvious weaknesses in the stiffness of the structure.

3.3. Determination of Rayleigh Damping

Rayleigh damping [31] is commonly used in engineering applications to approximate the damping of the structure. The damping matrix of the structure is assumed to be a linear combination of the mass and the stiffness matrices, as shown in Formula (1), where α and β are the mass damping coefficient and the stiffness damping coefficient, respectively. These coefficients can be obtained by Formulas (2) and (3), which involve the frequencies of the i-th and j-th modes (denoted by ω_i and ω_j), as well as the corresponding damping ratios (ζ_i and ζ_j). For space structures, a typical value of 0.02 is often used as the damping ratio [27]. Once α and β are obtained, the damping ratio of the k-th mode can be calculated from Formula (4).

In Formulas (2) and (3), the i-th mode is typically one of the first few modes that contribute significantly to the effective mass coefficient, while the j-th mode is the mode whose cumulative effective mass coefficient reaches 60–70% and has an effective mass coefficient exceeding 5%. Figure 5 shows the damping ratio of each mode when i is the 2nd mode, and j takes different values. Notably, the damping ratios of the modes between the 2nd and the j-th mode are typically lower than 0.02, while the damping ratios of the modes outside this range are greater than 0.02.

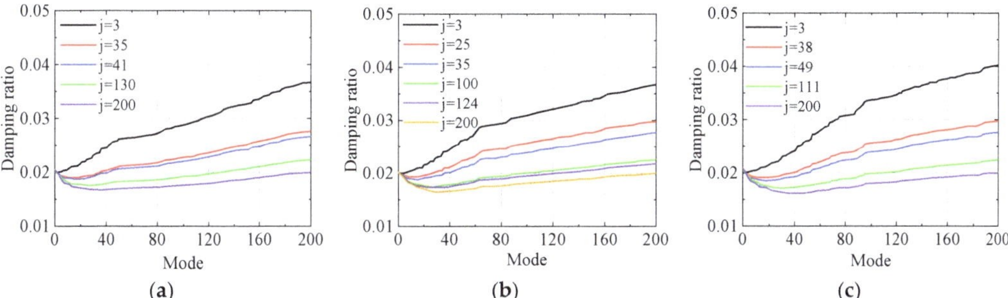

Figure 5. Damping ratios of the three types of spherical mega-latticed structure (**a**) Kiewitt-type; (**b**) Geodesic-type; and (**c**) Three-dimensional grid-type.

$$[C] = \alpha[M] + \beta[K] \tag{1}$$

$$\alpha = \left[2\omega_i\omega_j\left(\xi_i\omega_j - \xi_j\omega_i\right)\right] / \left(\omega_j^2 - \omega_i^2\right) \tag{2}$$

$$\beta = 2\left(\xi_j\omega_j - \xi_i\omega_i\right) / \left(\omega_j^2 - \omega_i^2\right) \tag{3}$$

$$\xi_k = \alpha / (2\omega_k) + (\beta\omega_k)/2 \tag{4}$$

Based on the results depicted in Figure 5, selecting the 130th, 100th, and 111th modes as the *j*-th mode yields more suitable Rayleigh damping coefficients for the Kiewitt-type, the Geodesic-type, and the Three-dimensional grid-type mega-latticed structures, respectively.

3.4. Earthquake Selection

To meet the requirements of seismic design, the earthquake selected should satisfy three characteristics: spectral characteristics, effective peak value, and duration. The spectral characteristics can be characterized by the seismic influence coefficient, and the designed response spectrum can be obtained from the seismic fortification intensity, the site category, and the design earthquake grouping [30]. When the seismic fortification intensity is degree 8 (The design basic ground motion acceleration is 0.02 times the gravitational acceleration), the site category is Class II (Cohesive soil with an allowable bearing capacity of foundation soil $[\sigma_0] > 150$ kPa), and the seismic design group is the first group, the design response spectrum of the structure can be determined.

The selected seismic records and earthquake information are shown in Figure 6. The seismic acceleration time history was amplitude modulated according to Formula (5), where $a'(t)$ is the amplitude modulated acceleration time history, $a(t)$ is the original acceleration time history, $A'_{\max}$ is the maximum value of the amplitude modulated acceleration time history obtained based on the specification [30], and $A_{\max}$ is the maximum value of the original acceleration time history. Additionally, the effective duration of the earthquake should generally be 5 to 10 times the fundamental period of the structure according to the specification [30]. For this study, a duration of 30 s has been selected.

$$a'(t) = \frac{A'_{\max}}{A_{\max}} a(t) \tag{5}$$

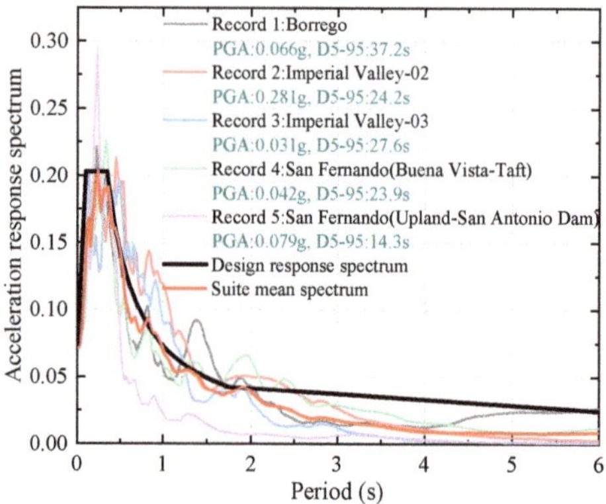

Figure 6. Acceleration response spectrum and earthquake information.

4. Effect of Seismic Direction on Structural Response

According to the Code for Seismic Design of Buildings (GB50011-2010) [30], vertical seismic should be considered for large-span structures with fortification degrees 8 and 9. Therefore, this study utilizes three-dimensional earthquakes for seismic analysis. To illustrate the impact of earthquake direction on the structural responses, taking the Three-dimensional grid-type mega-latticed structure as an example, the structural responses due to one-dimensional and two-dimensional earthquakes were calculated separately for comparison.

The structure was divided into five rings in the radial direction, designated as Ring 1 to Ring 5. In the ring direction, sectors were defined every 60°, starting with Sector 0 in the positive direction of the X-axis. The counterclockwise direction was considered positive, while the clockwise direction was negative, resulting in Sector ± 1, Sector ± 2, and Sector ± 3. Notably, Sector 3 and Sector − 3 were coincident, as shown in Figure 7.

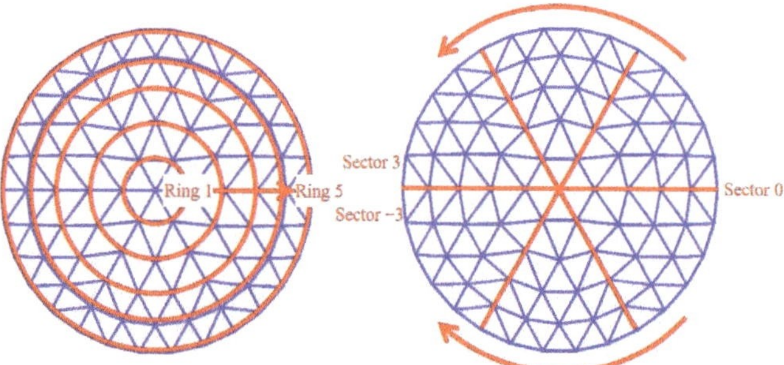

Figure 7. Schematic diagram of structural location.

Due to the large number of elements and nodes in the 800 m mega-latticed structure, only specific elements and nodes were analyzed. Specifically, the elements and nodes located at the pyramid position as well as the upper and lower chords at the mid-span, 1/4-span, and 3/4-span of each ring, were selected as the characteristic objects for analysis.

For the seismic analysis, Record 1 in Figure 6 was modulated to the effective amplitudes of 70 gal for frequent earthquakes in X, Y, Z, XY, and XZ directions and input to the structure. Subsequently, the time history analysis method was applied to assess the coupling degree of the structural responses when two horizontal earthquakes or horizontal and vertical earthquakes act simultaneously.

4.1. Effect of Seismic Direction on Displacement Response

4.1.1. Effect of X, Y, and XY Direction Seismic Action on Displacement Response

This section studies the distribution of the displacement increment relative to the static displacement of the characteristic nodes in the X, Y, and Z directions with the number of rings and sectors under the seismic action in X, Y, and XY directions. The results are presented in Figure 8.

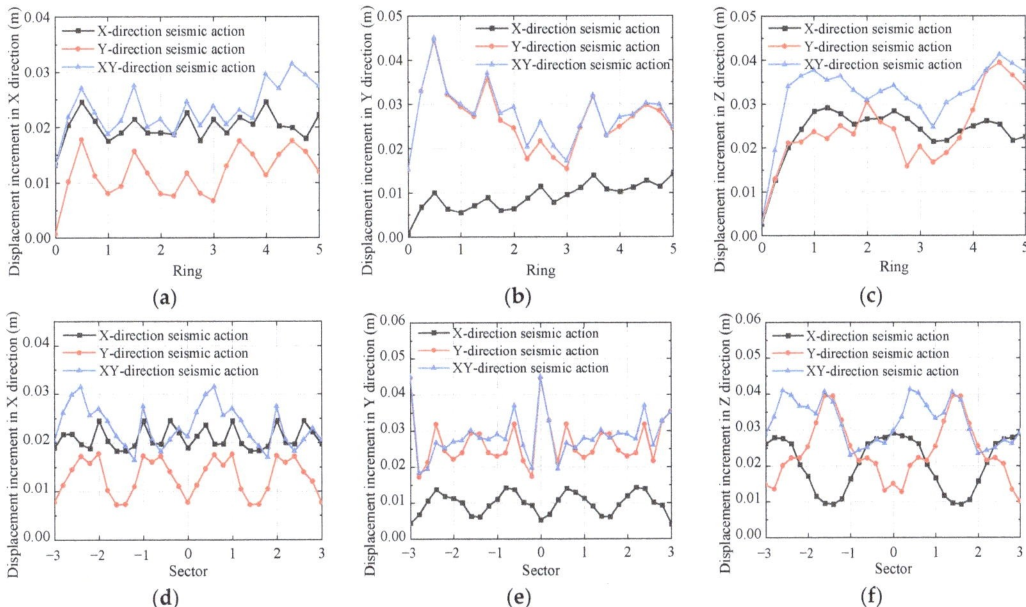

Figure 8. Distribution of displacement increment under the action of X, Y, and XY-direction earthquakes (**a**) Radial distribution in X-direction; (**b**) Radial distribution in Y-direction; (**c**) Radial distribution in Z-direction; (**d**) Circumferential distribution in X-direction; (**e**) Circumferential distribution in Y-direction; and (**f**) Circumferential distribution in Z-direction.

The displacement increments in the X, Y, and Z directions are of similar magnitudes, with the largest increment occurring in the Y direction. Figure 8a reveals that the seismic action in the Y direction contributes significantly to displacement increments in the X direction, especially at the 1.5 ring and the 4–5 ring, as well as the 0.5 sector and −2.5 sector in rotationally symmetric positions. It indicates a strong coupling between the X and Y directions. Furthermore, Figure 8c demonstrates that seismic action in the X and Y directions considerably contributes to displacement increments in the Z direction, suggesting a strong coupling across the three directions.

4.1.2. Effect of X, Z, and XZ Direction Seismic Action on Displacement Response

This section analyzes the coupling degree between horizontal and vertical seismic responses by comparing the displacement increments of the structure under X, Z, and XZ seismic action. The results, as shown in Figure 9, align with those observed in Section 4.1.1, wherein the displacement increments in the three directions are greater under XZ seismic action than under the one-dimensional seismic action. Furthermore, under vertical seismic action rather than horizontal seismic action, the displacement increment in the horizontal direction is smaller.

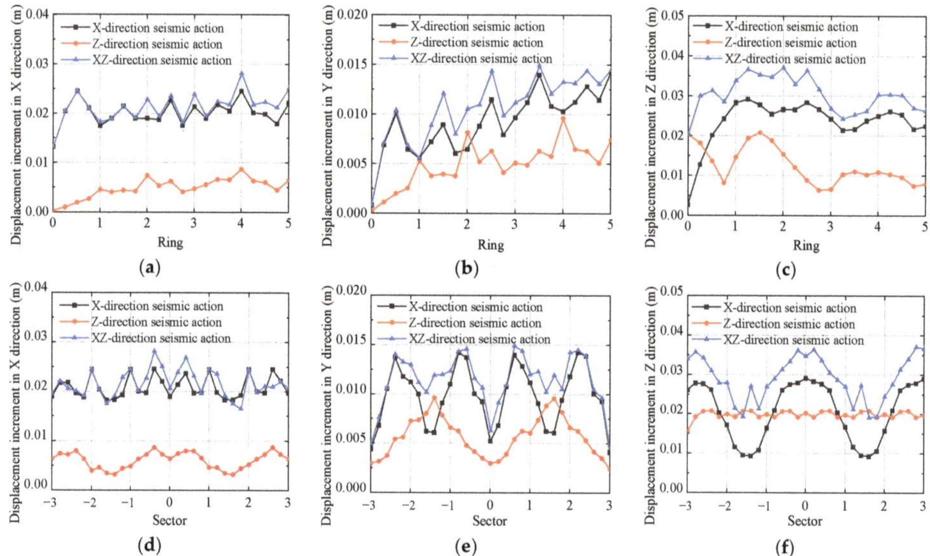

Figure 9. Distribution of displacement increment under the action of X, Z and XZ-direction earthquakes (**a**) Radial distribution in X-direction; (**b**) Radial distribution in Y-direction; (**c**) Radial distribution in Z-direction; (**d**) Circumferential distribution in X-direction; (**e**) Circumferential distribution in Y-direction; and (**f**) Circumferential distribution in Z-direction.

As seen in Figure 9c, the vertical displacement increment increases within the range of 0 to 0.75 ring under X-direction seismic action, whereas it decreases within the same range under Z-direction seismic action. After 0.5 rings, the vertical displacement increment under the X-direction seismic action exceeds that under the Z-direction seismic action. However, the distribution of Z-direction displacement increment under XZ-direction seismic action is consistent with that under X-direction seismic action, signifying that the influence of X-direction seismic action is dominant when considering the coupling of the earthquakes in the direction of X and Z.

4.2. Effect of Seismic Direction on Stress Response

To reflect the extent of the structural stress affected by seismic direction, the distributions of the equivalent stresses of the elements with the number of rings under various seismic action directions were compared, as shown in Figure 10.

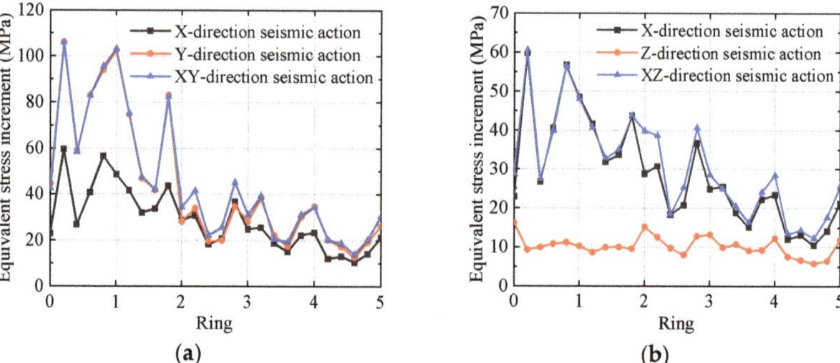

Figure 10. Effect of seismic direction on equivalent stress increment of structural members (**a**) Radial distribution of equivalent stress increment under horizontal earthquakes; and (**b**) Radial distribution of equivalent stress increment under horizontal and vertical earthquakes.

Equivalent stress increments in elements are larger than those observed under one-dimensional seismic action, however they nevertheless have a comparable magnitude, regardless of whether the earthquake action is XY or XZ. According to Figure 10b, the horizontal earthquake has a greater effect than the vertical earthquake in terms of the equivalent stress increment of the element. The stress response in the horizontal direction is more influenced by seismic action in the Y direction than in the X direction, which is consistent with the displacement response result. Because the X-direction seismic action passes longitudinally through one of the primary directions of the Three-dimensional grid-type mega-latticed structure, where the structural stiffness is comparatively high, while the Y-direction seismic action is transversely perpendicular to the truss, causing a more significant response from the structure.

Based on the observed effects of seismic action direction on structural displacements and stresses, it can be concluded that there is a high degree of coupling between the horizontal and vertical directions in the structures. Additionally, the results suggest that the vertical seismic action does not necessarily play a controlling role in the structural response; on the contrary, the horizontal seismic action may have a more significant impact. Therefore, seismic analysis should consider the joint action of horizontal and vertical earthquakes simultaneously, and the three-dimensional seismic input should be adopted to accurately capture the structural response.

5. Response Analysis of Structures under Frequent Earthquakes

The method used to analyze the displacement and stress response of structures subjected to frequent earthquakes involved the application of time history analysis. The selected seismic records were amplitude modulated to 70 gal before being input into the Kiewitt-type, Geodesic-type, and Three-dimensional grid-type mega-latticed structures, respectively. Since the selected seismic motions were chosen based on the structural spectral characteristics, each input motion can relatively accurately reflect the structures' response. Therefore, in the subsequent analysis, when displaying the results of a specific input motion, they correspond to the seismic action of Record 1 in Figure 6. The maximum envelope values at each position are displayed on the displacement and stress distribution curves.

5.1. Displacement Response under Frequent Earthquakes

Figure 11 illustrates the radial and circumferential displacement envelope of the Kiewitt-type, Geodesic-type, and Three-dimensional grid-type mega-latticed structure under the static force and Record 1 seismic action. Similar displacement distributions are observed under seismic and static loads across the three structures from the comparison of the displacement envelope. The maximum displacement in the X and Y horizontal directions increases and then decreases radially from the top of the dome to the supports. Due to the support constraints, the maximum displacement mostly occurs within Rings 3 to 4. The maximum vertical displacement occurs at the center of the dome and gradually decreases towards the supports. The distribution of the maximum displacement along the circumferential direction shows an opposing pattern in the X direction compared to the Y direction, as observed in Figure 11d,e.

In addition, the curves of the ratio of dynamic displacement to static displacement along the radial direction are drawn in Figure 12. The ratios of the two horizontal directions are relatively high within the 0–2 ring. The reason could be that the structural displacements within the 0–2 ring are relatively small in the static case, while the seismic action leads to a uniform increase in structural displacement. Conversely, the ratio in the Z direction is lower than that in horizontal directions, with no significant change observed along the radial direction.

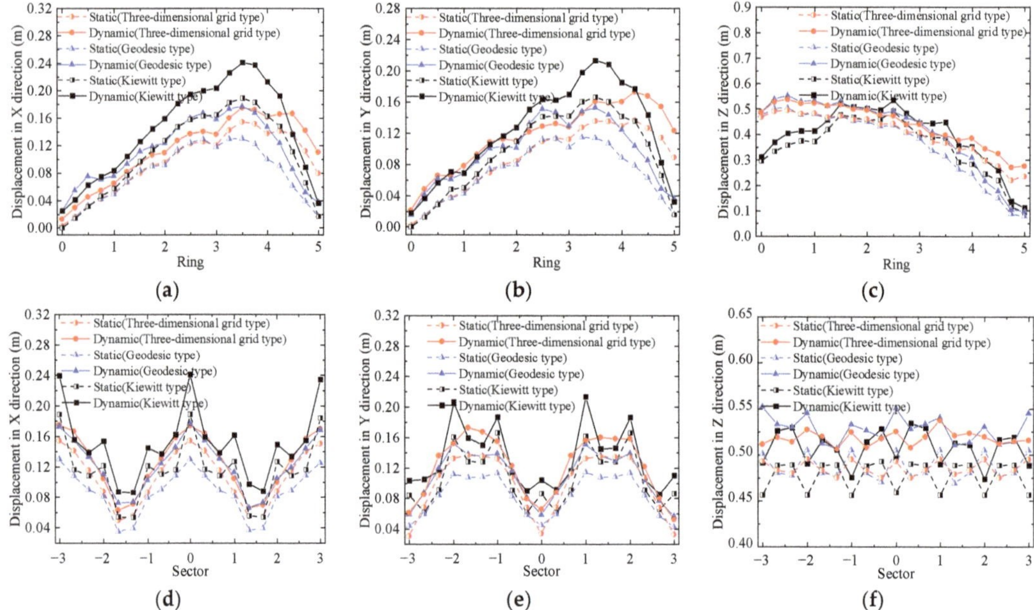

Figure 11. Displacement distribution of three spherical mega-latticed structures under static and seismic action: (**a**) Radial distribution in X-direction; (**b**) Radial distribution in Y-direction; (**c**) Radial distribution in Z-direction; (**d**) Circumferential distribution in X-direction; (**e**) Circumferential distribution in Y-direction; and (**f**) Circumferential distribution in Z-direction.

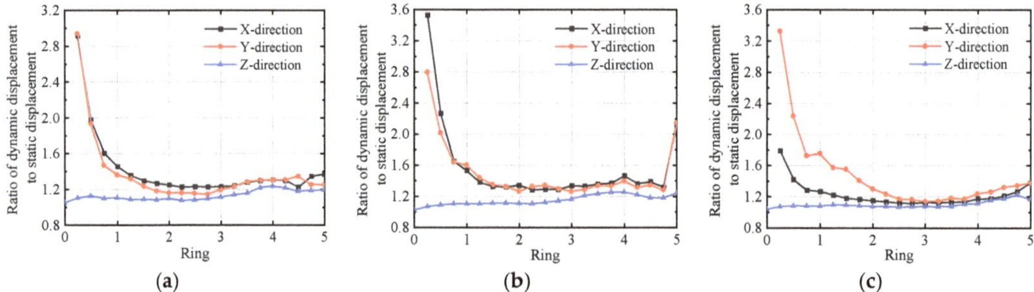

Figure 12. Radial distribution curves of dynamic displacement to static displacement ratio of structures: (**a**) Kiewitt-type; (**b**) Geodesic type; and (**c**) Three-dimensional grid-type.

Due to the unique characteristics of individual seismic records, different seismic records may result in varying responses from a structure. Therefore, the responses of the three mega-latticed structures were calculated separately under the action of the other four seismic records presented in Figure 6, and the statistical results are listed in Table 1. Among them, the displacements in the two horizontal directions of the Kiewitt-type mega-latticed structure were greater than those of the Geodesic-type and Three-dimensional grid-type. However, the vertical displacements of the three structures were comparable magnitudes.

Based on the average displacement response of the three structures under the action of five seismic records, the ratio of dynamic displacement to the static displacement of the structures was estimated to be approximately 1.10~1.50, with the Z direction value ranging between 1.10~1.15. It can be observed that the displacement ratio in all three directions of the Three-dimensional grid-type mega-latticed structure is relatively uniform, which could be attributed to the topology of the structure. The mega grids in the Three-dimensional grid-type structure are positively triangular in the horizontal projection plane, except for the outermost ring, resulting in an equal representation of behavior in all three directions.

Table 1. Average displacement responses under the action of five seismic records of three structures.

Structure Type	Average Maximum Dynamic Displacement/m			Maximum Static Displacement/m			Average Dynamic and Static Displacement Ratio		
	X	Y	Z	X	Y	Z	X	Y	Z
Kiewitt-type	0.243	0.224	0.543	0.189	0.166	0.493	1.29	1.35	1.10
Geodesic-type	0.177	0.165	0.566	0.131	0.112	0.505	1.35	1.47	1.12
Three-dimensional grid-type	0.169	0.171	0.557	0.155	0.137	0.469	1.09	1.25	1.12

5.2. Stress Response under Frequent Earthquake

The stress responses of the structural members were calculated based on the 8 types mentioned in Section 2.1, and the distribution laws of the maximum stress along the radial direction were investigated under the frequent seismic action. In consideration of the complex stresses of the structural members, the von Mises stress is utilized to describe the stresses levels.

Figure 13 presents the radial distributions of the maximum equivalent stress of the upper chords, the lower chords, the web members, the cross rods between upper chords, the diagonal rods between upper chords, and the members at the pyramid position in the three types of spherical mega-latticed structures under the static force and Record 1 seismic action. Under frequent earthquakes, the distribution of the maximum equivalent stress in the radial direction follows a similar pattern to that observed under the static force. The high equivalent stresses near the structural intersection are due to the high density of the rods at these locations, resulting in an increased stiffness. Moreover, Figure 13f demonstrates that the equivalent stresses of the pyramid members at the intersection position are greater than those of the other member types.

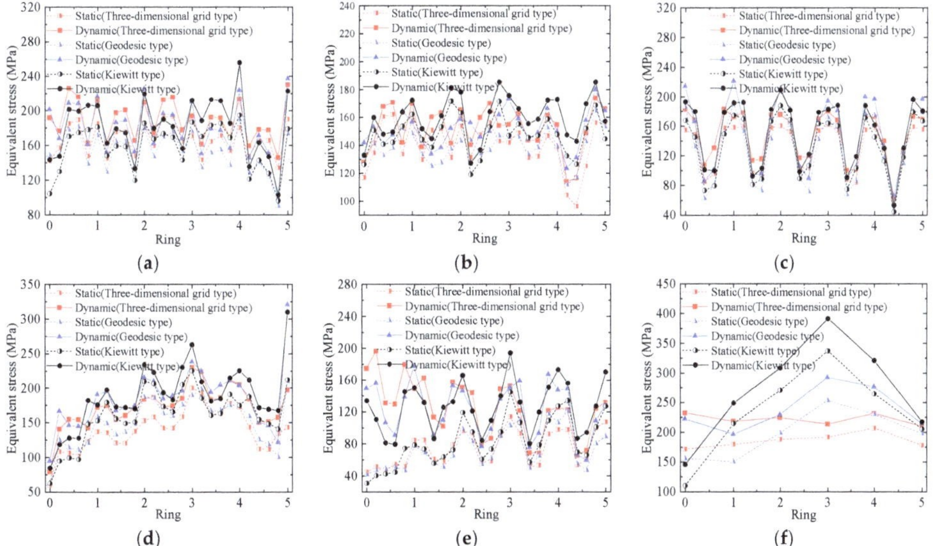

Figure 13. Stress distribution of three spherical mega-latticed structures under static and seismic action (**a**) Upper chord; (**b**) Lower chord; (**c**) Web member; (**d**) Cross rod between upper chords; (**e**) Diagonal rod between upper chords; and (**f**) Member at the pyramid position.

Furthermore, from Figure 13a,d, the equivalent stresses of the upper chord and the cross rod between upper chords at the outermost ring near the support position are higher than those of the nearby rods, which may be related to the separate classification of the lower chords and web members near the pyramid when selecting the member sections. However, when considering the overall magnitude of the equivalent stresses for each member type, the values are evenly distributed.

The radial distributions of the ratio of the maximum equivalent stresses under the seismic action to that under the static force for each member type in the three structures are drawn in Figure 14. The stress ratios of the diagonal rods between the upper chords fluctuate more strongly along the radial distribution compared to other types of members, indicating a more significant impact on the stresses of the diagonal rods at the top of the spherical mega-latticed structure.

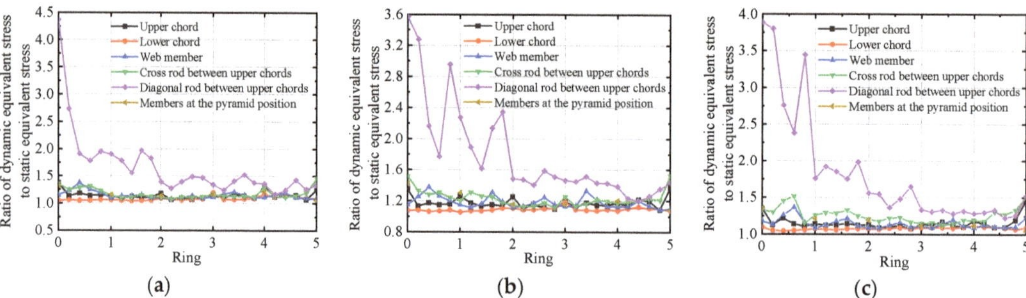

Figure 14. Radial distribution curve of dynamic equivalent stress to static equivalent stress ratio of structures: (**a**) Kiewitt-type; (**b**) Geodesic-type; and (**c**) Three-dimensional grid-type.

The stress responses of the three spherical mega-latticed structures under the action of the other four seismic records were also calculated, and the statistical results are listed in Table 2. The ratio of equivalent stress under seismic action to that under static force ranges from 1.10 to 1.85.

Table 2. Average stress responses under the action of five seismic records of three structures.

Type of Member	Kiewitt-Type			Geodesic-Type			Three-Dimensional Grid-Type		
	Average Maximum Dynamic Stress /MPa	Maximum Static Stress /MPa	Average Dynamic and Static Stress Ratio	Average Maximum Dynamic Stress /MPa	Maximum Static Stress /MPa	Average Dynamic and Static Stress Ratio	Average Maximum Dynamic Stress /MPa	Maximum Static Stress /MPa	Average Dynamic and Static Stress Ratio
Upper chord	250.92	194.91	1.29	247.79	190.15	1.30	247.80	195.68	1.27
Lower chord	189.98	171.84	1.11	186.19	164.56	1.13	197.62	180.66	1.09
Web member	213.51	188.16	1.13	228.84	192.51	1.19	185.91	162.17	1.15
Cross rod	293.79	225.59	1.30	310.67	210.12	1.48	236.15	200.51	1.18
Diagonal rod	188.36	144.91	1.30	221.18	121.01	1.83	199.11	114.21	1.74
Pyramid member	384.62	337.32	1.14	299.80	253.77	1.18	260.76	207.28	1.26

The stresses of the Kiewitt-type mega-latticed structure are significantly larger than those in the other two structures, indicating that the impact of the non-uniformity in the grid is amplified by dynamic loads, particularly in the pyramid position, where the maximum dynamic stresses have exceeded the stress ratio limit of 0.85 for the section selection in the static case, but have not yet reached the yield stress of the material.

Based on the above analysis of the displacement and stress response in the three 800 m-span spherical mega-latticed structures under the frequent earthquakes, and considering the steel consumption obtained from the section selection in the static case, it can be concluded that compared to the Geodesic-type and the Three-dimensional grid-type mega latticed structures, the Kiewitt-type structure has the higher steel consumption and member stresses under frequent earthquakes. Therefore, the Kiewitt-type structure may not be a suitable structural scheme for 800 m-span. Thus, only the Geodesic-type and the Three-dimensional grid-type mega-latticed structures will be studied in analyzing structures under rare earthquakes.

6. Response Analysis of Structure under Rare Earthquake

When the structure encounters a rare earthquake, properly designed structures should allow certain members to yield and develop into plasticity, and the stresses redistribute without causing damage to the structure [30]. The strategy sacrifices the secondary components of the structure to protect the primary ones. For instance, connecting beams in shear wall structures can have their stiffness deliberately reduced to absorb seismic energy and protect the primary structure. Similarly, in frame-core tube structures with outriggers, the outriggers can yield and dissipate energy while the core tube and outer frame remain elastic. After the 2022 M6.8 Luding earthquake in China, Qu [32] conducted field research in the disaster area and observed structures presenting this type of behavior. Therefore, this chapter performs an elastic–plastic analysis under the action of rare earthquakes using the Bilinear Kinematic Hardening (BKIN) model, which considers the Bauschinger effect. The model uses the Mises yield criterion and the kinematic hardening criterion to describe the stress–strain relationship of the material in two straight lines. The steel parameters are consistent with the previous analysis, and the tangent modulus is about 6300 MPa, which is 3% of the elastic modulus. Figure 15 depicts the stress–strain curve of the BKIN model. The seismic records in Figure 6 are amplitude-modulated to a peak effective acceleration of 400 gal for this analysis.

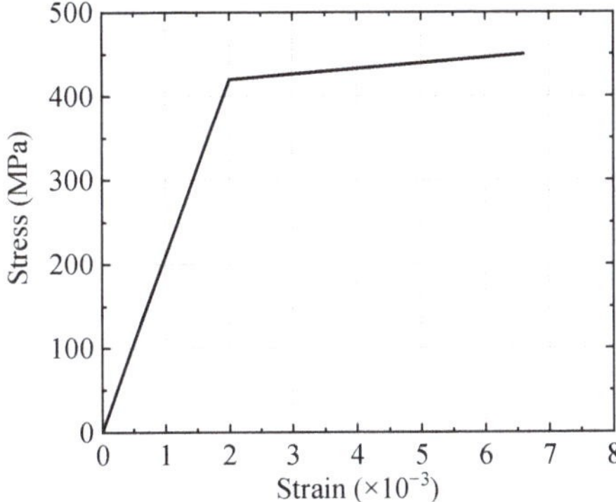

Figure 15. Stress–strain curve of the Bilinear Kinematic Hardening (BKIN).

6.1. Displacement Response under Rare Earthquake

Figure 16 illustrates the maximum dynamic displacements distributions in the X, Y and Z directions along the radial and circumferential directions for the Geodesic-type and Three-dimensional grid-type mega-latticed structures under the action of Record 1 seismic action.

In the case of rare earthquakes, the distribution of the maximum dynamic displacements along the circumferential direction no longer exhibits a symmetrical shape. In the radial direction, certain locations experience significantly higher maximum dynamic displacements in the horizontal direction than under frequent earthquakes. Consequently, the radial distribution no longer shows a relatively smooth curve but one with jagged fluctuations at some positions. Conversely, the distribution of the maximum vertical dynamic displacement from the top of the supports is similar to that under frequent earthquakes, displaying a gradually decreasing curve.

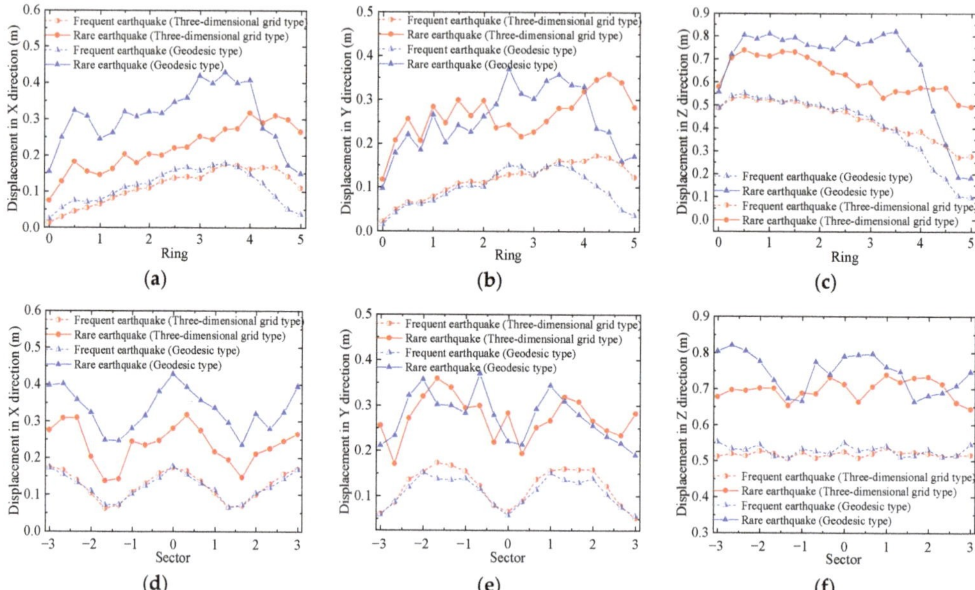

Figure 16. Displacement distribution of the Geodesic-type and the Three-dimensional grid-type mega-latticed structure under the action of frequent and rare earthquakes: (**a**) Radial distribution in X-direction; (**b**) Radial distribution in Y-direction; (**c**) Radial distribution in Z-direction; (**d**) Circumferential distribution in X-direction; (**e**) Circumferential distribution in Y-direction; and (**f**) Circumferential distribution in Z-direction.

To compare the radial variation of the maximum dynamic displacement ratios in the Geodesic-type and the Three-dimensional grid-type mega-latticed structures under rare and frequent earthquakes, α represents the ratio of elastic–plastic displacement under rare earthquakes to elastic displacement under frequent earthquakes. Meanwhile, α_{max} expresses the ratio of the maximum elastic–plastic displacement response to the maximum elastic displacement response. The curves of α_{max} along the radial direction are shown in Figure 17. In the horizontal direction, the values of α_{max} are relatively high at the top and the locations near the supports, indicating significant impact on nodal displacements at these positions with increasing seismic intensity. In contrast, the α_{max} in the vertical direction is smaller than in the horizontal directions, and the curves have slight fluctuations along the radial direction.

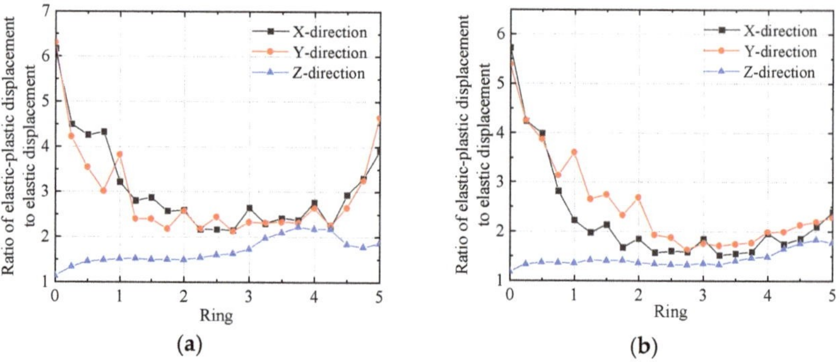

Figure 17. Radial distribution curves of elastic–plastic displacement to elastic displacement ratio of structures: (**a**) Geodesic type; and (**b**) Three-dimensional grid-type.

The displacement responses of the Geodesic-type and the Three-dimensional grid-type mega-latticed structures under four additional seismic records were also calculated, and the statistical results are listed in Table 3. Compared to the Geodesic-type mega-latticed structure, the Three-dimensional grid-type has a higher α_{max} in the horizontal direction and a lower α_{max} in the vertical direction. The significant difference between the values of α_{max} in horizontal and vertical directions implies greater impact on the structure's horizontal direction due to the seismic intensity. The average displacement response results of both structures under five seismic records indicate that the ratio of maximum elastic–plastic displacement to maximum elastic displacement is about 2.5 to 3.0 in the horizontal direction and approximately 1.5 to 2.0 in the vertical direction.

Table 3. Average displacement response results of the structures under the action of rare earthquakes.

Structure Type	Average Maximum Elastic–Plastic Displacement/m			Maximum Elastic Displacement/m			Average α_{max}		
	X	Y	Z	X	Y	Z	X	Y	Z
Geodesic-type	0.442	0.444	0.915	0.177	0.165	0.566	2.50	2.69	1.62
Three-dimensional grid-type	0.408	0.442	0.992	0.170	0.171	0.557	2.41	2.58	1.78

6.2. Stress Response under Rare Earthquake

When the Geodesic-type and the Three-dimensional grid-type mega-latticed structures are subjected to the rare earthquake of Record 1 in Figure 6, the radial distribution of the maximum stresses for six types of members mentioned in Section 2.1 are shown in Figure 18. The maximum equivalent stresses take place near the intersection of the structure under rare earthquakes. However, the equivalent stresses at the intersection of the upper chord and the intersection of the 3rd and 4th rings of the lower chord are particularly sensitive to the seismic intensity and increase significantly. Compared to the Geodesic-type, the Three-dimensional grid-type exhibits a more uniform increase in stresses, with greater increase in the equivalent stresses observed in the web members of the innermost ring.

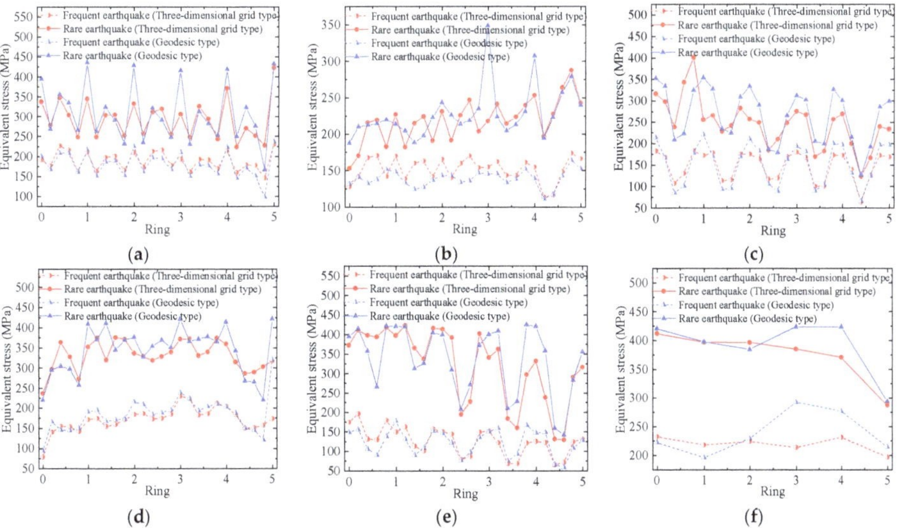

Figure 18. Equivalent stress distribution of the Geodesic-type and the Three-dimensional grid-type mega-latticed structure under the action of the frequent and rare earthquake (**a**) Upper chord; (**b**) Lower chord; (**c**) Web member; (**d**) Cross rod between upper chords; (**e**) Diagonal rod between upper chords; and (**f**) Member at the pyramid position.

The ratio of the elastic–plastic stress under the rare earthquakes is denoted by β, and the maximum ratio of elastic–plastic stress to elastic stress is represented as β_{max}. Figure 19 shows the radial distribution of β_{max} for each member type in the Geodesic-type and the Three-dimensional grid-type mega-latticed structures. The diagonal rod between the upper chords exhibits a larger β_{max} than other types of members. The distributions of β_{max} for the upper chord, the lower chord and the members at the pyramid position are relatively stable, while other member types demonstrate sharp fluctuations.

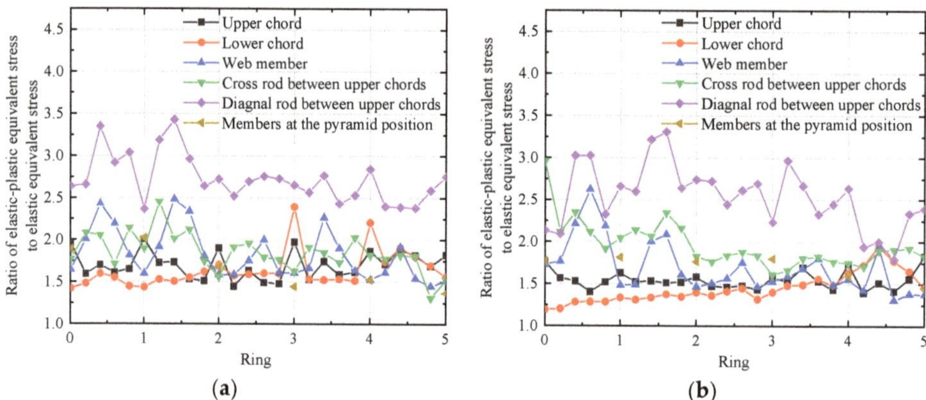

Figure 19. Radial distribution curves of elastic–plastic equivalent stress to elastic equivalent stress ratio of structures (**a**) Geodesic-type; and (**b**) Three-dimensional grid-type.

Table 4 provides the stress statistics of the Geodesic-type and the Three-dimensional grid-type mega-latticed structures under rare earthquakes. The Geodesic-type has more elements reaching the yield stress, whereas the Three-dimensional grid-type has fewer elements entering the elastic–plastic stage. For the 800 m-span mega-latticed structures of both types, the β_{max} of the structures can be taken to be 1.35 to 2.10.

Table 4. Average stress response results of the structures under the action of rare earthquake.

Type of Member	Geodesic-Type			Three-Dimensional Grid-Type		
	Average Maximum Elastic–Plastic Stress /MPa	Maximum Elastic Stress/MPa	Average β_{max}	Average Maximum Elastic–Plastic Stress/MPa	Maximum Elastic Stress/MPa	Average β_{max}
Upper chord	436.76	247.79	1.76	428.92	241.93	1.77
Lower chord	338.96	186.19	1.85	316.62	182.62	1.74
Web member	396.09	228.84	1.73	388.46	185.91	2.09
Cross rod	422.73	310.67	1.36	402.40	232.90	1.73
Diagonal rod	427.00	221.18	1.93	418.37	216.37	1.93
Pyramid member	426.24	299.80	1.42	416.07	272.46	1.53

6.3. Structural Plastic Development under Rare Earthquake

Under rare earthquakes, numerous members in the Geodesic-type and the Three-dimensional grid-type mega-latticed structures enter the plastic stage. Therefore, this section studies the plastic development in the structure, including exploration of the degree and process of plasticity in each member type to pinpoint the weak positions. The finite element model employs the PIPE20 element with 8 integration points denoted by 1P to 7P representing at least 1 to 7 of the 8 integration points yielding into the plastic state, while 8P indicates that the entire section has yielded into plasticity.

The extent and distribution of plasticity in members of the Geodesic-type and the Three-dimensional grid-type mega-latticed structures subjected to rare earthquakes were first analyzed. Colored circles represent elements with varying degrees of plasticity, where larger circles indicate more integral points yielding into plasticity. Figure 20 depicts the distribution of plasticity degree of each member type in both structures, along with the statistical results of the proportions of each plasticity status.

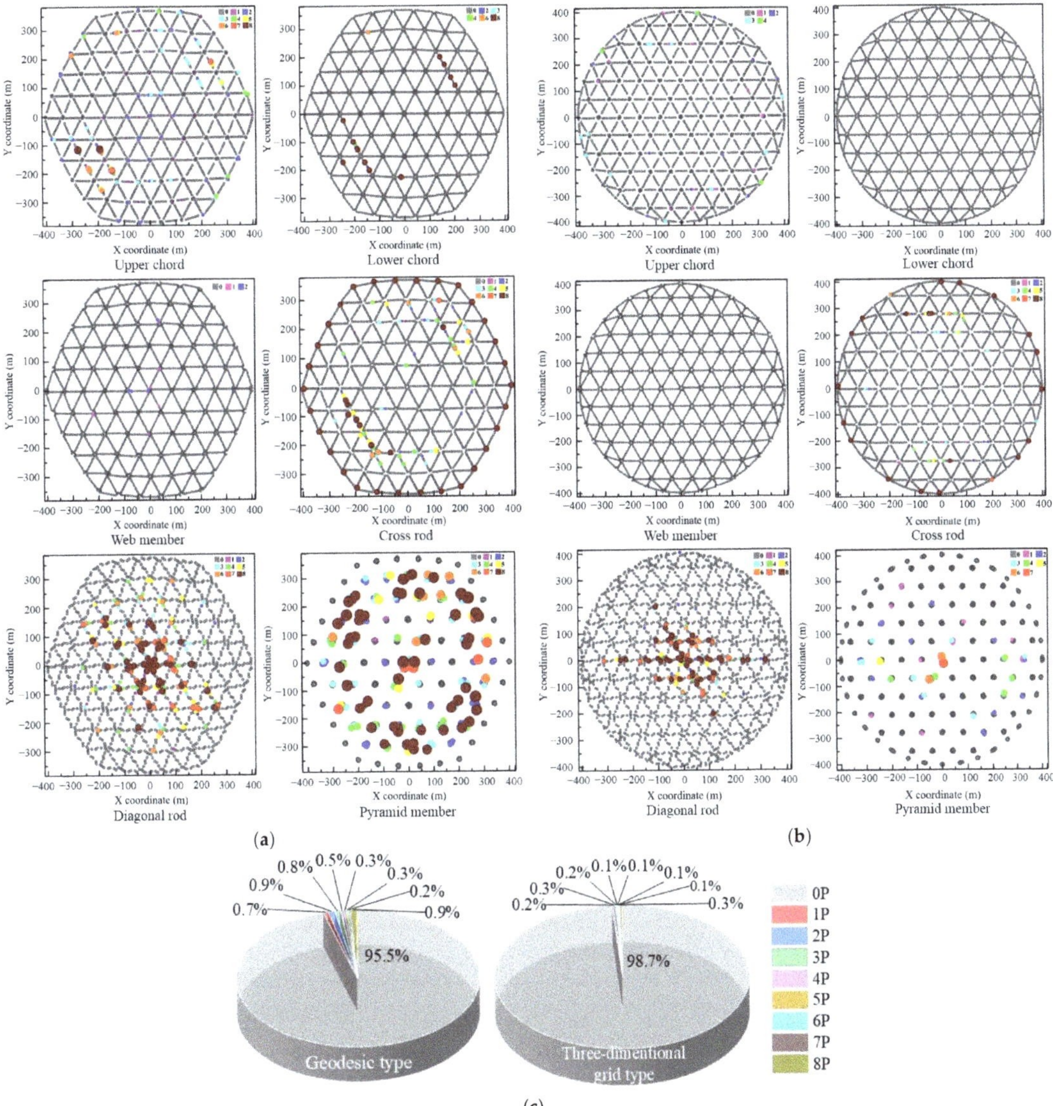

Figure 20. Plasticity distribution of the two structures and the percentage of each plasticity development degree (**a**) Distribution of plasticity of the Geodesic-type mega-latticed structure; (**b**) Distribution of plasticity of the Three-dimensional grid-type mega-latticed structure; and (**c**) Elements proportion of each plastic development degree.

The plastic development process of the Geodesic-type mega-latticed structure, which has a relatively higher degree of plasticity of the two structures, was also analyzed. Figure 21 shows the maximum degree of the plastic status at different moments. The cross rod between the upper chords at the outermost ring close to the support enters the plastic phase first, followed by the upper chords and the cross rods between the upper chords from near the supports to the inside of the structure. However, except for the cross rods between the upper chords at the outermost ring of the structure which quickly reach full-section yield, the plasticity of the remaining elements develops slowly and to a limited extent. The plastic development of the structure mainly occurs in the members at the pyramid position and the diagonal rods between the upper chords after 10 s of the action of the rare earthquake, progressing from the top of the structure towards the supports.

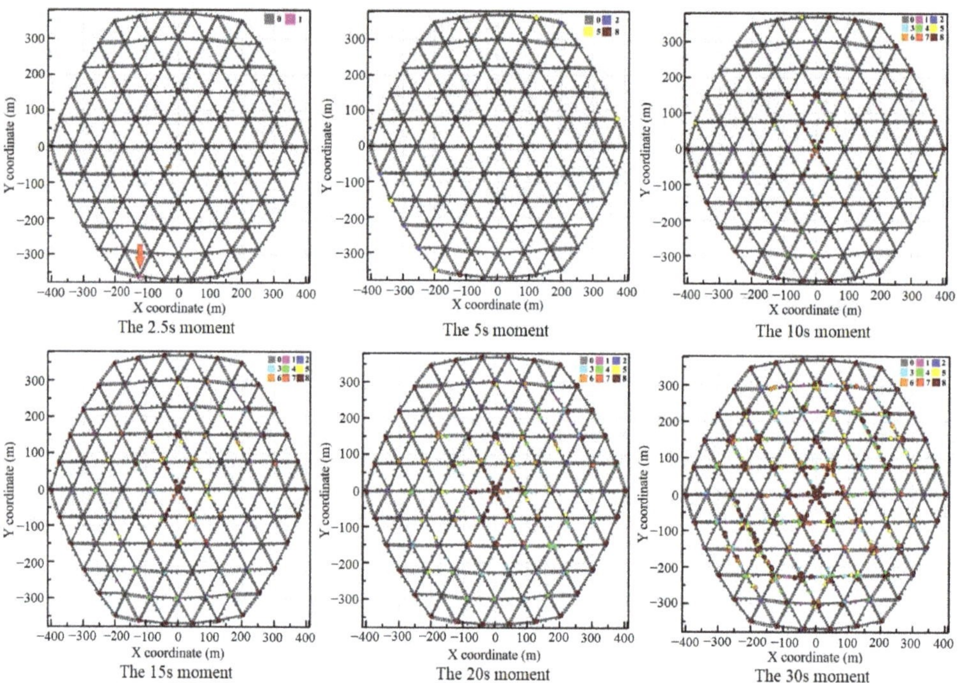

Figure 21. Plastic development process of the Geodesic-type mega-latticed structure.

Throughout the plastic development process of the structure, the pyramid members and the diagonal rods at the top of the structure, along with the pyramid members and the cross rods of the 3rd and 4th ring trusses, and the cross rods near the supports exhibit a high degree of plasticity, which are the weak parts of the structure. Hence, particular attention should be paid to the members at these positions in the design, and certain measures should be taken, such as increasing the cross-section of the rods to avoid the structure from being damaged by these positions during rear earthquakes.

7. Discussion and Conclusions

Following the results of the model selection based on the static performance of the structure [17], this study further investigated the seismic performance of the selected Kiewitt-type, Geodesic-type, and Three-dimensional grid-type mega-latticed structures at 800 m span by applying the time history analysis method. In this study, the method for seismic motion input in time history analysis was determined through seismic directionality analysis. The responses of the structure under seismic actions were calculated to determine the response amplification factor compared to static or frequent seismic actions. The study of the plastic development under rare earthquakes revealed the plastic development process and the weak areas of the structure, and gave conclusions on the model selection, which are of guidance for the practical application of this type of structure.

The following conclusions can be drawn:

1. The fundamental periods of the three studied spherical mega-latticed structures with 800 m spans are far from the characteristic period of the site, possessing low fundamental frequencies and a dense spectrum. The vertical vibration modes are primarily excited in all three structures, indicating that the vertical stiffness is smaller than the horizontal stiffness. Strong coupling exists among the three directions of the structures, thus multi-dimensional seismic motions should be input when calculating the structural response.

2. The response analysis under frequent seismic actions is studied in combination with static analysis. By comparing the maximum response of different positions of the structure with the static results, the amplification factor of the structure under frequent seismic actions were obtained, which can estimate the dynamic response results from the static results. The dynamic displacement ratio ranges from 1.10 to 1.50 and the dynamic stress ratio ranges from 1.10 to 1.85.

3. The amplification factor of the elastic–plastic response of the structure under rare seismic actions relative to the elastic response under frequent seismic actions can be taken as 2.50 to 3.00 for horizontal displacement ratios and 1.50 to 2.00 for vertical displacement ratios, while the stress ratios range from 1.35 to 2.10. The deformation of the structure is more affected by seismic intensity than stresses, and the horizontal displacement is more sensitive to seismic intensity.

4. Under rare seismic actions, only less than 5% of the elements enter the plastic stage, with higher plastic development degree observed in the cross rods, the diagonal rods, and the members at the pyramid positions. The cross rods close to the support first enter the plastic phase, and the top of the structure and the third and fourth rings from the top to the supports show a high degree of plastic development, which are the relatively weakness of the structure.

5. The Kiewitt-type mega-latticed structure requires a large amount of steel and experiences high stress under the seismic actions, making it not applicable to the 800 m span. Under the rare seismic actions, the number of elements entering plasticity and the degree of plastic development of the Geodesic-type mega-latticed structure are higher than those of the Three-dimensional grid-type. Therefore, when designing an 800 m span spherical mega-latticed structure, priority should be given to the Three-dimensional grid-type.

Author Contributions: Conceptualization, Y.Z.; methodology, Y.Z. and Z.Z.; software, Z.Z.; validation, Y.Z. and Z.Z.; formal analysis, Z.Z.; resources, Y.Z.; data curation, Z.Z.; writing—original draft preparation, Z.Z.; writing—review and editing, Y.Z.; supervision, Y.Z.; project administration, Y.Z.; funding acquisition, Y.Z. All authors have read and agreed to the published version of the manuscript.

Funding: This research was funded by National Natural Science Foundation of China, grant No. 51978207 and 51927813; National Science Fund for Distinguished Young Scholars, grant No. 51525802; Creative Research Groups of National Natural Science Foundation of China, grant No. 51921006; Heilongjiang Natural Science Foundation for Excellent Youth project, grant No. YQ2021E030.

Institutional Review Board Statement: Not applicable.

Informed Consent Statement: Not applicable.

Data Availability Statement: The data used to support the findings of this study are available from the corresponding author upon request.

Acknowledgments: This work was financially supported by the Funds for Creative Research Groups of National Natural Science Foundation of China (Grant No. 51921006), National Natural Science Foundation of China (Grant No. 51978207, 51927813), National Science Fund for Distinguished Young Scholars (Grant No. 51525802), and Heilongjiang Natural Science Foundation for Excellent Youth project (Grant No. YQ2021E030), which is gratefully acknowledged.

Conflicts of Interest: The authors declare no conflict of interest.

References

1. Mariangela, L. The Centennial Hall of Wroclaw: History of A Modern Architecture in Reinforced Concrete Classified World Heritage Site. *J. Archit. Conserv.* **2021**, *27*, 17–52.
2. Hladik, P.; Lewis, C.J. Singapore National Stadium Roof. *Int. J. Archit. Comput.* **2010**, *8*, 257–277. [CrossRef]
3. Wang, H.; Li, P.; Wang, J. Shape Optimization and Buckling Analysis of Novel Two-Way Aluminum Alloy Latticed Shells. *J. Build. Eng.* **2021**, *36*, 102100. [CrossRef]
4. El-Sheikh, A. Development of A New Space Truss System. *J. Constr. Steel Res.* **1996**, *37*, 205–227. [CrossRef]
5. Shen, S.Z.; Lan, T.T. A Review of the Development of Spatial Structures in China. *Int. J. Space Struct.* **2001**, *16*, 157–172. [CrossRef]
6. Kron, W.; Löw, P.; Kundzewicz, Z.W. Changes in Risk of Extreme Weather Events in Europe. *Environ. Sci. Policy* **2019**, *100*, 74–83. [CrossRef]
7. Balling, R.J. Tall buildings + skybridges + envelope + green = greenplex: A sustainable urban paradigm for the 21st century. In Proceedings of the Council on Tall Building and Urban Habitat 2011 Seoul Conference, Seoul, Republic of Korea, 10–12 October 2011; TS 19-03. pp. 340–347.
8. Liddell, I.; Westbury, P. Design and Construction of the Millennium Dome, UK. *Struct. Eng. Int.* **1999**, *9*, 172–175. [CrossRef]
9. Ren, X. Architecture and Nation Building in the Age of Globalization: Construction of the National Stadium of Beijing for the 2008 Olympics. *J. Urban Aff.* **2008**, *30*, 175–190. [CrossRef]
10. Zhou, G.P.; Li, A.Q.; Li, N.; Li, J.H. Design and analysis of the health monitoring system for the steel arch of the main stadium roof in Nanjing Olympic sports center. In *Applied Mechanics and Materials*; Trans Tech Publications Ltd.: Stafa-Zurich, Switzerland, 2017; Volume 858, pp. 3–9.
11. He, Y.J.; Zhou, X.H.; Liu, Y.J.; Dong, S.L.; Li, J. Super-Span Reticulated Mega-Structure. *J. Archit. Civ. Eng.* **2005**, *22*, 25–29. (In Chinese)
12. He, Y.J.; Zhou, X.H. Static Properties and Stability of Cylindrical ILTDBS Reticulated Mega-Structure with Double-Layer Grid Substructures. *J. Constr. Steel Res.* **2007**, *63*, 1580–1589. [CrossRef]
13. Zhou, X.H.; He, Y.J.; Xu, L. Formation and Stability of a Cylindrical ILTDBS Reticulated Mega-Structure Braced with Single-Layer Latticed Membranous Shell Substructures. *Thin-Walled Struct.* **2009**, *47*, 537–546. [CrossRef]
14. Zhou, X.H.; He, Y.J.; Xu, L. Stability of A Cylindrical ILTDBS Reticulated Mega-Structure with Single-Layer LICS Substructures. *J. Constr. Steel Res.* **2009**, *65*, 159–168. [CrossRef]
15. He, Y.J.; Zhou, X.H. Formation of the Spherical Reticulated Mega-Structure and Its Stabilities in Construction. *Thin-Walled Struct.* **2011**, *49*, 1151–1159. [CrossRef]
16. Zhang, Q.; Zhang, Y.; Yao, L.; Fan, F.; Shen, S. Finite Element Analysis of the Static Properties and Stability of A 800 m Kiewitt Type Mega-Latticed Structure. *J. Constr. Steel Res.* **2017**, *137*, 201–210. [CrossRef]
17. Zhang, Q.; An, Y.; Zhao, Z.; Fan, F.; Shen, S. Model Selection for Super-Long Span Mega-Latticed Structures. *J. Constr. Steel Res.* **2019**, *154*, 1–13. [CrossRef]
18. Wang, B.; Jiang, H.J.; Lu, X.L. Experimental and Numerical Investigations on Seismic Behavior of Steel Truss Reinforced Concrete Core Walls. *Eng. Struct.* **2017**, *140*, 164–176. [CrossRef]
19. Zhang, D.W.; Li, Y.Q.; Huang, B.H. Investigation on Seismic Performance of the Fully Assembled Steel Frame Applying Beam-Column Joints with Replaceable Energy-Dissipating Elements. *Sustainability* **2023**, *15*, 5488. [CrossRef]
20. Barkhordari, M.S.; Tehranizadeh, M. Ranking Passive Seismic Control Systems by Their Effectiveness in Reducing Responses of High-Rise Buildings with Concrete Shear Walls Using Multiple-Criteria Decision Making. *Int. J. Eng.* **2020**, *33*, 1479–1490.
21. Abdalla, J.A.; Mohammed, A.S. Dynamic characteristics of large reinforced concrete domes. In Proceedings of the 14th World Conference on Earthquake Engineering, Beijing, China, 12–17 October 2008.
22. Feizolahbeigi, A.; Lourenço, P.B.; Golabchi, M.; Ortega, J.; Rezazadeh, M. Discussion of the Role of Geometry, Proportion and Construction Techniques in the Seismic Behavior of 16th to 18th century bulbous discontinuous double shell domes in central Iran. *J. Build. Eng.* **2021**, *33*, 101575. [CrossRef]
23. Zhang, H.D.; Liang, X.; Gao, Z.Y.; Zhu, X.Q. Seismic Performance Analysis of a Large-Scale Single-Layer Latticed Dome with A Hybrid Three-Directional Seismic Isolation System. *Eng. Struct.* **2020**, *214*, 110627. [CrossRef]
24. Fan, F.; Wang, M.; Cao, Z.G.; Shen, S. Seismic Behavior and Design of Spherical Reticulated Shells with Semi-Rigid Joint System. *China Civ. Eng. J.* **2010**, *43*, 8–15. (In Chinese)
25. Sun, J.; Zhang, Q. Seismic Performance of Long-Span Space Reticulate Shells and Its Analysis. *J. Nat. Disasters* **2011**, *20*, 193–198. (In Chinese)
26. ANSYS, Inc. *Theory Manual. 001369*, 12th ed.; SAS IP, Inc.: Cheyenne, WY, USA, 2009.

27. *JGJ 7-2010*; Technical Specification for Space Frame Structures. China Architecture & Building Press: Beijing, China, 2010. (In Chinese)
28. Fan, F.; Cao, Z.G.; Ma, H.H.; Yan, J.C. *Elasto-Plastic Stability of Reticulated Shells*; Science Press: Beijing, China, 2015; p. 16. (In Chinese)
29. Resapu, R.R.; Perumahanthi, L.R. Numerical Study of Bilinear Isotropic & Kinematic Elastic–Plastic Response under Cyclic Loading. *Mater. Today Proc.* **2021**, *39*, 1647–1654.
30. *GB50011-2010*; Code for Seismic Design of Buildings. China Architecture & Building Press: Beijing, China, 2010. (In Chinese)
31. Dasgupta, S.P. Computation of Rayleigh Damping Coefficients for Large Systems. *Electron. J. Geotech. Eng.* **2003**, *8*, 83–88.
32. Qu, Z.; Zhu, B.J.; Cao, Y.T.; Fu, H.R. Rapid Report of Seismic Damage to Buildings in the 2022 M6.8 Luding Earthquake, China. *Earthq. Res. Adv.* **2023**, *3*, 100180. [CrossRef]

sustainability

Article

Research on the Bearing Capacity and Sustainable Construction of a Vacuum Drainage Pipe Pile

Wei-Kang Lin [1,2], Xiao-Wu Tang [1,2,*], Yuan Zou [1,2], Jia-Xin Liang [1,2] and Ke-Yi Li [1,2]

[1] Research Center of Coastal and Urban Geotechnical Engineering, Department of Civil Engineering, Zhejiang University, Hangzhou 310058, China; 11912043@zju.edu.cn (W.-K.L.); 18401606407@163.com (Y.Z.); jaynaliang@zju.edu.cn (J.-X.L.); 20lky@zju.edu.cn (K.-Y.L.)

[2] Engineering Research Center of Urban Underground Space Development of Zhejiang Province, Hangzhou 310058, China

* Correspondence: tangxiaowu@zju.edu.cn; Tel.: +86-139-5809-1045

Abstract: The vacuum drainage pipe (VDP) pile is a new type of pipe pile on which the current research is mainly focused on laboratory tests. There is little research on bearing characteristics and carbon emissions in practical engineering. To further explore the bearing capacity and sustainable construction of vacuum drainage pipe piles, static load tests were conducted to investigate the single-pile bearing capacity of ordinary pipe piles and vacuum drainage pipe piles, as well as soil settlement monitoring around the piles. Then, the Q-S curves of the two piles, the pile-side friction resistance under different pile top loads, and the development law of pile end resistance were compared and analyzed. Finally, based on the guidelines of the IPCC, the energy-saving and emission-reduction effects of VDP piles in practical engineering were estimated. The results indicate that, after vacuum consolidation, the VDP pile basically eliminates the phenomenon of soil compaction and does not cause excessive relative displacement of the pile and soil. VDP piles have increased lateral friction resistance, and compared to traditional piles, their ultimate bearing capacity is increased by 17.6%. Compared with traditional methods, the VDP pile method can reduce carbon emissions by 31.4%. This study provides guidance for the production and design of future VDP piles and demonstrates the potential of VDP piles for energy conservation and emission reduction in comparison to traditional methods.

Keywords: pile foundation; sustainable construction; soft soil; compressive bearing capacity; carbon emission

Citation: Lin, W.-K.; Tang, X.-W.; Zou, Y.; Liang, J.-X.; Li, K.-Y. Research on the Bearing Capacity and Sustainable Construction of a Vacuum Drainage Pipe Pile. *Sustainability* **2023**, *15*, 7555. https://doi.org/10.3390/su15097555

Academic Editors: Jurgita Antuchevičienė and Noradin Ghadimi

Received: 14 March 2023
Revised: 25 April 2023
Accepted: 3 May 2023
Published: 4 May 2023

1. Introduction

Many large-scale railway, highway, port, and airport infrastructures, as well as major industrial bases and logistics centers, are distributed in coastal soft soil areas. At present, the commonly used treatment method is the plastic drainage board and PHC (prestressed high-strength concrete pile) piles combined method [1,2]. This method has the following disadvantages: (1) the consolidation process takes a long time; (2) the plastic drainage board does not degrade easily, which is not conducive to sustainable development; and (3) the pile driving after the foundation has hardened increases the energy consumption during construction. Therefore, there is an urgent need to study low-carbon and sustainable foundation treatment methods for coastal soft clay.

In the context of global "carbon neutrality", traditional civil engineering is actively implementing the concept of sustainable development [3,4]. To reduce construction costs and improve the sustainability of their projects, domestic and foreign scholars have explored the feasibility of integrating foundation treatment and pile foundation engineering, with examples including permeable concrete piles [5], perforated piles [6], drainage plate combination piles [7,8], geotextile-encased stone columns [9], etc. However, their efforts are limited by the following deficiencies: The pile body material is filled with voids, and the pile body strength is low, so it cannot be directly used as an engineering pile [10]. Small holes open up in the pile body, and the water and soil are not separated, resulting in silting and blocking, thus preventing water from being effectively drained for a long time. The combination of the pile body and drainage board reduces the friction area and friction coefficient, and the bearing capacity decreases. The energy consumption is similar to that of common construction methods. The strength of the pile depends on the confining pressure. Once the geomembrane is broken, the overall strength will significantly decrease [11].

To sum up, there is still a large research gap regarding the integration of foundation treatment and pile foundation engineering. In order to further promote the green, low-carbon, and sustainable development of foundation treatment, Tang et al. [12–14] proposed the vacuum drainage pipe pile (VDP pile). The pile body is uniformly arranged with small holes, and the pile body is covered with degradable geotextile for reverse filtration to prevent the small holes from silting up and ensure the long-term stability of the drainage channel. The VDP piles are driven when the soil is soft, and the excess pore water pressure can be dissipated through the drainage channel. After vacuum consolidation, they can be directly used as engineering piles. No plastic drainage boards are used for drainage, which reduces costs and environmental pollution.

The production and construction of pile foundation engineering generate large amounts of solid waste and greenhouse gases. Currently, most research on building carbon footprints focuses on the operational stage rather than the production and construction stages [15]. However, during the production and construction stages, a large amount of material is consumed, and construction machinery and transportation equipment consume a large amount of energy, resulting in the generation of a large amount of carbon dioxide [16]. Therefore, it is necessary to conduct more accurate and comprehensive calculations of the carbon emissions for each stage [17]. Many studies have been conducted on the life cycle of buildings abroad, accumulating a large amount of raw data. However, research in this field is still in its early stages and lacks a unified standard database and evaluation model. Moreover, the extreme differences in energy composition between China and foreign countries make it impossible for foreign raw data to be directly used in China [18]. Therefore, this article combines foreign raw data and the latest research conducted in China to adjust the means of carbon emission factors in pile foundation engineering in order to better understand carbon emissions and determine carbon emission trends.

Thus far, most of the new types of piles [6,19,20] used for soft foundations have achieved good results in experimental research, but they have few applications in practical engineering. This is also true for VDP piles, and the current research mainly focuses on laboratory tests. The production and construction processes of the VDP pile are not clear, and the improvement of the bearing capacity of the vacuum drainage pipe pile has not been verified in an actual project. To further promote the sustainable construction and application of VDP piles, field tests were carried out to compare and analyze the ultimate single-pile bearing capacity, pile side friction, and pile end resistance of ordinary piles and vacuum drainage pipe piles. Here, the effects of the VDP pile on energy conservation and emission reduction in practical projects are discussed.

2. Sustainable Construction of the Vacuum Drainage Pipe Pile in Soft Soil

2.1. Production Process of the Vacuum Drainage Pipe Pile

The PHC pile is the most common pipe pile used in Chinese pile foundation engineering. Due to its simple production, strong load capacity, small settlement deformation, and high efficiency, it has been widely used in the field of construction engineering. Most comparative studies of new types of piles use PHC piles as a reference. Therefore, the ordinary piles used in this study were PHC piles. The main material of the VDP piles used in the test was concrete, and the concrete strength of the pile was C80 (the compressive strength of the cubic concrete block was 80 MPa). The outer diameter of the pipe pile was 500 mm, the inner diameter was 250 mm, and the pile length was 9 m. The number of holes in a single layer was 2, the spacing between the layers of holes was 1 m, and the hole diameter was 30 mm, giving 14 holes in total. The production process of VDP piles is the same as that of ordinary pipe piles (PHC piles), with the following differences: To ensure the integrity of the holes, after the reinforcement cage is placed into the mold, a PVC pipe with a length of 50 cm and an outer diameter of 3 cm is bound to the reinforcement cage using fine steel wires to form a new mold. After concrete curing, the PVC pipes are knocked out to form holes. A metal mesh with a filter membrane is inserted into the holes. Airtight glue is used to fit the metal mesh with the pile body to prevent air leakage in the vacuum process, as shown in Figure 1.

Figure 1. Production process of the vacuum drainage pipe pile.

2.2. Construction Process of the Vacuum Drainage Pipe Pile

Unlike the plastic drainage board and pile foundation combined method, the construction process of the VDP piles is as follows:

(1) When the soil is soft, the universal prefabricated pipe pile driver is used to drive the vacuum drainage pipe pile. At this time, the energy consumption required for driving the pile is low;
(2) One uses the pile driving disturbance and its own drainage channel to reduce the soil-squeezing effect, and the other connects an external vacuum machine to accelerate drainage and consolidation;
(3) After the soil hardens, the pile composite foundation is formed so as to jointly bear the upper load and directly serve as the engineering pile;
(4) The plastic drainage board is not used in the process, which is environmentally friendly.

The construction and service of the vacuum drainage pipe piles are integrated, which improves the overall bearing capacity and greatly reduces the construction period, with good economic benefits, as shown in Figure 2.

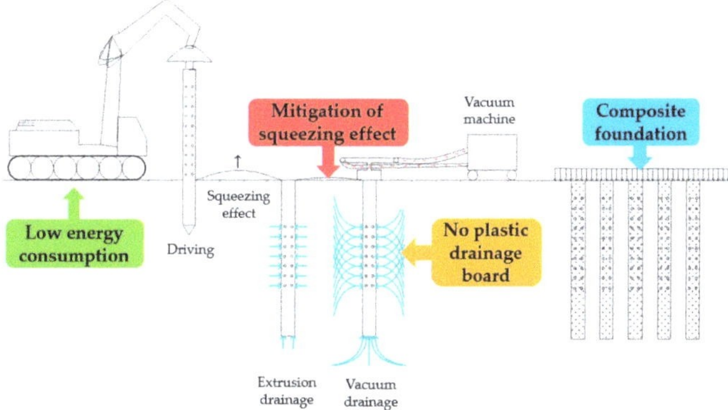

Figure 2. The construction and service of the vacuum drainage pipe pile.

3. Field Test

3.1. Test Site and Pile Description

The soil parameters, vacuum method, and strain gauge distribution are shown in Figure 3. The field tests were carried out on the project site of a highway to be built in Hangzhou, Zhejiang Province. The soil layer of the site is mainly composed of plain fill, silt, and silty clay, with a groundwater-level depth of 1.8 m. To ensure the progress of the pile foundation project, the vacuum consolidation time was set at 10 days for this test, and the vacuuming time was 18 h per day for a total of 180 h. During the vacuum interval, the water inside the pile was pumped into the water storage bucket. During the vacuum period, the vacuum degree had an atmosphere of approximately 0.5~0.7. FBG (fiber Bragg grating) and had the advantages of high precision and real-time performance. A fiber grating strain gauge and fiber grating thermometer from Zhixing Technology Nantong Co., Ltd. were used in this test. The strain gauges were arranged 500 mm below the pile top with a spacing of 1000 mm. The thermometer was buried 5000 mm below the pile top.

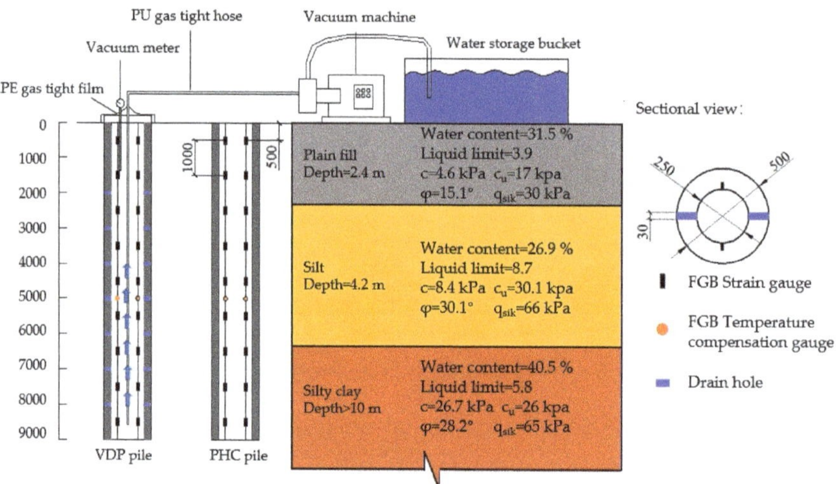

Figure 3. Soil parameters and pile parameters.

3.2. Test Process

A single-pile static load compression test was carried out for one PHC pile and one VDP pile. For the PHC pile, the first-level load was 300 kN, and then the load for each following level was 150 kN. For the VDP pile, the first-level load was 400 kN, and the load for each following level was 200 kN. The slow maintenance load method was adopted in the test [21]. The specific process is shown in Figure 4: (1) After applying the first-level load, one measures and records the pile top displacement at 5 min, 15 min, 30 min, 45 min, and 60 min, respectively, and once every 30 min thereafter. (2) The next level of load can be applied if the displacement does not exceed 0.1 mm within two consecutive hours. (3) The loading can be terminated when the total settlement of the pile top exceeds 40 mm. The single-pile compressive static load test device is mainly composed of a surcharge, buttress, hydraulic jack, steel beam, and displacement meter, as shown in Figure 5.

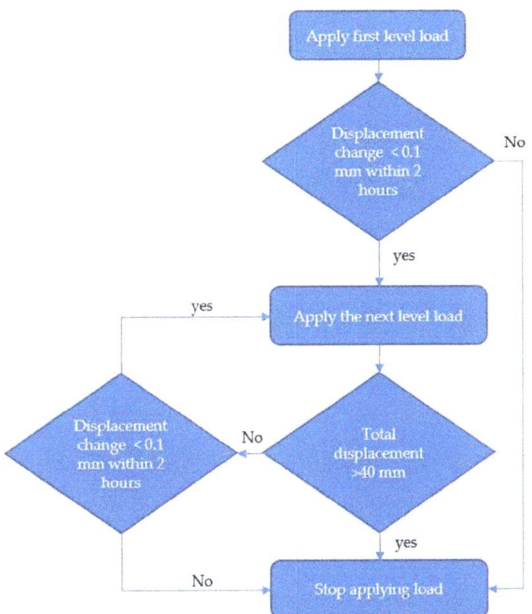

Figure 4. The flow chart of the slow maintenance load method.

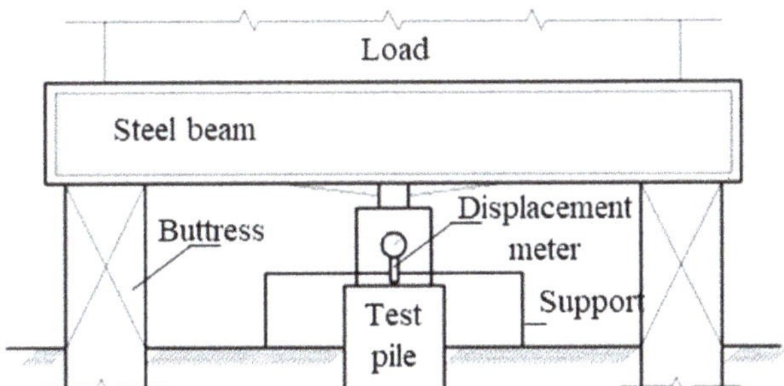

Figure 5. Overview of the static load tests.

To verify that the VDP pile will not produce a large relative displacement of the pile and soil during vacuum consolidation, it is necessary to monitor the surface displacement and pile top displacement of the soil around the test pile. One uses a precision total station to observe the ground settlement around the pile. The layout of the measuring points is shown in Figure 6. A total of 6 measuring points are uniformly arranged at a distance of 0 mm~1000 mm from the pile side, and the same measuring points are arranged in four vertical directions. Finally, the average displacement is taken as the surface displacement of the soil around the test pile.

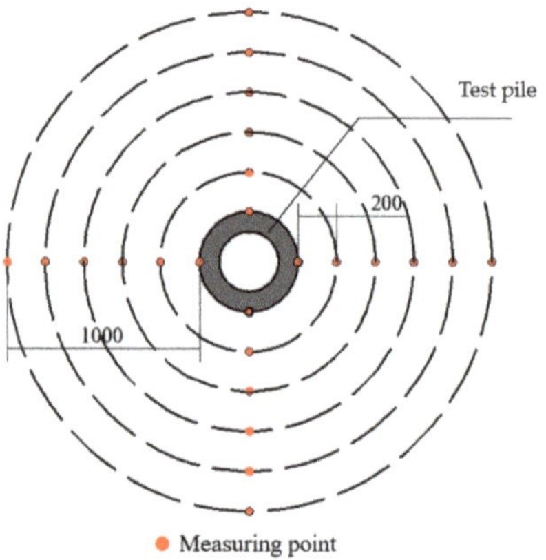

Figure 6. The layout of the measuring points (unit: mm).

3.3. Test Results

3.3.1. Surface Displacement of Soil around the Piles

Figures 7 and 8 show the surface displacement curves of the soil around the PHC pile and the VDP pile, respectively. The soil surface uplift was the highest at 20 cm from the side of the PHC pile, and there was still an uplift after 10 days. This shows that the soil-squeezing effect caused by pile driving has not completely disappeared. On the 4th day, the VDP pile basically eliminated the soil squeeze and even caused a settlement of approximately 1.5 mm at the 60 cm side of the pile. From the 5th day to the 7th day, the settlement velocity of the pile-side soil of the two types of piles decreased, and the reverse arch phenomenon was observed. This is mainly due to the heavy rain on these two days, which increased the amount of pore water among the soil particles. Over time, the soil around the two types of piles continued to consolidate, and the surface uplift gradually disappeared. This development trend of soil settlement around ordinary pipe piles is basically consistent with the results of existing research; thus, the monitoring method explored in this article is effective [22,23].

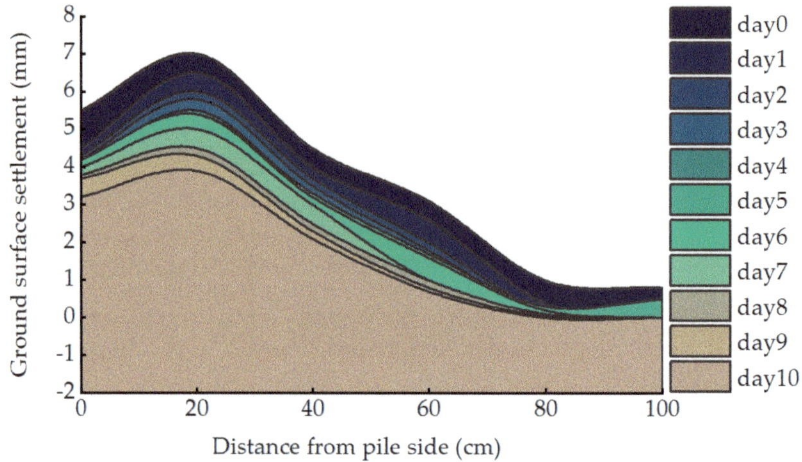

Figure 7. Surface displacement of the soil around the PHC pile.

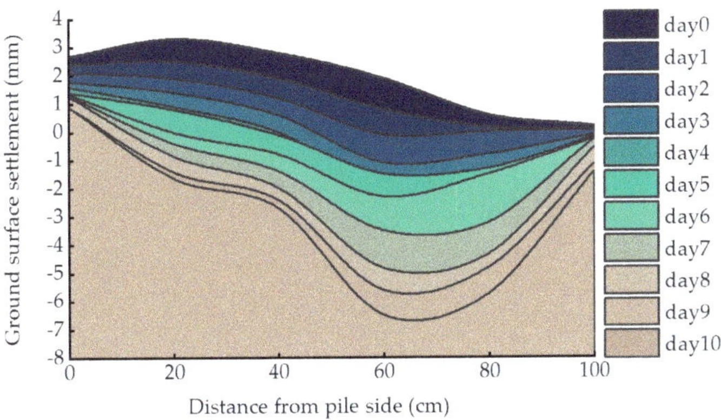

Figure 8. Surface displacement of the soil around the VDP pile.

If the settlement of the soil around the pile is greater than the settlement of the pile itself, this will cause negative frictional resistance. The displacement of the top of the VDP pile and the ground displacement on the 0 cm side of the pile were drawn in the same diagram, and the relative displacement curve was added (as shown in Figure 9). It can be seen from the relative displacement curve that the maximum difference between the two was 0.18 mm~−0.12 mm. The relative displacement of the pile and soil caused by the vacuum effect is almost negligible. Therefore, vacuum consolidation will not lead to excessive pile-soil displacement or negative frictional resistance.

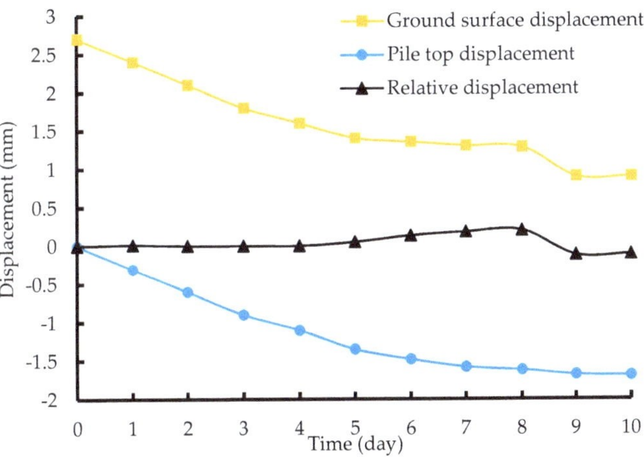

Figure 9. Relative displacement curve.

3.3.2. Relationship between Pile Head Displacement and Load

The load-settlement curves of the PHC pile and the VDP pile are slowly changing types, as shown in Figure 10. The slope of the VDP pile curve is comparatively smaller. Throughout different stages, the VDP pile demonstrated a larger bearing capacity and anti-deformation capacity than the PHC pile. To determine the vertical ultimate bearing capacity of a single pile, the method outlined in the technical code [21] was employed. The load value corresponding to a 40 mm settlement of the pile top was considered the ultimate compressive bearing capacity. Using linear interpolation, the ultimate compressive bearing capacities of the PHC pile and the VDP pile were calculated to be 1482 kN and 1743 kN, respectively. The single-pile ultimate bearing capacity of the VDP pile was approximately 17.6% larger than that of the PHC pile.

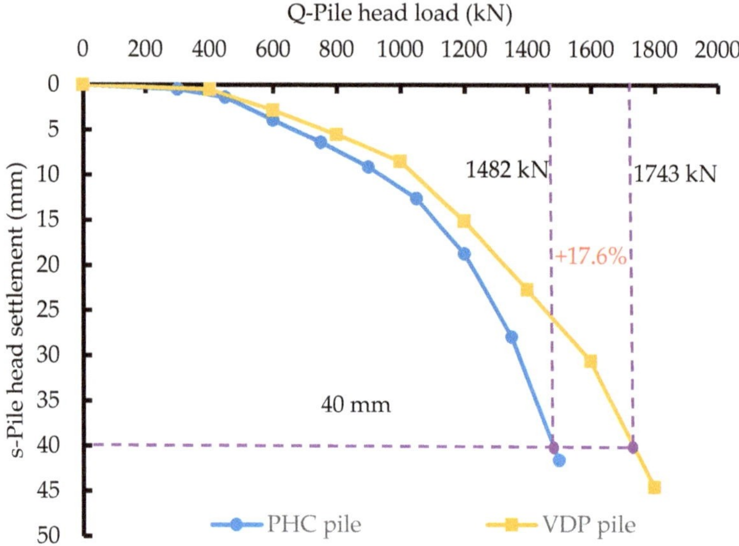

Figure 10. Load-settlement curve of pile.

3.3.3. Analysis of the Pile Side Frictional Resistance

The pile-side friction resistance can be calculated by dividing the difference between the axial forces of two sections by the pile-side area of the section. The average pile side friction resistance of the PHC pile and VDP pile under different pile top loads is shown in Figures 11–13. At a depth of 1 m~2 m, as the pile top load increased from 600 kN to 1200 kN, the friction resistance of the PHC pile and the VDP pile increased from 30 kPa~35 kPa to 55 kPa~60 kPa, but the gap between the two did not further expand. At a depth of 3 m~6 m, the friction resistance of the two piles increased from 35 kPa~40 kPa to 65 kPa~75 kPa, and the difference between them increased slightly. At a depth of 7 m~8 m, the friction resistance of the PHC pile increased from 20 kPa~25 kPa to 30 kPa~35 kPa, while the friction resistance of the VDP pile increased from 30 kPa~35 kPa to 60 kPa~65 kPa, and the difference between the two increased from 40.40~55.32% to 81.79~98.29%.

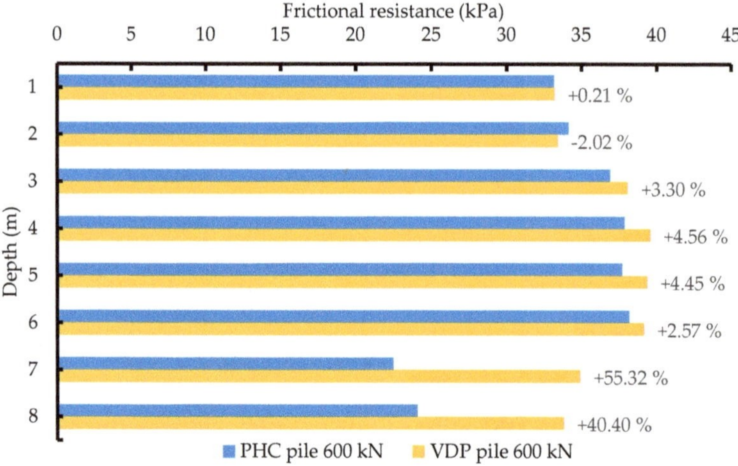

Figure 11. Distribution of pile side frictional resistance when the pile top load is 600 kN.

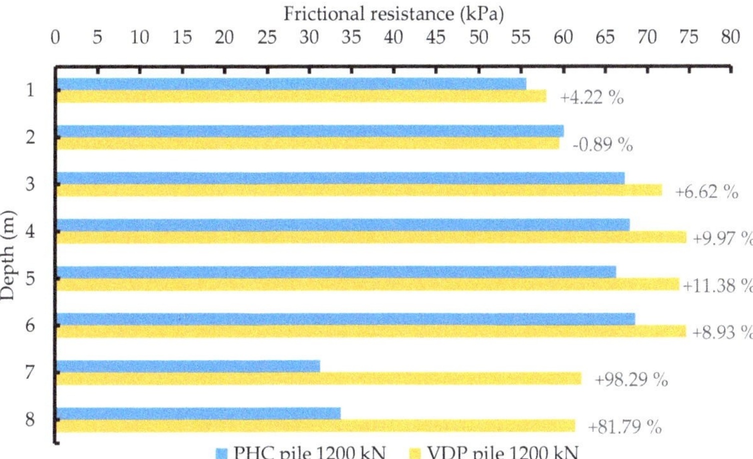

Figure 12. Distribution of pile side frictional resistance when the pile top load is 1200 kN.

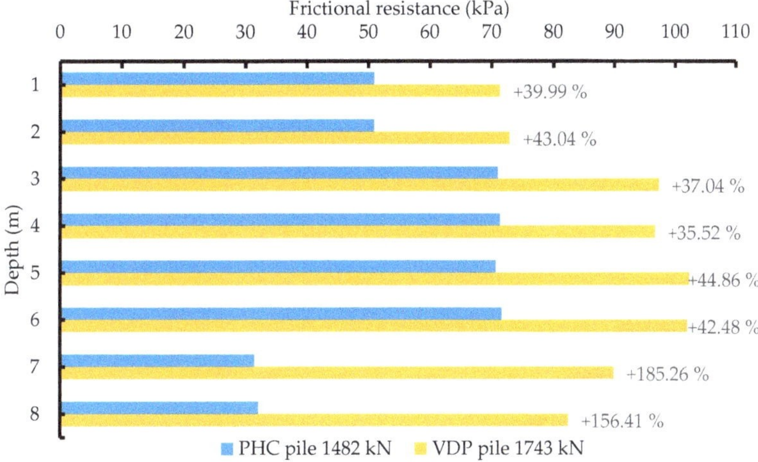

Figure 13. Distribution of pile side frictional resistance when the pile top load is the ultimate load.

It is worth noting that when the depth was 2 m, regardless of whether the pile top load was 600 kN or 1200 kN, the lateral friction resistance of the PHC piles was higher than that of the VDP piles. This may be because groundwater has a greater impact on VDP piles compared to PHC piles, as VDP piles can be permeable. According to the geological survey report, the groundwater level was located at a depth of 1.8 m, and the frequent flow of groundwater reduced the compaction of the soil around the pile, thereby reducing the friction resistance at this depth. However, at a depth of 2 m, when the top load of the pile reached its limit, the lateral friction resistance of the VDP pile was significantly higher than that of the PHC pile. This is mainly because the ultimate load of the former was greater than that of the latter, and the pile-soil displacement was greater. Therefore, at a depth of 2 m, the soil around the pile provided more lateral friction resistance. Overall, the groundwater interface will reduce the lateral friction resistance of vacuum drainage pipe piles, but as the pile top load gradually increases to the ultimate load, this impact will gradually decrease.

The distribution of pile-side frictional resistance under the ultimate load is shown in Figure 13. The frictional resistance of the VDP pile mainly acts on the middle and lower parts of the pile, and the distribution is more uniform than that of the PHC pile. For the PHC pile, at depths of 7 m~8 m, the pile side frictional resistance is only 30 kPa~35 kPa, while the lateral friction resistance of the VDP pile can reach 80 kPa~90 kPa. This indicates that vacuum consolidation can greatly enhance the lateral friction resistance near the bottom of the pile, greatly improving the load transfer efficiency of the pile body.

To further analyze the influence of the soil layers on the side friction resistance, the average side friction resistance of different soil layers was calculated using the difference method. The depth of the plain fill was 0 m~2.4 m, the depth of the silt was 2.4 m~6.6 m, and the depth of the silty clay was more than 6.6 m. The distribution of pile side friction resistance for the two piles according to the soil layer is shown in Figure 14. It can be seen that compared with the PHC pile, the pile side friction resistance of the VDP pile to the silty clay layer is greatly increased, with an increase of 170.69%. For the other two soil layers, there is also an increase of nearly 40%. Therefore, the vacuum consolidation effect of the VDP pile can improve the pile side friction resistance, and the lifting effect is more apparent for the soil layer with a worse structure.

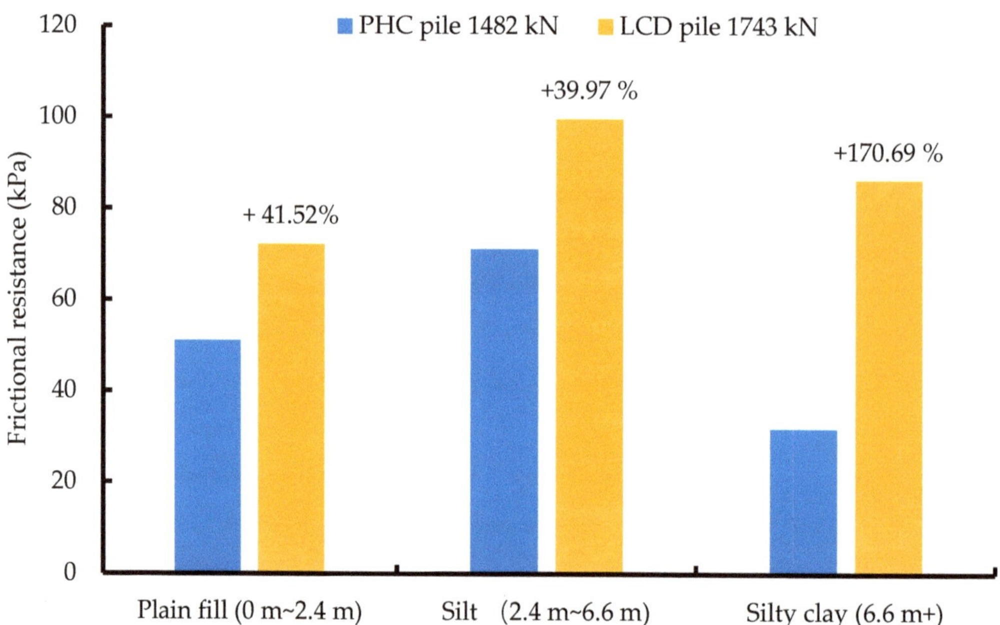

Figure 14. Distribution of pile side frictional resistance according to the soil layer.

3.3.4. Analysis of the Pile end Resistance

The pile end resistance ratio can be defined as the ratio of the pile bottom resistance to the top pile load, which is used to analyze the bearing characteristics of the pile foundation and determine the pile type classification. In this field test, no pressure sensor was installed at the pile end, and the pile end resistance was approximately calculated by subtracting the total side friction from the pile top load. The relationship between the pile end resistance ratios of the PHC and VDP is shown in Figure 15.

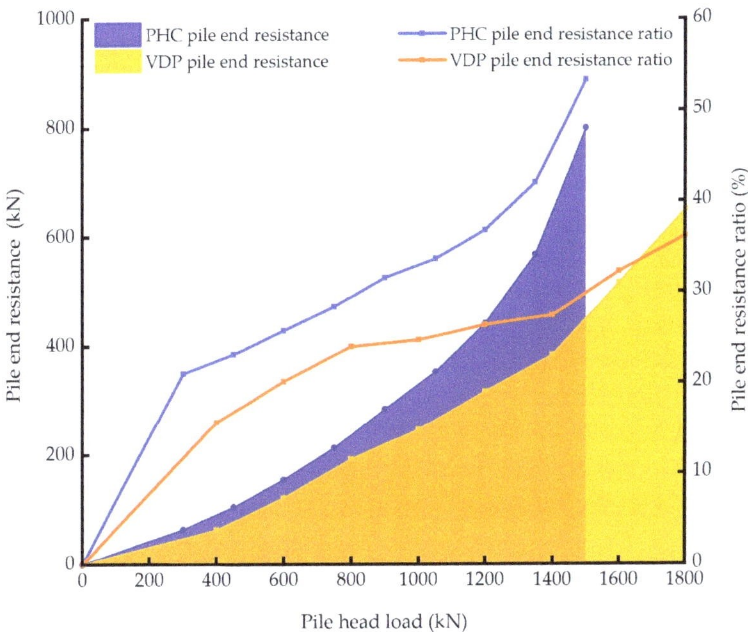

Figure 15. The pile end resistance of the piles.

In the initial stage of pile top load loading, the rising slopes of the two pile types are basically the same. When the load on the top of the pile exceeds 750 kN, the increase in the pile-end resistance of the VDP pile levels off. At this time, further pile-side frictional resistance develops so as to bear the load. The slope of the PHC pile is basically unchanged, and the load is borne mostly by the pile bottom. When approaching the ultimate load, both slopes increase, the pile end resistance of PHC changes abruptly, and the pile end resistance ratio exceeds 50% at the last level of loading, which can be determined as the frictional end bearing pile, dominated by end bearing. On the contrary, the slope of the VDP pile is increased, but it is basically consistent with the initial stage of load loading. There is no sudden change, and the final pile-end resistance ratio is less than 40%. It can be determined as an end-bearing friction pile, dominated by pile-side frictional resistance.

4. Carbon Emission Estimation for the Whole Process of the VDP Pile

The carbon emissions of pile foundation projects refer to the total amount of greenhouse gas emitted into the external environment as building materials and energy are consumed throughout the whole life cycle, from production and transportation to the construction of piles. The calculation method is the total consumption of materials or energy multiplied by the carbon emission factor, which can be estimated according to Equation (1). The carbon emission factors used in this paper refer to the IPCC Guidelines for the National Greenhouse Gas Emission Inventory and the Chinese Academy of Engineering.

$$C = \sum_{i=1}^{m} M_i \times F(M_i) + \sum_{j=1}^{n} E_i \times F(E_i) \tag{1}$$

where

C is the total carbon emission (kg);

M_i is the total consumption of the i-th material in engineering (kg);

$F(M_i)$ is the carbon emission factor of the i-th material (kg/kg);

E_i is the total consumption of the j-th energy in engineering (kg, kW·h);

$F(E_j)$ is the carbon emission factor of the j-th energy (kg/unit).

The material of the pile body is mainly composed of concrete and steel; the material of the plastic drainage board is mostly PVC; and the metal mesh for the VDP piles is steel. The energy data required for transportation is taken from the *China Transportation Yearbook 2021* [24]. The materials and energy consumption required for construction are calculated according to the Budget Quota of Electric Power Construction Project [25] and the Quota of Construction Machinery Shift Cost of Electric Power Construction Project [26].

Taking the working conditions of this field test as an example, for the treatment of a 10,000 m^2 soft soil foundation, if the combined method of common piles and drainage boards is used, the pile spacing is 2.5 m, 1600 piles with a diameter of 0.5 m and a length of 9 m are required, and 1600 plastic drainage boards with a length of 15 m are added. If VDP piles are used, the pile spacing is 3 m, and 1111 piles with a diameter of 0.5 m and a length of 9 m are required. The bearing capacity of VDP piles is higher than that of PHC piles; thus, the pile spacing can be appropriately increased. For soft soil, when the pile spacing exceeds a diameter of 5, the pile group effect coefficient will be approximately 1.3 [27].

The carbon emission factor refers to the amount of greenhouse gas generated when consuming a unit mass of substance and is an important parameter that characterizes the greenhouse gas emission characteristics of a certain substance [28]. In the estimation of carbon emissions, the selection and determination of carbon emission factors are crucial. The carbon emissions of pile foundation engineering materials mainly derive from two sources: one is direct or indirect emissions generated through the use of energy, and the second is direct emissions generated through the chemical reactions of raw materials. The carbon emissions from the first source can be calculated based on production energy consumption statistics obtained by multiplying the energy usage by the carbon emission factor. The carbon emissions from the second source can be estimated by comprehensively considering the carbon content of the material and the chemical reaction process.

In the production of PHC piles, the steel bars and spiral bars are made of cold-drawn low-carbon steel wire; and the circular bars, the end plates, and the pile sleeves are made of Q235 steel. Considering the process and technical level of domestic steel production, the carbon emission factor of the steel used for PHC piles can be taken as 2.0 (kg/kg) [29].

The concrete strength of PHC piles is generally C80, and currently, there is no carbon emission factor for C80 concrete in the data. Therefore, according to the law of carbon emission factors for C20, C30, C40, and C50, the carbon emission factor for C80 can be recursively calculated to be 470 (kg/m^3) [30].

The carbon emission factor of energy includes the total carbon emissions per unit mass of energy in various stages of acquisition, processing, and use. The main types of energy consumed in pile foundation engineering are coal, electricity, and diesel. The carbon emission factors for energy are as follows: coal has a carbon emission factor of 0.73 (kg/kg), electricity has a carbon emission factor of 0.28 (kg/kW·h), and diesel has a carbon emission factor of 0.59 (kg/kg) [31].

The results regarding the carbon emissions for the common piles and drainage boards throughout the whole process of the combined method are shown in Table 1, and the results regarding the carbon emissions of the VDP piles throughout the whole process are shown in Table 2. Regarding total carbon emissions, the VDP pile method can lead to a reduction of 31.4%, and the main factor for reducing carbon emissions is the reduced use of piles. In addition, the energy consumption of the whole process is also decreased, with 6.8 tons of diesel oil, 13.9 tons of coal, and 8678 kw·h of electricity. Therefore, compared with traditional methods, the VDP pile treatment of soft soil foundations reduces the use of pile foundations, which conforms to the development principles and models of a circular and low-carbon economy. It reduces carbon emissions and helps to protect the ecological environment. In addition, it reduces energy consumption and promotes the sustainable development of the project.

Table 1. Carbon emissions of PHC piles and plastic drainage boards.

Project	Unit	Material or Energy Consumption				Emission Factor (kg/Unit)	Carbon Emissions (kg)
		Fabrication	Transport	Construction	Total		
Concrete	m^3	2128	\	\	2128	470	1,000,160
Steel	kg	252,160	\	1942	254,102	2	508,204
PVC	kg	3000	\	\	3000	6.79	20,370
Diesel oil	kg	\	5704	12,862	18,566	0.59	10,953
Electricity	kw·h	19,420	\	8580	28,000	0.28	7840
Coal	kg	45,431	\	\	45,431	0.73	33,164
		Total carbon emissions for the whole process					1,580,693

Table 2. Carbon emissions of VDP piles.

Project	Unit	Material or Energy Consumption				Emission Factor (kg/Unit)	Carbon Emissions (kg)
		Fabrication	Transport	Construction	Total		
Concrete	m^3	1478	\	\	1478	470	694,660
Steel	kg	175,405	\	1942	177,347	2	354,694
Diesel oil	kg	\	3960	7761	11,721	0.59	6915
Electricity	kw·h	13,480	\	5842	19,322	0.28	5410
Coal	kg	31,537	\	\	31,537	0.73	23,022
		Total carbon emissions in the whole process					1,084,701

5. Conclusions

To further explore the bearing capacity and sustainable construction of vacuum drainage pipe (VDP) piles, a VDP pile and a PHC pile with a pile length of 9 m and a pile diameter of 500 mm were fabricated. The field test was carried out on a site composed of fill, silt, and silty clay. The ultimate single-pile bearing capacity, pile side friction, and pile end resistance of the PHC piles and VDP piles were compared and analyzed. The effects of the VDP pile on energy conservation and emission reduction in practical projects were discussed. This study thus provides guidance for the production and design of VDP piles in the future. The following conclusions were drawn:

1. The monitoring results of the total station showed that, in the vacuum consolidation stage, the relative displacement of the pile and soil for the VDP pile was less than 0.5 mm, which is almost negligible. After 4 days, the squeezing effect could basically be eliminated, while the PHC pile still had a protrusion of 3~4 mm on the 10th day. This indicates that VDP pipe piles can alleviate the squeezing effect;

2. In this field test, the ultimate bearing capacity of a single VDP pile was 1743 kN, and that of the PHC pile was 1482 kN, an increase of 17.6%. The VDP pile can improve the bearing capacity of a single pile. This is mainly because the VDP pile enhances the side friction of the pile;

3. Compared with the PHC pile, the pile side friction resistance of the VDP pile to the silty clay layer was increased by 170.69%, and the vacuum consolidation effect was more apparent for the soil layer with a poor structure. When the load on the pile top reached the limit, the pile end resistance ratio of the VDP pile was less than that of the PHC pile, and the bearing capacity was mainly provided by friction. Vacuum consolidation can compact the pile and soil, which increases the frictional resistance at the pile-soil interface;

4. Regarding total carbon emissions, the VDP pile method can lead to a reduction of 31.4% compared with the traditional method, and the main factor influencing this reduction is the reduced use of piles. The energy consumption is also less than that of the traditional method, which conforms to the principle of low-carbon sustainable development.

Author Contributions: Conceptualization, W.-K.L. and X.-W.T.; methodology, W.-K.L.; software, W.-K.L. and K.-Y.L.; validation, W.-K.L., X.-W.T., and Y.Z.; formal analysis, J.-X.L.; investigation, W.-K.L.; resources, W.-K.L.; data curation, Y.Z.; writing—original draft preparation, W.-K.L.; writing—review and editing, X.-W.T.; visualization, Y.Z.; supervision, X.-W.T.; project administration, X.-W.T.; funding acquisition, X.-W.T. All authors have read and agreed to the published version of the manuscript.

Funding: This work was funded by the Zhejiang Province Public Welfare Technology Application Research Project (LGG22E080002), the Key Water Science and 477 Technology Project of Zhejiang Province (RB2027), and the Cultural Heritage Bureau of Zhejiang Province (2023006).

Institutional Review Board Statement: Not applicable.

Informed Consent Statement: Not applicable.

Data Availability Statement: Not applicable.

Acknowledgments: The authors express their gratitude to the construction department of Hongyang Construction Co., Ltd. for contributing to the construction of the test piles and assistance with the static load tests and to the vacuum equipment production department of Jiangsu Xintai Rock Technology Co., Ltd. for contributing to the vacuum equipment supply and installation. The authors also express their gratitude to the pile production department of Zhejiang Xinye Pipe Pile Co., Ltd. for contributing to the production and transportation of the pipe piles and to the sensor production department of Zhixing Technology Nantong Co., Ltd. for contributing to the sensor installation and technical guidance.

Conflicts of Interest: The authors declare no conflict of interest.

Abbreviations: Notation

C	Total carbon emissions (kg);
M_i	Total consumption of the i-th material in engineering (kg);
$F(M_i)$	The carbon emission factor of the i-th material (kg/kg);
E_i	Total consumption of the j-th energy in engineering (kg, kW·h);
$F(E_j)$	The Carbon emission factor of the j-th energy (kg/unit).

References

1. Mesri, G. Discussion of "Field study of pile—Prefabricated vertical drain (PVD) interaction in soft clay". *Can. Geotech. J.* **2021**, *58*, 747. [CrossRef]
2. Wu, Y.; Fang, L.; Li, X.; Hu, S. Technical discussion on tube pile combined with prefabricated strip drain to soft soil treatment. *Chin. J. Rock Mech. Eng.* **2006**, *25*, 3572–3576. (In Chinese). Available online: https://www.cnki.com.cn/Article/CJFDTOTAL-YSLX2006S2035.htm (accessed on 1 October 2006).
3. Cakiroglu, C.; Islam, K.; Bekdaş, G.; Kim, S.; Geem, Z.W. CO_2 Emission Optimization of Concrete-Filled Steel Tubular Rectangular Stub Columns Using Metaheuristic Algorithms. *Sustainability* **2021**, *13*, 10981. [CrossRef]
4. Zavadskas, E.K.; Antucheviciene, J.; Vilutiene, T.; Adeli, H. Sustainable Decision-Making in Civil Engineering, Construction and Building Technology. *Sustainability* **2018**, *10*, 14. [CrossRef]
5. Umaiyan, U.; Muthukkumaran, M. Improved radial consolidation in soft clay using pervious concrete piles. *Civil Eng.-Geotech.* **2022**. [CrossRef]
6. Ni, P.; Mangalathu, S.; Mei, G. Compressive and flexural behaviour of reinforced concrete permeable piles. *Eng. Struct.* **2017**, *147*, 316–327. [CrossRef]
7. Chen, Z.; Wang, B.; Wang, C.; Wang, Y.; Xiao, P.; Li, K. Performance of a Subgrade-Embankment-Seawall System Reinforced by Drainage PCC Piles and Ordinary Piles Subjected to Lateral Spreading. *Geofluids* **2023**, *2023*, 4489478. [CrossRef]
8. Deng, Y.; Zhang, R.; Sun, J. Novel technology of statical-drill and rooted drainage pile and its large-scale model test. *J. Build. Struct* **2022**, *43*, 293–302. (In Chinese) [CrossRef]
9. Miranda, M.; Da, C.; Castro, J. Influence of geotextile encasement on the behaviour of stone columns: Laboratory study. *Geotext. Geomembr.* **2017**, *45*, 14–22. [CrossRef]
10. Poornachandra, V.; Venkataraman, P.; Krishna, P.B. Discrete element method to investigate flexural strength of pervious concrete. *Constr. Build. Mater.* **2022**, *323*, 126477. [CrossRef]
11. Tan, X.; Cai, M.; Feng, L. Numerical Study on Mechanical Behaviors of Geotextile-wrrapped Stone Column. *China J. Highw. Transp.* **2020**, *33*, 136–145. (In Chinese) [CrossRef]
12. Tang, X.; Yu, Y.; Zhou, L. A Prefabricated Pipe Pile with Drainage and Enlarged Friction Resistance and Its Construction Method. CN Patent 104846809A, 19 August 2015. (In Chinese).

13. Tang, X.; Zou, Y.; Lin, W. Model experiment on improving bearing capacity of perforated pipe piles by vacuum consolidation. *J. Zhejiang Univ. (Eng. Sci.)* **2022**, *56*, 1320–1327. (In Chinese) [CrossRef]
14. Tang, X.; Lin, W.; Zou, Y. Experimental study of the bearing capacity of a drainage pipe pile under vacuum consolidation. *J. Zhejiang Univ.-Sci. A (Appl. Phys. Eng.)* **2022**, *23*, 639–651. [CrossRef]
15. Li, X.J.; Zheng, Y.D. Using LCA to research carbon footprint for precast concrete piles during the building construction stage: A China study. *J. Clean. Prod.* **2019**, *245*, 118754. [CrossRef]
16. Kwok, K.Y.G.; Kim, J.; Chong, W.K.; Ariaratnam, S.T. Structuring a comprehensive carbon-emission framework for the whole lifecycle of building, operation, and construction. *J. Archit. Eng.* **2016**, *22*, 04016006. [CrossRef]
17. Du, Q.; Chen, Q.; Yang, R. Forecast carbon emissions of provinces in China based on logistic model. *Resour. Environ. Yangtze Basin.* **2003**, *22*, 143–150. [CrossRef]
18. Li, X.; Wang, C.; Kassem, M.A.; Wu, S.-Y.; Wei, T.-B. Case Study on Carbon Footprint Life-Cycle Assessment for Construction Delivery Stage in China. *Sustainability* **2022**, *14*, 5180. [CrossRef]
19. Ni, P.; Mangalathu, S.; Mei, G.; Zhao, Y. Permeable piles: An alternative to improve the performance of driven piles. *Comput. Geotech.* **2017**, *84*, 78–87. [CrossRef]
20. Ni, P.; Mangalathu, S.; Mei, G.; Zhao, Y. Laboratory investigation of pore pressure dissipation in clay around permeable piles. *Can. Geotech. J.* **2018**, *55*, 1257–1267. [CrossRef]
21. *JGJ106-2014*; Technical Code for Testing of Building Foundation Piles. Ministry of Housing and Urban-Rural Construction of the People's Republic of China: Beijing, China, 2014. (In Chinese)
22. Sun, Y.; Li, Z. Study on Design and Deformation Law of Pile-Anchor Support System in Deep Foundation Pit. *Sustainability* **2022**, *14*, 12190. [CrossRef]
23. Le, X.; Cui, X.; Zhang, M.; Xu, Z.; Dou, L. Behavior Investigation of Necking Pile with Caps Assisted with Transparent Soil Technology. *Sustainability* **2022**, *14*, 8681. [CrossRef]
24. Ministry of Transport of People's Republic of China. *China Transportation Yearbook 2021*; China Communications Press: Beijing, China, 2021.
25. National Energy Administration. *Budget Quota of Electric Power Construction Project*; China Electric Power Press: Beijing, China, 2018.
26. China Electricity Council. *Quota of Construction Machinery Shift Cost of Electric Power Construction Project*; China Electric Power Press: Beijing, China, 2020.
27. Liu, J. Group Effects and Some Problems on the Concept Design of Pile Group Foundation Under Vertical Load. *China Civ. Eng. J.* **2004**, *37*, 78–83. [CrossRef]
28. Gurvich, A.; Creamer, G.G. Overallocation and Correction of Carbon Emissions in the Evaluation of Carbon Footprint. *Sustainability* **2021**, *13*, 13613. [CrossRef]
29. Zhang, X.; Wang, F. Assessment of embodied carbon emissions for building construction in China: Comparative case studies using alternative methods. *Energy Build.* **2016**, *130*, 330–340. [CrossRef]
30. Roh, S.; Tae, S.; Kim, R.; Park, S. Probabilistic Analysis of Major Construction Materials in the Life Cycle Embodied Environmental Cost of Korean Apartment Buildings. *Sustainability* **2019**, *11*, 846. [CrossRef]
31. Chen, K.H. Research on Carbon Emission Accounting in the Construction Delivery Stage of Building Engineering. Ph.D. Thesis, Guangdong University of Technology, Guangzhou, China, 2014.

MDPI
St. Alban-Anlage 66
4052 Basel
Switzerland
Tel. +41 61 683 77 34
Fax +41 61 302 89 18
www.mdpi.com

Sustainability Editorial Office
E-mail: sustainability@mdpi.com
www.mdpi.com/journal/sustainability